THE
NINTH REVOLUTION
Transforming Food Systems for Good

THE
NINTH REVOLUTION
Transforming Food Systems for Good

Sayed Nader Azam-Ali

World Scientific

NEW JERSEY · LONDON · SINGAPORE · BEIJING · SHANGHAI · HONG KONG · TAIPEI · CHENNAI · TOKYO

Published by

World Scientific Publishing Co. Pte. Ltd.

5 Toh Tuck Link, Singapore 596224

USA office: 27 Warren Street, Suite 401-402, Hackensack, NJ 07601

UK office: 57 Shelton Street, Covent Garden, London WC2H 9HE

Library of Congress Cataloging-in-Publication Data
Names: Azam-Ali, S. N. (Sayed N.) author.
Title: The ninth revolution : transforming food systems for good / Sayed Nader Azam-Ali.
Description: Hackensack, NJ : World Scientific, [2021] |
    Includes bibliographical references and index.
Identifiers: LCCN 2021009911 | ISBN 9789811236440 (hardcover) |
    ISBN 9789811236457 (ebook) | ISBN 9789811236464 (ebook other)
Subjects: LCSH: Agriculture--History. | Agrobiodiversity. | Food crops. | Agricultural diversification.
Classification: LCC S439 .A935 2021 | DDC 631.5/8--dc23
LC record available at https://lccn.loc.gov/2021009911

British Library Cataloguing-in-Publication Data
A catalogue record for this book is available from the British Library.

First published 2021 (hardcover)
Reprinted 2022 (in paperback edition)
ISBN 9789811250101 (pbk)

For any available supplementary material, please visit
https://www.worldscientific.com/worldscibooks/10.1142/12262#t=suppl

Printed in Singapore

This remarkable book, outstandingly-well written, represents a bold *tour de force* through the several thousand years of human history that have seen agriculture develop from humble beginnings of small plots of diverse domesticated local foodplants to where it is today — a vast worldwide industry almost entirely based on just four major crops — wheat, maize, rice and soy bean, largely grown in monocultures. Interpreting this transformation through a series of nine revolutionary steps, the author elegantly, and uniquely in terms of virtually all other recent books on food security, weaves together history, culture, religion, belief systems, languages, strategic and applied science and technological innovation, and how these have represented drivers for agricultural change. The underlying emphasis throughout the work is the critical importance of conserving and utilizing agrobiodiversity, and to highlight that the current global system of reliance on relatively very few staple crops grown in monocultures is totally unsustainable and will fail to provide a large majority of 10 billion people by 2050 with adequate food and nutritional security in the face of extremes of climate change, major losses of biodiversity and erosion of other natural resources including water and soil. This book is eminently readable, unique in terms of its scope, and suitable for a very wide audience from the general public and students to researchers and policy makers and shapers.

George Rothschild
Emeritus Professor
Natural Resources Institute
University of Greenwich, UK
Former Director-General
International Rice Research Institute (IRRI)

The Ninth Revolution is Professor Sayed Azam-Ali's brilliant and important explanation of how human civilizations have moved through eight phases of agricultural revolution, each transforming our world for better and, often unnoticed, for worse. It lays bare the perilous risks to humanity from a dangerous and unsustainable food system, and concisely identifies crucial steps for a Ninth Revolution of agriculture.

Azam-Ali passionately and unflinchingly explains how our dependence today on only a handful of mainstream crops is killing people through malnutrition; crushing cultures; helping to warm the planet and destroying biological diversity. Perhaps most critically, with an inexorable rise in global population that is expected to reach around 10 billion by 2050, he confronts us with realization that our present approach to modern agriculture will be unable to provide nutritional security for millions of people on a hotter, drier planet.

In this transcending and highly original work, Azam-Ali argues that hope for our future lies in a Ninth Revolution of food systems that will rediscover and cultivate crops that are currently neglected or forgotten. He eschews romantic nostalgia for pre-industrial eras and firmly recognizes that science and technology are necessary enablers for a complex, sophisticated food system of the future. This book is destined to become one of the great manifestos of human adaptation for the 21st century, which is shaping up perhaps to be the most dangerous and challenging ever for human civilization.

Max Herriman
CEO of the international consultancy firm,
Sea Resources Management

Sayed Azam-Ali has not only written a history of agriculture, but has proposed practical approaches to increase the food supply for poor people, what he deems a Ninth Revolution in agriculture. This is not a technical treatise for geneticists or food scientists, but after examining an historical context, he proposes imaginative solutions to improve the food supply and nutritional security of humanity in an era of changing climates. This is a book that should find its way to think tanks and students of international relations, agricultural development, natural resources, ecology, biological conservation, and policy studies, as well as the wider public.

Seth Beckerman
Editor and Writer (International Development), Pennsylvania, USA
s.beckerman@vip.cgnet.com

Professor Sayed Azam-Ali's The Ninth Revolution is a must-read, especially for anyone interested in understanding the underlying contexts and intricacies behind the world's most pressing issues in food and agricultural security. In fact, we are adding it into the official curriculum of my organization's world-leading food and agriculture learning and startup acceleration program, which comprises a global community of 60,000 millennial and gen-z leaders in 175 countries.

Through his multispectral approach that combines perspectives from anthropology to deep technical innovation, Professor Azam-Ali provides an excellent overview of how we got to where we are, and the next steps we need to take to revolutionize how we feed and nourish the planet. This is especially important for our world's young people who, as digital natives, are drawn to the technological aspects of modern food and agriculture innovations, but also require an in-depth exploration of the historical background underpinning the previous eight agricultural revolutions.

The young people we work with all over the world are deeply interested in getting involved in solving the issue of feeding 10 billion people by 2050 and Azam-Ali's compelling outlook can inform their actions locally and on-the-ground, right up to national and international advocacy.

The global challenges we collectively face — from climate change, biodiversity threats and intensified soil erosion — strongly call for actions that are built on inherently next-generation values and approaches such as circular economy, sustainable use of agrobiodiversity, and inclusive policies for food security and nutrition. Azam-Ali's excellent book is approachable, informative, and thought-provoking for all readers — from expert through to novice — and is essential reading for anyone serious about making necessary changes to our food and agricultural systems.

Christine Gould
Founder & CEO of the World's Largest Innovation Engine
for Food and Agriculture
Thought For Food Foundation

for Sue, Sami and Jahan — the World

FOREWORD

Sayed and I share a love of growing plants and a desire to see that the whole world is properly fed. Neither of us grew up on a farm or even in the countryside. My interest was sparked by my father and the grandfather of my best friend who encouraged me to grow flowers and vegetables, while Sayed's inspiration came from his uncle who taught him how to grow runner beans and radishes in the garden of their semi-detached house in South London. His love of food was inspired by his mother who introduced him to the traditional cuisines and food cultures of her family background in Iran, the Indian sub-continent and Southeast Asia. With no familial links to agriculture, it was somewhat of a surprise when, in 1976, Sayed chose to study Agricultural Botany at university in North Wales. While his degree focussed mainly on plant sciences, it was the lectures in Agricultural Meteorology that most interested him. He was especially inspired by the emerging discipline of Environmental Physics that was being pioneered by Professor John Monteith and his multidisciplinary team at the University of Nottingham. I studied Soil Science at Reading

xi

and first met Sayed in 1979 when he came to Nottingham as a young postgraduate student and joined the Monteith team of which I was a part. We got to know each other well when we worked together on experiments with pearl millet especially on a study in the Sahel region of West Africa. We have been friends ever since.

At the time of his PhD, agricultural sciences focused mainly on the breeding of high yielding varieties of major staple crops such as wheat, rice and maize that formed the basis of what became known as the *Green Revolution*. Whilst agronomic management of mainstream crops were part of the green revolution package, an understanding of how the physical environment impacted on crops was of less interest. The argument was that inputs such as irrigation and fertilizers could adequately replace any physical shortages of water and nutrients. Environmental physicists took a rather different approach and considered that the way in which crops were able to capture and convert resources such as light, water and nutrients determined their actual productivity within the limits set by their genetics. Monteith's team included physicists, biologists and engineers who worked together to study complex challenges resulting from environmental stresses such as drought and high temperatures on crops growing in Africa and India linked with controlled environment studies. This foundation in multidisciplinary research and work in environments that were academically, physically and culturally different to his own provided the diversity of ideas, people and plants that gave Sayed the grounding and inspiration for his own career — the fabric of this book.

The question that has most intrigued Sayed Azam-Ali for over four decades is why humanity depends on so few crops and animals for most of its food. This simple question has guided his teaching, research and leadership in Europe, Africa, the Indian sub-continent and Southeast Asia. The thread that connects these activities and the chapters of this book has been Sayed's fascination with hitherto underutilized crops in agricultural and food systems. Like his mentor, John Monteith, Sayed Azam-Ali has been an outsider to the traditional worlds of agriculture and food. This has allowed him to take a more complete and holistic perspective of food systems and a longer-term view of the purposes and values of research and education. Chapters 4 and 5 provide an insight into these perspectives and explain his concerns about the consolidation of the global food system and the research that underpins it. These concerns and their consequences for humanity and the planet will help the reader appreciate why the author calls for a Ninth Revolution to transform food systems. Further chapters describe Sayed's introduction to underutilized crops and the opportunities that they offer to diversify agricultural production, rediscover ingredients that have been part of our food systems for millennia, and how their wider adoption can provide important tools for the transformation of food systems. In Chapter 7, Sayed describes his leadership of the 'Crops For the Future Research Centre' in Malaysia in which he played a pivotal role and to which I was pleased to contribute. The full potential of this ground-breaking work has yet to be fully realized, but I believe that history will recognize its significance and, as with Monteith's Nottingham group, appreciate that it was ahead of its time.

The final chapter provides a powerful justification for the 'Ninth Revolution'. The author argues that business-as-usual will not be enough for humanity and our food systems to navigate the multiple challenges of climate change and variability, further pandemics and political instability. He provides a coherent case for why we must urgently diversify food systems beyond the products of a narrow range of elite species. He highlights the exciting initiatives that are now being developed beyond farms, fields and conventional research and education and by practitioners, often young and women, without a formal background in conventional agriculture. He includes nine 'Collective Actions' that form the basis of the revolution that he proposes. Through these, he argues that communities must be at the heart of innovations in agriculture and food, and that embracing the diversity of people, species and thinking is the best chance that humanity has to meet the challenges ahead.

In 1962, Rachel Carson's book 'Silent Spring' led to the banning of toxic agricultural chemicals such as DDT and the introduction of legislation to protect planetary ecosystems. These actions did not come about through the publications of scientists or the concerns of the agricultural industry but through the demands of a wider public when presented with evidence in plain language. Sayed Azam-Ali's book similarly provides a compelling case to repurpose our food systems as a global public good that goes beyond the current 'yield for profit' paradigm to one that places heathy people and a healthy planet with human values at the heart of the food system.

Professor Peter Gregory

Former Chief Executive of Scottish Crop Research Institute and East Malling Research and Emeritus Professor of Global Food Security at the University of Reading, UK

CONTENTS

Foreword   xi
Introduction   xxiii

1 The First Agricultural Revolution   **1**

  Context   1
  Agriculture — A Brief History   2
    The First Agricultural Systems   4
  Genetic Resources   6
    Conservation of Genetic Diversity   6
    Nikolai Vavilov   9
  Knowledge   14
    Technologies   14
    Written (Codified) Knowledge   15
    Spoken (Vernacular) Knowledge   16
    English — the Killer Language   18
    Language Diversity, Biodiversity, and Agriculture   19
  Final Thoughts   23

2 Empires of Power   **27**

Context   28
Agriculture, Civilizations, and Empires   29
     Hydraulic Empires   29
     Akkadian Empire (Revolution Number 2)   30
     Roman Empire (Revolution Number 3)   32
     Muslim Empire (Revolution Number 4)   37
     British Empire (Revolution Number 5)   40
     English Agricultural Revolution   43
     Industrial Revolution   45
     Markets   46
     Capitalist Farmers   47
     Trade and Empire   48
     Cotton: Nexus of Agricultural Revolution, Industrial
         Revolution, and British Empire   49
     The Plantation   51
Revolution Numbers 2, 3, and 4 Revisited   52
     Akkadian, Roman, and Muslim Empires: Fractured Legacies   52
     British Empire: An Enduring Legacy   54
Final Thoughts   56

3 The Green Revolution   **59**

Context   60
Agricultural Biodiversity   60
     Norman Borlaug — A Breeding Approach to Agricultural
         Biodiversity   61
     CIMMYT (Centro Internacional de Mejoramiento
         de Maiz y Trigo)   62
     Nobel Prize   62
The Green Revolution   63
     Plant Breeding   66
     High-Yielding Varieties (HYVs)   66
     Fertilizers   66
     John Bennet Lawes   67
     Haber-Bosch   67

Crop Health   69
  Herbicides   69
  Pesticides   71
Irrigation   71
Mechanization   72
The Green Revolution in Perspective   73
  A Success Story   73
  Borlaug's Legacy   74
  Politics   75
    Mexico   75
    India   77
    Africa   79
  People   80
  Plants   81
    Nutritional Diversity   83
    Knowledge Diversity   83
  Planet   84
Final Thoughts   86

4 The Titanic Agrifood System   **89**

Context   90
Cheap Food Paradigm   91
  Industrial Agrifood System   92
Cost of Cheap Food   95
Environment   97
Society   98
Food Insecurity   99
  Undernourishment   101
  Overnourishment   102
  Triple Burden of Malnutrition   102
  Malnourished Children   104
  Stunting   104
  Wasting   105
  Obesity   105
Climate Crisis   106

Role of Agriculture in Climate Change 108
Impacts of Climate Change on Agriculture 108
    Likely Impacts on Agricultural Production 109
    Effect on Quality 111
    Insect Pests, Diseases, Weeds 112
    Food Prices 113
Weathering the Perfect Storm 113
    Climate Resilient Cropping Systems 114
        Multicropping: A Model for Climate-Resilient Cropping
            Systems 115
        Climate Resilient Crops 119
    Climate-Resilient Nutrition 121
        Nutritionism 121
        Diversifying Diets 125
    Climate Resilient Food Systems 128
        Diversifying Value Chains 128
Final Thoughts 129
    Nature of Food 129

5 Global Agricultural Research and Education **133**

Context 134
A Global Model for Higher Education 134
    Emergence of Global Rankings 138
    Modularization 139
    Agricultural Sciences and Modularization 140
    Research Assessment and Agricultural Sciences 141
        Quality 144
        Quantity 144
        Consequences 145
Agricultural Research Systems 146
    Colonial Agricultural Research Systems 148
    National Agricultural Research Systems 150
    Official Development Assistance 151
    Global Agencies 153

The United Nations Food and Agriculture
    Organization (FAO)   154
    Achievements of FAO   156
    Criticisms of FAO   158
International Agricultural Research Centres (IARCs)   160
    Consultative Group on International Agricultural
        Research (CGIAR)   161
    CGIAR Programmes   163
    Successes of CGIAR   165
    Criticisms of CGIAR   167
    Association of International Research and Development
        Centers for Agriculture (AIRCA)   170
Revolution Number 8 Revisited   172
    Winners   173
    Losers   174
Is the Agricultural Research System Fit for the Future?   174
Final Thoughts   177

6 Underutilized Crops   **181**

Context   182
John Monteith — A Mechanistic Approach to Biological
    Systems   183
    ODA Tropical Microclimatology Unit   188
    Keith Scott   191
    Tropical Crops Research Unit (TCRU)   192
    Starting on Underutilized Crops   194
Bambara Groundnut   196
    Early Faculty Work at Nottingham   197
        Initial Comparisons with Groundnut   199
        Daylength Enigma   199
    Out of Africa   201
        Mapping Global Potential   203
    Bambara Groundnut — The Groundnut of the Women   205
    EU Support for Research on Bambara Groundnut   206
        Operational Method of Hybridization   208

First Genetic Linkage Map of Bambara Groundnut   209
Working in Teams   209
Underutilized Crops Revisited   210
Groundnut — Case Study of a Major Crop   211
An American Crop   212
Emergency Food Source   214
The East African Groundnut Scheme — A Case
Study in Hubris   214
Catalogue of Errors   216
Serious Flaws Ignored   218
Bambara Groundnut — Towards a Research
Framework for Underutilized Crops   220
Tenacity   221
Where Does the Framework Fail?   222
Belief Systems in Rural Malawi — A Case Study   223
Final Thoughts   226
Why Is Groundnut a Major Crop and Bambara Groundnut
Is Not? A Researcher's Perspective   227

7 Crops for the Future   **231**

Context   232
Plan A — The Case for a Special Plants Only Agrifood
System   233
Plan B — The Case for a More Biodiverse Agrifood System   234
Funding Crop Research   237
Conducting Crop Research   238
Major Crops   238
Underutilized Crops   240
Crops For the Future Research Centre (CFFRC) — An
Experiment in Plan B   242
Research Value Chain   243
Five Negative Assumptions About Underutilized Crops   245
No Demand for these Crops or Markets for their Products   246
FoodPLUS — Novel Foods and Functional Ingredients
from Underutilized Crops   246

FishPLUS — Insect Meal Using Underutilized Crops as
a Sustainable Source of Aquaculture Feed 247
Forgotten Foods — The Future is History 249
If a Market Exists, There is No Supply Chain to Get
Products to Supermarkets 251
If a Supply Chain Exists, Quality Seed and New Varieties
Are Needed to Make Products 253
Even With Products and Supply Chains, There is
No Research Expertise on These Crops 256
We Can't Decide Their Uses and Where to Grow Them
Without Reliable Evidence 262
What Crop Can I Grow at My Location under Current
and Future Climates? 263
What Is the Composition and Nutritional Value of
Ingredients and Products? 264
A Food System Approach 264
Crops for the Future — Work in Progress 265
Final Thoughts 273

8 Revolution Number 9 **275**

Context 276
Ode to Joy (All men shall be brothers) 276
Transforming Energy, for Good 277
Energy Sector 277
Negative Externalities 277
Positive Externalities 280
Transforming Food Systems, for Good 284
Global Agrifood System 284
Negative Externalities 284
Business as Usual 286
Climate Changes Everything 289
Climate Change, Food, and Population 289
How Did We Get Here? 292
Complex Systems and Complicated Shadows 294
The Climate Crisis Requires Complex Solutions 295

Why the Yield-for-Profit Paradigm Won't Work in the
   Future   296
Why We Need to Transform Food Systems for Good   299
Sustainable Development and the Climate Crisis   301
The Ninth Revolution — Transforming Food Systems
   for Good   304
We Need New 'Light Bulbs' for the Food System   304
The Light Bulb is Diversity: Nine Collective Actions
   to Transform Food Systems for Good   307
Collective Action 1: A Global Evidence-Base
   for Agricultural Diversification   309
Collective Action 2: Community of Practice for Research
   on Agricultural Diversification   310
Collective Action 3: Food and Sustainable Development
   Curricula for Schools   311
Collective Action 4: Decision Support for
   Climate-Resilient Agroecosystems   312
Collective Action 5: Decision Support System
   for Climate-Resilient Species   313
Collective Action 6: Digital Tools for Traceable
   Agrobiodiverse Value Chains   314
Collective Action 7: Genetic Resources to
   Breed and Improve Agricultural Species   315
Collective Action 8: National Guidelines for Dietary
   Diversification   316
Collective Action 9: Global Action Plan for Agricultural
   Diversification (GAPAD)   317
Final Thoughts   319
This Is Not a Drill   320
Food is Joy   325

Acknowledgements   **329**

Endnotes   **333**

Index   **373**

INTRODUCTION

Evolution is the change process for all forms of life. It occurs gradually because many generations are required for a particular species to adopt a new characteristic. In contrast, *revolutions* involve rapid, transformational changes in society. Because they occur suddenly, revolutions mark a turning point in the history of a nation, region, or even the world. They are often accompanied by a period of upheaval during which an existing order is replaced by new cultural norms and economic and political systems. Unlike human evolution that has taken millions of years, human history is brief. We only began to settle in communities about 10,000 years ago, and our oldest historical records can be traced for less than half that time. Since then, human history has been turbulent and punctuated by numerous revolutions, many of which involved agriculture. Some aspired to control the resources to grow food, make and store its products, and maintain authority over its distribution, availability, and price. Others aspired to control the research, education and knowledge that underpin food systems and the values associated with it.

In this book, I trace eight revolutions in human history that are associated with agriculture, from the onset of farming through to our modern globalized and industrialized agrifood system. I include the empires associated with these revolutions from the Akkadian Empire that first secured control over water for agriculture; the rise and fall of the Roman, Muslim, and British empires that gained and then lost control over land, resources, and technologies to produce food; to the Empire of the Market that now determines which and at what price communities have access to food. In this sequence, the genesis and legacy of the *Green Revolution* is considered as the template for our current agrifood system and the research and educational systems that underpin it. Using examples from my own experience, I argue that we cannot simply rely on market forces to determine the cost of food nor do we have time for incremental improvements to our current model. Instead, a revolution is proposed - in thinking and actions, as well as social, cultural, and moral values to repurpose the food system so that it can nourish more people on a hotter planet without destroying the biodiversity and ecosystems on which we all depend. At its core this requires us to recognize nutrition as a human right rather than food merely as a means of sustenance and profit. The book concludes with an urgent call to action for a *Ninth Revolution* whose aim is nothing less than to transform agriculture and food systems for good.

1

THE FIRST AGRICULTURAL REVOLUTION

Revolution Number 1

Context

Around 10,000 years ago, our ancestors gave up capturing and collecting food and began to grow it.[1] A shift to sedentary agriculture, however primitive, can be traced to that transition. The division between those who farmed food and those who only consumed it gave an opportunity for human groups to move around the globe and transport their crops and animals. It also led to the exchange of ideas, technologies, knowledge and beliefs among groups involved with cultivating, processing, cooking and consuming food. It was the first step in globalization. It was Revolution Number 1.

Revolution Number 1 has made us who we are today. Without farming, we cannot imagine how those freed from the daily burden of capturing food could discover new lands, build armies, wage wars, establish empires, and subjugate and pacify colonies. Without Sedentism there would be no nation states, written languages, tools, extractive industries, pollution, sport, or global pandemics spread by travellers. Without Revolution Number 1, the Beatles could not have sung Revolution Number 1 nor could Beethoven have written his Symphony Number 9.

Ten thousand years later, Revolution Number 1 still has a huge legacy for our agrifood system. Since we began farming, our food intake has declined from over 30,000 edible plants to fewer than 30 plant species[2] of which a handful of staple crops now provide most of the food consumed by 7.8 billion people.[3] From the tens of thousands of languages and dialects spoken at the beginning of farming, just six are now spoken by around half of humanity.[4] Of the countless agricultural systems practiced around the world, the monoculture of individual crops is now the global norm.[5] And from the many political, social, and economic systems devised by human beings, one — the neoliberal market model — now holds sway across most of the world. This chapter looks at the legacy of Revolution Number 1 (the First Agricultural Revolution), the diverse agricultural, knowledge and food systems that we have inherited over millennia and the consequences of their marginalization, denigration and demise in pursuit of a single global model.

Agriculture — A Brief History

The history of agriculture is brief — at least compared to our own time on the planet. Bipedal primates — our ancestors — have evolved over the last

5 to 7 million years. The *hominid* or primate family includes all our past and modern relatives and can be divided into the *gorilla* or ape line, the *pan* or chimpanzee line, and the *homo* (hominin) or human line from which we are derived. Even if we only consider our most recent — *Homo sapiens* — model, our genome (the sum of our genetic material) has been on the planet for at least 250,000 to 300,000 years. For almost all of this time, our predecessors were hunter-gatherers, killing their prey and collecting plant foods from the wild. In contrast, sedentary agriculture, or *farming*, has been practiced for about 10,000 years and only intensively for the last few decades. Assuming that all our predecessors ate, the cultivation of crops and management of livestock represent a meagre fraction of our homonin past — a blip in our history but not in its consequences for us and our planet.

A major change in climate around 11,000 BCE, which coincided with the end of the last ice age, probably led to the start of agriculture.[6] Rather than an equable climate that suited hunting and gathering throughout the year, much of the earth experienced periods of drought. This seasonality favoured the growth of annual plants whose storable seeds or tubers allowed hunter-gatherers to establish the first settled villages. Climate change provided two drivers for agriculture. First, the need for sedentary lifestyles because there was insufficient food to hunt throughout the year, and, second, preference for organized farming to cultivate crops in designated spaces and whose stored products allowed communities to survive the dry season. The rest, literally, is history.

Because climate change had differential effects around the globe, it imposed new geographical limits (*seasons*) for the cultivation of different plants. Diverse natural vegetation and local climates meant that agriculture could start in disparate parts of the globe with different crops and growing seasons. Early farmers will have observed that some plants growing in their local habitat yielded products that were both edible and could also be stored and planted for future harvests. There were several regions where agriculture evolved independently by local, settled communities. However, we shouldn't assume that there was an immediate or complete switch to sedentism. Not all hunter-gatherers were constantly on the move and not all sedentary communities stayed put — communities could do both simultaneously. Nevertheless, there was a gradual transition to settled agricultural communities in which the seeds of wild stands of plants, often grasses or cereals, started to be sown, harvested, and consumed alongside the hunting and gathering of other foods. As the choice and cultivation methods of these *crops* improved, the need to hunt and gather food declined, though such practices still exist today in different parts of the world.

For most of hominin history our ancestors lived variously as true hunters and consumers of animals and/or gatherers of roots, grains, fruits, and seeds. Over millions of years, changes in global and local climates and transmigration to new climates indicate that the diet and nutrition of hominins was altered and varied. More importantly, no single universal diet would have been consumed by all hominins.[7] Instead, distinct and isolated groups subsisted on very different diets determined by local geography, ecology, climate, and the particularities of each community.

Diets may have been different from each other, but certain features of all early hominin diets would have been similar. For example, the lack of tools and cooking would have meant that any wild plants or animal foods were unprocessed. Also, an inability to store food meant that food reserves could not be distributed to ensure a more constant energy supply throughout the year. Under such circumstances, dietary intake and composition varied between and within years in response to the availability of food items, which themselves were a consequence of local environmental conditions. Much like their ancestors, the diets of modern hunter-gatherer populations are largely determined by variations in the availability of food sources set by climate and season. For example, in tropical and mid-latitudes, modern hunter-gatherers subsist mainly on roots, fruits, flowers, and leaves supplemented by animal protein. The proportion of animal protein increases with latitude because climate and season limit the growth of edible vegetation — meat contributes between only 5 and 20 per cent of the diet of tropical hunter-gatherers, yet it accounts for more than 70 per cent of the winter diet of sub-Arctic and Arctic hunters.[8]

The First Agricultural Systems

The transition from hunting and gathering in a natural *ecosystem* to farming in an *agroecosystem* imposed limits on the use and timing of resources by the communities that *farmed* them. At least initially, farming mimicked nature in that it was diverse, complex, and dynamic in the use of space, time, and people. The first farmers began using flora (plants) and fauna (animals) from the natural ecosystem for their own benefit. This could have been, for example, from forest gardens cleared within natural vegetation or through shifting cultivation in which areas of forest were cleared to grow plants for a few years and then abandoned and cultivation moved to a different area. As favoured plants became more productive they would have been moved to dedicated locations, *farms*, for their cultivation and to nearby villages for their consumption. These farms would not have been familiar in the modern sense but rather a transition from a natural to an agroecosystem. The first agroecosystems were complex assemblages of annual and perennial species, and animals.[9] These *agrisilvopastoral* (agriculture/tree/animal) systems were gradually replaced by distinct animal and cropping systems.[10] The cropping systems themselves would have included diverse forms, from *agroforestry* (crops planted within tree canopies) and mixtures of annual crops (*intercrops*), through to uniform *monocultures* (single crops of one species). In our early history, human populations were sparse, and hunter-gatherers and nomadic lifestyles allowed natural ecosystems to recover from human effects. By becoming farmers, humans increased the frequency of harvesting plants from the same land and reduced the fallow periods when the agroecosystem could recover.

Wild cereals produce small grains that are hard to harvest and need processing to make them digestible. Before agriculture, there was no previous evolutionary experience of these species in the hominin genome because they were never part of the hunter-gatherer diet. However, sedentary agriculture allowed the domestication of some cereals whose small seeds could be protected, stored, and processed into different products. These species, first domesticated by Neolithic farmers 10,000 years ago, still form the basis of the global human diet today.

At the start of farming, goats and sheep may have been the first animals to be domesticated for food. In Southeast Asia chickens were also domesticated at about the same time, followed by the domestication of larger animals such as oxen or horses for ploughing and transport. Animal systems would have included various forms of *pastoralism* in which livestock were herded across the landscape in search of pasture, fodder, and water.

In regions where plant cultivation proved difficult because soils were poor or climates too hostile for crops, animals were herded by pastoralists whose lifestyle differed from that of hunter-gatherers in that they did not rely exclusively on naturally occurring resources. They kept animals for dairy products and wove their wool into textiles that they could trade with farming societies. As humans began to experiment with farming, they started domesticating and breeding animals whose labour helped make more intensive farming possible and whose milk and meat provided nutrition for increasingly stable and dependent populations.

Genetic Resources

Agriculture uses the genetic resources of plants and animals to benefit humans. These genetic resources exist in *germplasm* that can be stored in *gene banks* for later use or modification. Conventionally, global genetic resources are divided into six *kingdoms*; animals, plants, bacteria, archaebacteria, fungi, and protozoa. Each kingdom is subdivided into six levels or *taxa*; Phylum, Class, Order, Family, Genus and Species. Each species includes *varieties* of a group of living organisms that are capable of exchanging genes or *breeding*. Germplasm encompasses all living genetic resources within each kingdom.

Conservation of Genetic Diversity

For agriculture, seeds, tissues and other genetic resources can be used for plant and animal breeding, conservation of genetic material, and research. These resources include seed collections stored in dedicated seed banks, living plants growing in protected spaces, or nurseries and animals maintained in animal breeding programmes. Germplasm collections can range from stored material of wild species to dedicated breeding lines and elite genotypes that have undergone deliberate human selection or genetic modification. Germplasm collection, protection, and conservation are critical to maintaining biological diversity and agricultural gene banks have been established for many decades. For example, the Millennium Seed Bank at the Royal Botanic Gardens, in Kew, London, has over

96,000 seed collections, representing nearly 40,000 species, over 6,100 genera and 349 plant families.[11]

Despite its significance as a repository of genetic diversity, the Royal Botanic Gardens in Kew illustrate the conflicting interest in conserving genetic resources for commercial benefit, their uses for recreation and leisure, and their protection for posterity and the planet. Kew Gardens were founded in the 1760s by Princess Augusta, elder sister of King George III who ruled Britain during the American Revolutionary War. In 1838, after years of official neglect, a Royal Commission reviewed the future of the gardens and decided to develop and expand their activities. At that time, the growing affluence and leisure time of Britain's emerging middle classes coincided with an increasing interest in gardening and horticulture and the gardens offered an ideal location for recreation linked with a curious interest in plants from across Britain's expanding empire. As well as their recreational value, Kew Gardens became an important agent of the British Empire. The third director of Kew, William Thistleton-Dyer, advocated that it should be a "botanical clearing-house or exchange for the empire".[12] From the 1840s, Kew received large public support because scientific discoveries and colonial activity were priorities for the government and the plants of empire served both objectives. Successive directors of Kew ensured that the gardens coordinated a network of colonial botanical gardens from where Kew-trained specialist botanists were sent to Britain's dominions. Their primary purpose was *economic botany* — to discover plants of commercial value for the empire.

Two examples illustrate how economic botany served the empire.[13] During the 19th century, Malaria was a major killer of British soldiers serving in tropical regions of the empire. Quinine, which is used in the treatment of malaria, is extracted from the bark of the *Cinchona* tree that is native to South America. The British government had no option but to purchase cinchona for troops in India. After the Great Rebellion in 1857, the government of India decided to introduce cinchona to support the needs of an expanding British military presence in the region to quell further uprisings. Botanists from Kew were sent to South America to secure cinchona seeds to establish plantations in southern India, and as a result, quinine became cheaply available in India at the expense of the economies of the newly independent South American countries. This wiped out Latin American trade in cinchona but boosted the reputation of Kew and its activities in economic botany that demonstrated the power of the British Empire.

The second example is rubber, which also is native to South America.[14] In 1876, Kew again under the direction of the India Office sent botanists to South America to smuggle 70,000 rubber seeds out of Brazil, which at the time was able to supply the world's needs for rubber. Once in Kew's custody, the seeds were planted in Ceylon (now Sri Lanka), and Malaya (now Malaysia) where they thrived and have since formed the basis of a hugely successful rubber industry. By 1938, colonial plantations met almost all the world's demand for rubber at the expense of Brazil, its centre of genetic diversity.

At the end of World War Two and the embers of empire, Kew found a new purpose and switched its focus from economic botany to the conservation of plant genetic resources. It forged close links with global plant conservation efforts and established visionary initiatives such as the Millennium Seed Bank Partnership which, working with nearly 100 countries, is the largest *ex situ* (off site) plant conservation programme in the world. Kew is also responsible for world-leading research that is focused on rescuing plants that are at greatest risk from climate change and the impacts of human activities. As a result, Kew has successfully transformed itself from its colonial past of *exploiting* plant genetic resources for commercial benefit to *protecting* them from extinction. Its conservation efforts are not restricted to commercial plants but also include botanical and mycological collections that represent around 95 per cent of vascular plant genera and 60 per cent of fungal genera. Kew also hosts one of the largest repositories of botanical publications in the world that include the systematics and taxonomy of wild plants and collections of herbaria in microform.[15]

At present, the 11 gene banks hosted by the centres of what was the Consultative Group on International Agricultural Research (CGIAR) and is now called the CGIAR System Organization are located around the world and store the genetic resources of many of the world's major crops.[16] These are often stored as plant *accessions* from individual species that have been collected at a single time from a specific location. In addition, the remarkable Svalbard Global Seed Vault in Norway, in collaboration with CGIAR gene banks, stores over one million accessions that represent almost all of our agricultural history.[17] These samples are destined to be hosted at Svalbard as a long-term backup for the 1,750 seed banks that protect the world's treasure trove of agricultural biodiversity. Donated accessions from 53 international gene banks include 156 distinct crop genera. Through funding by the Norwegian government, the Svalbard Seed Vault (the 'Doomsday Vault') was built on the island of Spitsbergen in Norway's remote Arctic Svalbard archipelago. It was established to store duplicates of seed samples that were held in international gene banks to protect global agricultural biodiversity in the event of regional or global crises that would risk the safety of seeds in other gene banks. While safe from political upheavals, even locations as remote as Spitsbergen, well inside the Arctic Circle and at a latitude of nearly 80 degrees north, are not immune to climate change. In October 2016, the seed vault experienced a large water intrusion at the tunnel entrance due to higher than average temperatures in the Arctic and heavy rainfall. The water didn't intrude far enough to damage seeds within the vault, but required improvements, including waterproofing the tunnel walls and digging exterior drainage ditches.

In addition to gene banks (living genetic resources), there is a vast fossil record of animal and plant history that allows us to trace organisms that were once living. Together, gene banks and fossils provide a comprehensive record of plant and animal distribution, evolution, and significance both for global biodiversity and agricultural biodiversity.

Nikolai Vavilov

Agriculture started independently in many locations, but the plant species that were first cultivated (the first *crops*) came from a range of natural ecosystems. In the 1930s, the Russian scientist Nikolai Vavilov identified centres of origin for many of the world's major crops. His life's work cost him his life.[18] Nikolai Ivanovich Vavilov was born in Moscow on 25 November 1887. His younger brother, Sergey Ivanovich Vavilov, was a renowned physicist who founded the Soviet School of Physical Optics and with Pavel Cherenkov discovered the Vavilov-Cherenkov effect of electromagnetic radiation, for which Cherenkov received the Nobel Prize in physics. He also co-developed the Kasha-Vavilov rule of luminescence quantum yields.

Unlike his brother, Nikolai Vavilov was obsessed with farming and the frequent famines that bedevilled his native Russia. In 1913 he travelled in Europe to study plant diseases with William Bateson, the British scientist who was an early pioneer in the science of genetics. From 1924 to 1935 Vavilov directed the Lenin All-Union Academy of Agricultural Sciences at Leningrad. He was also a member of the USSR Central Executive Committee, president of the All-Union Geographical Society, and in 1926 was awarded the Lenin Prize. Vavilov was heavily influenced by Charles Darwin whose theory of evolution proposed that all forms of life have descended from common ancestors. From Darwin's ideas, Vavilov proposed that plants evolve over time and that the theory of evolution provides a basis to search for where certain types of plants originated and evolved into different forms through a process of natural and later human selection and breeding. Vavilov built scientific evidence around testable hypotheses, not just empirical descriptions of what he saw and collected, reliance on blind faith, or political ambition.

In contrast with Vavilov's scientific approach, the Russian agronomist Trofim Lysenko launched a political campaign against genetics and science-based agriculture. *Lysenkoism* rejected natural selection and favoured Lamarckism, which claimed that physical changes in organisms during their lifetime could be transmitted to their offspring. For example, this theory proposed that the length of a giraffe's neck was associated with successive generations of giraffes reaching for leaves and thereby developing longer necks which they passed on to the next generation.

By 1936 Lysenko had become director of the Soviet Union's Lenin All-Union Academy of Agricultural Sciences. During this time, he secured the endorsement of Joseph Stalin for his views and rejection of the scientific approach of Vavilov. As a result, over 3,000 scientists were dismissed, imprisoned, or executed. On 6 August, 1940, Nikolai Vavilov was arrested during an expedition to Ukraine and in July 1941 sentenced to death. In 1942, even though his life sentence had been commuted to 20 years imprisonment, he died of starvation while still in prison. On 7 August 1948, at a session organized by Lysenko and with the approval of Stalin, the V.I. Lenin Academy of Agricultural Sciences proclaimed that Lysenkoism would henceforth be considered as "the only correct theory". Soviet scientists had to disown any work that contradicted Lysenkoism. Critics of Lysenko were declared to be bourgeois or fascist, and similar anti-bourgeois theories were encouraged in matters relating to linguistics and the arts in Soviet Russia.

During World War Two, the Leningrad seed bank, established by Vavilov as the largest seed bank in the world, was protected and its collections carefully preserved throughout the siege of the city which lasted for over two years. While the Soviets had evacuated art collections from Leningrad, they were unconcerned about the fate of the 250,000 seed, root, and fruit samples that were kept in the seed bank. Scientists at the Vavilov Institute moved the seed collections to the basement for their protection. Nine of those who had been guarding the seed collections had starved to death by the time the siege was lifted in the spring of 1944. In 1955, Vavilov was officially pardoned as part of a de-Stalinization process. By the 1960s he was formally declared a hero of Soviet science. The N.I. Vavilov Institute of Plant Industry in St. Petersburg (what was Leningrad) still hosts one of the world's largest collections of plant genetic resources. On its 75th anniversary in 1968, the institute was renamed after Vavilov. In 1977, the Soviet astronomer Nikolai Stepanovich Chernykh discovered a minor planet which he named after Vavilov and his brother Sergey Ivanovich. The crater Vavilov on the far side of the Moon is similarly named after them. However, Nikolai Vavilov's greatest contribution was rather closer to Earth — identifying the centres of global crop diversity that are named after him.

Vavilov recognized that humans had been domesticating and selectively improving crops in their own locations for millennia. He argued that plants were not domesticated at random, but in *centres of origin* where there was already high diversity of a crop's wild relatives. These centres were distinct and independent of each other, but each could be the centre of diversity for a number of crop species. Now armed with a general hypothesis, Vavilov began to look for scientific evidence to identify those centres of genetic diversity through expeditions, plant evaluation at each location, and the knowledge of local agriculturalists. Vavilov was careful to include visits to marginal environments such as mountainous regions near deserts or semi-deserts where local farmers had improved wild species to produce local *landraces* of crops of particular interest to them.[19] (A landrace is a type of a plant or animal species that has developed through adaptation to its natural environment and the cultural influences of local farmers and pastoralists).

Vavilov identified his first centres of diversity in 1924.[20] This initial list was refined and expanded, and there are now eight generally recognized Vavilov centres of diversity.

I. **East Asian (central and western China, Korea, Japan, and Taiwan).** The native region for soybean, millet, many vegetable crops, and fruits. Vavilov estimated that 20 per cent of the world's cultivated flora originate from eastern Asia.

II. **Hindustani (tropical India, Indochina, southern China, and southeast Asia).** The native region of rice, sugar cane, tropical fruit, and vegetables. More than 30 per cent of the world's cultivated plants originate here.

III. **Inter-Asiatic (Asia Minor, Iran, Syria, Palestine, Jordan, Afghanistan, inner Asia, and north-western India).** Native crops include the origin of wheat, rye, and fruit trees. About 15 per cent of the important cultivated plants for the world originated here.

IV. **Caucasian (Armenia, Azerbaijan, Georgia, and Russia).** Includes original species of temperate fruit trees. Additional species of wheat and rye are also found here.

V. **Mediterranean (countries bordering the Mediterranean).** Includes vegetables, forage crops, and olive. About 10 per cent of the species of cultivated crops originated here.

VI. **Abyssinian (Ethiopia).** Includes teff, niger seed, ensete, and coffee. Probably about 4 per cent of the world crops originated here.

VII. **Central American (southern North America, Mexico, and the West Indies).** Includes maize, cotton, beans, pumpkin, cocoa, avocado, and subtropical fruits. About 8 per cent of the important crops of the world originated here.

VIII. **Andean (Venezuela, Colombia, Ecuador, Peru, Bolivia, Argentina, Brazil and Chile).** Includes many tuberous crops such as potato, the quinine tree, and the coca bush.

Vavilov was a genius. His life's work, political persecution and tragic death are vividly described in Peter Pringle's excellent biography.[21] Not only did he successfully identify the centres of diversity for many of the world's major crops, he also established two other theories that underpin modern plant genetics, breeding, and conservation — the *Law of Homologous Series in Variation*[22] and the concept of *genetic erosion*.[23] Vavilov's law introduced the concept of *parallel variation* among species — if a characteristic is observed in one crop species it is also likely to exist in its related species. This means that as research continues on a more favoured crop its outputs can be *translated* to other less favoured crops without starting all over again for each crop. The saving in time, effort, and cost in such translation is potentially huge. On the basis of this law and novel genetic technologies, plant and animal breeders can systematically seek and find desired characters in different species in a variety of cultivated plants and domestic animals and their wild relatives. This search for desirable traits can be used in conjunction with Vavilov's centres of diversity for cultivated plants in which there is likely to be a reservoir of genetic material in wild ancestors of modern crops. Vavilov's law of homologous series has become a powerful stimulus for breeding new varieties of cultivated plants and the establishment of a scientific basis to introduce and acclimatize these varieties to changing and changed environments.

The concept of genetic erosion was also first introduced by Vavilov. Genetic erosion occurs when an organism dies without the chance to breed and pass on its genes to the next generation. Endangered species, by definition, have few individuals and are especially vulnerable to genetic erosion. The danger of genetic erosion in wild relatives of crop plants has enormous consequences for agriculture. When plant breeders incorporate desirable characters into a single variety, they may discard existing varieties or ancient relatives that do not have that character. Once discarded, these traits can be lost forever unless they are conserved in seed banks such as Vavilov's in St. Petersburg. Genetic erosion includes loss of genetic diversity in local landraces of domesticated plants and animals that have become adapted to the natural environment in which they originated and may be suitable for unpredicted future environments. The maintenance of genetic diversity offers potential options stored within landraces and wild relatives that may include nutritional and health properties, resilience to hostile or fragile environments, and tolerance to stresses such as pests, pollution, cold, heat, drought, and salinity.

Throughout his career, Vavilov recognized that genetic resources have to be conserved for the benefit of humanity. He realized not only that the genetic diversity of wild plants contained the basis of crop improvement for climate and disease challenges, but that changes in environment, either naturally or through human activities, could erode that genetic base. He saw how genetic erosion led to the loss of wild plant diversity and local crop landraces and the consequences of their loss on the world's food security. His theories have become guiding principles for advanced plant breeding, molecular genetics, and plant conservation. His single-minded dedication to conserving genetic resources, and his commitment to truth and scientific methods remain an inspiration and a role model for scientists and societies.

Knowledge

Technologies

Sedentary agriculture may have evolved only about 10,000 years ago in the Neolithic or New Stone Age, but the tools and implements used by humans can be traced back to about 3.3 million years ago in the Palaeolithic or Old Stone Age.[24] This is when our hominid antecedents first used stone tools, sharpened sticks, hammer stones, choppers, cleavers, spears, and even harpoons for hunting and gathering. The first Neolithic farmers inherited such tools, and over time, fashioned their own implements specifically for agriculture. Many of these early farming implements are still in use today. For example, the *shadoof* was an early crane-like water-lifting device used in irrigation from around 3,000 BCE in Mesopotamia and is still used to water small crop areas in prepared basins in tropical regions. At the start of agriculture, soil cultivation involved using simple digging sticks and hoes that were used to produce furrows in which seeds could be planted into prepared soil. Digging sticks, hoes, and mattocks were developed in many regions where agriculture was practised. Hoe-farming remains a traditional method of soil tillage in much of the tropics and mechanized versions or derivatives are in use throughout the world. The ability to harvest the products of cultivation, usually as seeds or tubers, led to the development of storage facilities, techniques to preserve crops, and processing into readily usable forms for easy cooking. Again, we can see that many of these technologies are still in use now with little modification by location or generation. Where traditional agricultural and food processing implements are no longer in use, their archaeological record is preserved from many locations around the world should we wish to rediscover their functions.

The beginnings of agriculture represent perhaps the biggest disruption in human history. We can, however, trace the genetic and archaeological record of plant, animal, and technological resources that relate to agriculture over the past 10,000 years. This allows us not only to conserve and protect these resources, but also to use or modify them for modern agricultural purposes — for example, by using wild relatives of crops and rare breeds of animals in breeding programmes. However, for agriculture to evolve, agricultural *knowledge* must also pass from one generation of farmers to the next or to intermediaries who can translate that knowledge. Uniquely, human beings can transfer knowledge through diverse spoken and sign languages, codify knowledge through written texts, or more recently, store, redefine, and reaggregate that knowledge through computer databases, knowledge systems, and imagery. At least for genetic and technological resources, the historical record from the beginning of agriculture is clear, traceable, and verifiable, but the vast store of knowledge and belief systems associated with agriculture since its inception is not. For this, we would need an unbroken record of codified (written) and vernacular (spoken) knowledge that relate to agriculture over this period; we have neither.

Written (Codified) Knowledge

If we take codified knowledge, the first evidence of script (cuneiform) began only about 5,000 years ago. The oldest known written language, Sumerian, dates back to around 3,500 BCE.[25] The earliest evidence of written Sumerian is the Kish tablet, which was found in Iraq.[26] This and other texts are not exclusively about agriculture. The first texts that relate specifically to agriculture are probably inventories, ledgers, and accounts of exchanges of commodities and not specific knowledge of crop and animal management. A torrent of textual information has been generated since then, but there remains the issue of verification. Without verification, information is of limited value; *fake news* is not a purely modern phenomenon. Nevertheless, much of the evidence accumulated and codified over the last 3,500 or so years is still available for us to make decisions on its value and validity for current agricultural and social systems.

Spoken (Vernacular) Knowledge

There is no written record of agriculture between about 10,000 and 3,500 years ago. Those from the beginning of written texts are patchy and none are easily verifiable. At best, written evidence represents only about half of the period in which agriculture has been practiced and when much of the world was illiterate, i.e., unable to write in any language. For most of our agricultural history, knowledge has been conveyed through speech or sign language both within and across generations. The main mechanism for the transfer of *vernacular* knowledge of agriculture (that held in people's heads) is through language.

Anthropologists disagree when language was first used. Proto-language — the use of words without syntax — may have existed for over 2 million years. However, what we now consider as structured language may have co-evolved with Homo sapiens. Languages can be considered as the bases of cultural diversity, and as such, allow the world's people to be systematically placed into different knowledge groups. The Ethnologue Languages of the World Project[27] estimates that there are 7117 living languages of which most are spoken by relatively few people in a few countries. Only eight countries account for over half of the world's spoken languages and just two (Papua New Guinea (840 languages) and Indonesia (711 languages)) speak four times as many languages as all of Europe. In contrast, a very small number of languages are spoken by vast populations — only six languages (English, Mandarin Chinese, Hindi, Spanish, French and Standard Arabic) are spoken by around half of the global human population and over 88 per cent of humanity speaks one of only 200 languages as their native tongue, and hundreds of millions more speak them as second languages.

The smaller the population that speaks a language, the greater the likelihood of that language being unrecorded — unwritten or its speakers being illiterate — and therefore endangered or lost. Roughly 40 per cent of languages are now classed as endangered, often with less than 1,000 speakers remaining.[28] In 1992 a seminal paper[29] by the Alaskan linguist Michael Krauss, *The World's Languages in Crisis* estimated that only 600 oral languages would still be spoken in 2100, an astonishing loss of around 50 spoken languages per year.

The Endangered Languages Project (ELP)[30] is a web-based platform for the world's endangered languages. Its predicted rates for language extinction range from Krauss's worst-case scenario, where 95 per cent of all the world's languages will be extinct or doomed within 100 years, to less catastrophic rates at which one language dies about every three months.[31] Irrespective of their exact accuracy, these predictions represent a catastrophic loss not just of languages but of the knowledge stored within them. Their death also means the loss of cultural or ethnic identity,[32] knowledge of prehistory that, without a written record, has been conveyed through the spoken word,[33] and a decline in the sum of knowledge curated by human beings for millennia.[34,35]

Languages spoken by similar numbers of people are not accorded similar worth. The eminent Finnish linguist Tove Skutnabb-Kangas coined the term *linguicism* to refer to discrimination based on language. She defined linguicism as the;

"ideologies and structures which are used to legitimate, effectuate, and reproduce unequal division of power and resources (both material and non-material) between groups which are defined on the basis of language".[36]

The next step from discrimination against a specific language is to destroy it. When the United Nations prepared for what became the 1948 International Convention for the Prevention and Punishment of the Crime of Genocide, the issue of linguistic genocide was discussed along with cultural and physical genocide as a serious crime against humanity.[37] The *ad hoc* committee responsible for the statement specified the following types of acts as examples of cultural genocide in Article III;

"Any deliberate act committed with intent to destroy the language, religion or culture of a national, racial or religious group on grounds of national or

racial origin or religious belief, such as (1) Prohibiting the use of the language of the group in daily intercourse or in schools, or the printing and circulation of publications in the language of the group; and (2) Destroying or preventing the use of libraries, museums, schools, historical monuments, places of worship or other cultural institutions and objects of the group".

Eighteen countries, including the UK, US and, oddly, Denmark, voted against the full definition of linguistic genocide to be included in the 1948 convention, successfully reducing article 1 to;

"Prohibiting the use of the language of the group in daily intercourse or in schools, or the printing and circulation of publications in the language of the group".

Nevertheless, linguistic genocide as originally defined by the UN in 1948 is practiced throughout the world and is indeed accelerating because of what the Singaporean linguist Anne Pakir has described as *killer* languages that have the power to eliminate others.[38]

English — the Killer Language

Over half of humanity now speaks one of only eight *killer* languages, of which English is the pre-eminent culprit. Information technologies, the advent of the World Wide Web, and the publication of books increasingly mean that the English language is rapidly becoming the *lingua franca* (!) of global knowledge, in general, and scientific investigation in particular. When linguistic diversity disappears, the speakers of a *dying* native language are assimilated into the domains of other *living* adopted languages. This may allow native speakers to *speak* a new language by adopting its words, but it does not mean that they can transfer the knowledge and experience associated with their original language. Where such knowledge is vernacular (unwritten), its loss is permanent.

Colonialism often accelerates language loss through linguicism that denigrates, marginalizes, or even bans certain languages. This is evident through *educational support for development programmes*, especially in the former colonies of the countries that speak killer languages. For example, in much of Africa and Asia the dominant role of English in secondary and higher education and its association with upward social mobility and professional development is strengthened by the mechanism and rules of aid from British and American donors. As a result, there is a growing dichotomy between the language (or languages) used in local society and the language used in education, research, government, trade, and markets. Of course, linguicism doesn't need outward colonialism as its agency. For example, linguicism explains the decline of the native languages of Wales, Scotland, and Ireland and their replacement not just by English but by an accent known as Upper Received Pronunciation associated with the English aristocracy, its ruling class, and elite educational institutions.

So what if languages disappear? Surely loss of language is a natural process that allows more and more people to communicate with each other, exchange ideas, and share experiences in a common idiom? Wouldn't life be easier if all of us spoke the same language, or at least only a few common languages? Does it matter if all but the killer languages disappear? In terms of agriculture, yes — for a number of reasons.

Language Diversity, Biodiversity, and Agriculture

The conservation and protection of the world's natural biodiversity are now recognized as global priorities. The loss of genetic resources in each of the seven kingdoms of biodiversity continues at an alarming rate, and many efforts now seek to arrest or even reverse this decline. For example, the *Convention on Biological Diversity* (CBD)[39] is one of the most comprehensive international agreements ever adopted. Its primary objectives are;

> *"the conservation of biological diversity, the sustainable use of its components and the fair and equitable sharing of the benefits arising out of the utilisation of genetic resources"* (Article 1);

where;

> *"biological diversity means the variability among living organisms from all sources including, inter alia, terrestrial, marine, and other aquatic ecosystems and the ecological complexes of which they are part; this includes diversity within species, between species and of ecosystems"* (Article 2).

In essence, the CBD is a critical international legal instrument with a focus on the conservation of diverse forms of life at the genetic, population, species, habitat, and ecosystem level. Its origins stem from the 1972 United Nations Conference on Human Environment held in Stockholm, which identified biodiversity conservation as a specific priority. The first session of the Governing Council of the United Nations Environment Programme (UNEP) identified the *"conservation of nature, wildlife and genetic resources"* as a priority. In the same decade numerous international treaties were completed — the *Convention on Wetlands of International Importance Especially as Waterfowl Habitat* (known commonly as the Ramsar Convention on Wetlands) (1971); the *Convention Concerning the Protection of the World Cultural and Natural Heritage* (World Heritage Convention) (1972); the *Convention on International Trade in Endangered Species of Wild Fauna and Flora* (CITES) (1973); and the *Convention on the Conservation of Migratory Species of Wild Animals* (Convention on Migratory Species) (1979).

In the early 1980s the *World Conservation Strategy*[40] and the *World Charter for Nature*[41] were adopted. In 1983, the United Nations General Assembly approved the establishment of the *World Commission on Environment and Development,* which in 1987 published the Brundtland Report[42] that linked environment with development and proposed strategies for sustainable development. In 1992, the *United Nations Conference on Environment and Development* (UNCED, the Earth Summit), was held in Rio de Janeiro, Brazil.[43] At this critical meeting, Agenda 21 (the Programme of Action for Sustainable Development), the Rio Declaration on Environment and Development, and the Statement of Forest Principles were adopted, and both the *United Nations Framework Convention on Climate Change* and the *Convention on Biological Diversity* were opened for signature.

We can see from the above examples and international commitments that the loss of biodiversity has received enormous and deserved global attention. However, few mention the loss of linguistic diversity. Although the 1948 *Universal Declaration of Human Rights* (UDHR)[44] identifies language as a category for equal rights, it doesn't provide a definition of linguistic rights. Even where there have been declarations and rules to protect specific languages and their rights, no binding document was produced that referred to linguistic rights. In 1996, the *Universal Declaration of Linguistic Rights* (UDLR) proposed policies to respect linguistic rights regardless of whether they were official, non-official, majority, local, regional, or minority languages. Later that year at the *World Conference on Linguistic Rights* in Barcelona, Spain, the UDLR was delivered to a representative of the UNESCO Director General. However, the declaration is not constitutional, and unlike the UDHR, has not been passed by the UN General Assembly nor adopted by UNESCO.

There is a real danger that linguistic rights will continue to be ignored until the human rights of minority peoples (e.g., their physical survival) and the conservation of biodiversity have been secured. Meanwhile, linguistic diversity is disappearing — indeed much faster than biological diversity. Some estimates suggest that the percentage of languages that will disappear in the next century is larger than the percentage of all biological species that will be lost over the same time.[45] Twenty per cent of biological species are estimated to become extinct in 100 years, and possibly more than 45 per cent of languages may be dead or moribund (that is, no longer learned by children) over the same period.[46]

There is evidence that biological and linguistic diversity are linked, with a clear overlap between countries with high number of endemic languages and their endemic biodiversity. For example, when comparing the 25 countries with the most endemic languages against the 25 that have the greatest number of higher vertebrates, as many as 16 countries are on both lists.[47] The number of languages and biological species are at their greatest in equatorial regions and decline when moving toward the poles, a remarkable overlap between global areas of biological megadiversity and those of high linguistic diversity.[48] Overlap is similar between less diverse cultural systems and low biodiversity. The links between linguistic and cultural diversity and biodiversity not only correlate but may also be causal. Recent theories of *human-environment coevolution* include the proposition that cultural diversity might enhance biodiversity or vice versa.[49]

Linguistic and cultural diversity are as important as biodiversity to humanity, and in the case of agricultural biodiversity, are indivisible. Endangered languages are the repositories of much of our knowledge of biological resources in the most marginal and diverse environments in the world. Where these resources include food crops, knowledge of their management is stored in the heads and tongues of the growers themselves. It has needed centuries, if not millennia, for people to understand their environments and the complex ecological relationships that are critical for maintaining biodiversity. Consistent with a phenomenon foretold by George Orwell in his novel *1984*, when indigenous peoples lose their languages, much of this knowledge is also lost. If this knowledge is not recorded and translated, dominant languages cannot absorb the biological, medical, and nutritional vocabulary of indigenous people. Their loss is our loss.

Where the language that contains important knowledge is unwritten, the decline of indigenous agricultural knowledge is generational. For example, the death of a parent means that a repository of centuries of knowledge on the growth and management of a particular species is unavailable to children who no longer speak the language of their ancestors and/or may well move to urban centres where specialist agricultural knowledge is not required. Loss of linguistic diversity doesn't simply mean replacing one language with another. The American anthropologist-linguist, Edward Sapir, observed "The world in which different societies live are distinct worlds, not merely the same world with different labels attached".[50] Nor is its consequence solely the loss of knowledge stored in one language that is not transferred to another. It also means a diminished value of *indigenous* or *vernacular* knowledge, which are seen as inferior to the scientific and technical information that is stored in the English language in publications and reports.

We can see that loss of linguistic diversity means losing knowledge in its translation and diminishing the value of such knowledge in its transfer. Worse still is the corruption and possible misappropriation of the knowledge that is transferred. As an illustrative example, imagine a child growing up in a village in East Africa. Throughout her childhood she might learn the experiences and belief systems of elders through the vernacular language of her family — literally her mother tongue. She might then move to a city for further education in which the medium of instruction is Swahili, and then move on to a bachelor's degree that is taught in English. On graduating, she might win a scholarship for PhD studies in perhaps the UK or USA, where she chooses to develop a computer model of an agricultural challenge that affects her own village.

In this process, usually without intent, the knowledge of her own ancestors will be corrupted and devalued through conversion from its vernacular form through Swahili, then English, and finally into a computer language and simulation modelling of the original condition. The final result might still be of value, but it will be the poorer for its loss in translation and transfer. The final irony is that the vernacular language in which the treasure trove of local agriculture is curated is no longer spoken by our student's generation. In Africa, at least, much of the knowledge of indigenous species and systems is in the heads of the farmers who maintain them. There is an African saying: *every time a farmer dies a library goes with her*. Effectively, by losing its language the 'family library' has been burned to the ground before anyone could rescue its books.

Language is not only the repository of knowledge about how to grow individual species, but also of how to store, process, and use their products as foods, medicine, or for social customs such as weddings and funerals. Cuisine, customs, taboos, and belief systems are all associated with specific crops and animals and their combinations into specific products. We can see that the loss of language means not only the loss of agricultural knowledge, but also the customs and practices associated within that language. Together, the loss of a language and the movement of its speakers ruptures the association that people have had with their crops and environments for countless generations. The speed of this loss allows no time to adjust and adapt — with consequences for future generations. We can (and do) maintain the essential genetic material contained within indigenous species through seed banks and stock centres, but the loss of indigenous knowledge in languages that are not codified is terminal; we can only lose an oral language once.

Final Thoughts

Throughout that very brief and turbulent part of human existence represented by agriculture, there has been a tendency away from diversity toward monocultures of species, knowledge systems and societies. These trends have been encouraged and often enforced through linguistic, imperialist, or political hegemony. Each has had lasting effects on the world's biodiversity and our understanding of it. It is difficult to see how these impacts can be reversed but ours is probably the last generation that can try.

Kew Gardens is the home of the world's largest botanical and myco-logical collections.[51] It is one of London's top tourist attractions and is a World Heritage Site. It represents a treasure trove of global biodiversity and the knowledge associated with it. One of its most significant impacts, however, is the role that it played in Britain's economic hegemony in the 17th and 18th centuries and how it helped to reframe its global collection into an English setting. The display of foreign plants within an English environment (albeit inside glasshouses) demonstrated the process of natu-ralization by empire. Flora collected from colonial voyages were brought back, reborn at the heart of the British Empire, and re-organized into the global microcosm of Kew as biological specimens as if with no previous cultural identity. This separation between nature and society enabled land-scapes to be constructed outside their political and cultural context and selective histories to be reconstructed to emphasize the history of Britain rather than that of its colonies. As the biological nerve centre of the colo-nial network, Kew stimulated the monocultural standardization of colonial landscapes throughout the British Empire.

Nikolai Ivanovich Vavilov, the scientists' scientist, dreamt of ending hunger and famine by applying the emerging science of genetics as a *mis-sion for all humanity*. His vision was that the best way to understand the potential for the cultivation of plant species was to identify their centres of diversity as natural laboratories where they evolved until the most suit-able ones developed. At least initially, Vavilov's dream fitted perfectly into the ambitions of the young Soviet state. Lenin endorsed the brilliant Vavilov and his search for genetic treasure on five continents. In what is now St. Petersburg, Vavilov built the first global seed bank with over 250,000 specimens as a living repository of plant diversity.

Vavilov was from the social class that was most detested by the Bolsheviks and the new generation of Soviet scientists and Stalin's secret police charged him with trumped up charges of sabotage and espionage. Meanwhile, Stalin's disastrous collectivization of farms so damaged Soviet food production that millions died of starvation in frequent fam-ines. Vavilov's scientific vision of agriculture, which was designed to work over decades, did not fit into Stalin's political demands for quick wins. Russian geneticists were marginalized in favour of the fraudulent claims and untested theories of Trofim Lysenko.

Vavilov's *Five Continents*[52] is one of the most important books ever on the origins of plants, with the potential to transform our understanding and cultivation of important crops. But Stalin rejected Vavilov's dedication to science. Instead, he promoted Lysenko's claim that by changing the conditions in which a plant is grown, a new species would be created that passed on its new characteristics to its offspring. Lysenko argued that crops that grew in warm climates could be quickly adapted to cold environments and therefore help to secure the Soviet food supply — an alluring prospect. The Lamarckist theory that characteristics acquired during the lifetime of an organism can be passed on to its offspring had been debunked by Darwinism. But in Russia, it was hailed as the new *Soviet genetics* and Lysenko became its most influential advocate. Lysenko dismissed genetics as dangerous nonsense and Stalin developed a five-year plan to collectivize all farms based on Lysenko's theories which, whilst *lacking supporting evidence*, fitted Stalin's view that the Soviet state could control everything. When the collectivization experiment inevitably failed, a scapegoat other than Lysenko had to be found. Between 1934 and 1940, 18 of Vavilov's colleagues were arrested, others denounced him, and he was charged with being an anti-Soviet spy responsible for sabotaging crop production, wasting state funds, creating seed shortages, and disrupting the collectivist plan. Vavilov was imprisoned and died of starvation in 1943, but until Lysenko's demise, in 1954, Soviet agriculture continued to decline in the absence of an independent, outward-looking, vibrant and diverse scientific community. This purge of Russian science set the country back for decades and ended Vavilov's remarkable career, destroyed by demagoguery, blind ideology that rejected scientific method and evidence in favour of political expedience.

Quite rightly, our concern is to maintain the genetic diversity of plants and animals that contribute to global biodiversity. We now have the tools to conserve and protect germplasm, especially the world's cultivated species. We increasingly recognize the importance of conserving the characteristics stored in wild relatives and landraces of modern crop species and halting genetic erosion. However, we have yet to afford the same importance, dignity or effort to halting linguistic erosion. The loss of language brings with it loss of knowledge stored in that language. Unlike genetic resources that we can store and rediscover at a later date, where linguistic resources are vernacular (in people's heads), their loss is irreversible. The last generation to speak a language is the last generation that can transfer perhaps thousands of years of experience to its successors. We may or may not agree on whether the world's languages are a global public good and deserve protection as a human right, but for agriculture they represent the only link in knowledge that we have with over 10,000 years of farming. They may also contain the wisdom and evidence that we will all need to survive the pandemics and climates of now and the future.

Using the metaphor of killer languages may seem alarmist. However, language (written or spoken) underpins the market forces and knowledge systems that govern agriculture. Where there are unifying or *free market* forces, these tend toward centralization, homogenization, standardization, efficiencies of scale, marketing, and consumer behaviour — usually through a single medium (English). The consequences of globalization for linguistic (and cultural) diversity are dire. Globalization brings with it uniformity of processes, products, and exchange in which the English language becomes the only medium of communication, the culture of aspiration, the symbol of modernism, and the metric of progress. The consequence of this monolingual and monocultural tsunami is that we diminish the diversity of agricultural species, their products, and methods of cultivation that we have available for future generations, and crucially, the knowledge and belief systems of the communities associated with them. In the empire of the market, those markets *force* the loss of traditional knowledge. From many thousands of edible plants and spoken languages, humanity now depends on a handful of crops and the knowledge related to them contained in a few written languages. Taken to its logical conclusion, is it impossible to conceive that we could retreat to one primary crop, one language, and one agrifood system. It is not.

Monocultures, whether in their societal or biological form, are inherently unstable and xenophobia and political expedience, whether in their personal or institutional form, are the enemies of science and progress. The rest of this book attempts to demonstrate why diversity, whether genetic, cultural or societal is our greatest ally in the troubled times ahead.

2

EMPIRES OF POWER

Revolution Numbers 2–5

Context

This chapter is about two forms of agricultural power. The first relates to the *physical* power of technologies that reduced the burden of farming on humans, their draught animals, and beasts of burden. The second relates to the *economic* power exerted by humans freed from the physical constraints of farming. Together, both forms of power have enabled conditions for civilizations and empires that were built on revolutions in agriculture. A *civilization* represents a complex form of society that is characterized by urban development, social hierarchy, a distinct form of government, and a common form of communication. A *revolution* represents a fundamental and sudden change in society — one from which a new civilization can emerge. Here we describe examples of civilizations that were built on agricultural revolutions and went on to become *empires* of power.

In much of the world, agriculture gradually evolved from hunter-gathering to sedentary farming, but its advance was sometimes so distinct, drastic, and disruptive that it could be described as a revolution. The term *Agricultural Revolution* often refers to a period during the 18th and 19th centuries when new technologies transformed agriculture, first in England, then in Europe and then elsewhere.[1] While it had a distinct character, the English Agricultural Revolution was built on others that occurred earlier and over longer timescales. It is the consequences of each of these agricultural revolutions that have helped shape the modern world. To a large extent, they have determined who we are, where we are, what we are, and the plants and animals that we now farm, consume and trade, irrespective of our and their origins. This chapter is not a comprehensive analysis of the many revolutions and civilizations that have been built throughout human history. Rather, it uses examples of four revolutions where changes in agriculture provided the catalyst for transformational changes both in the societies responsible for the revolution and, through imperial power, the societies impacted by it. Later chapters consider how the legacies of these revolutions have determined our current agrifood system and lessons from them that can guide those of the future.

Agriculture, Civilizations, and Empires

Hydraulic Empires

The preconditions to establish the first civilizations meant that agricultural communities had to consistently produce enough surplus food to provide the resources for further production increases, the onset of trade, and the stratification of society. Soon after adopting sedentism, human societies recognized that agriculture had to focus on locations and *seasons* of the year when temperatures and rainfall were suitable for specific crops. Altering the seasonal temperature at any one place was not a realistic prospect until much later in human history (for example using glasshouses), but supplementing rainfall was possible if there was access to water for irrigation. The first *foundational* civilizations all appeared where floodplains containing rich soil and their river sources supplied enough water to irrigate crops, provide drinking water, and transport people and their products. Each of these early civilizations developed its own unique characteristics and structures, and all emerged where the geography was suitable for agriculture and water was available for irrigation.

The German scholar Karl August Wittfogel described these early civilizations as *hydraulic empires* in which power was exerted through exclusive control over access to water.[2] By controlling the access of their societies to water, governments could impose centralized coordination of its management through a specialized bureaucracy and a hierarchy based on class or caste. Control over resources (such as food, water and energy) and its enforcement (through the threat of physical force) are vital for the maintenance of centralized control. Wittfogel maintained that government could exert control over these hydraulic empires because river water was critical for economic processes and was limited by the environment. This allowed for a monopoly over supply and demand and prevented the use of other resources to compensate — there is no alternative to water in biological systems. According to Wittfogel, hydraulic hierarchies led to permanent institutions of government, in other words, the control of water justified the formalization of the state with laws, rewards, and penalties. Popular revolution in such a state would have been difficult without the use of force to secure the control of water. Any new regime would resemble the old one since it would use control over the same resource to impose authority. Hydraulic empires usually collapsed either through internal divisions or were destroyed by foreign conquerors with access to greater force than their vanquished foes.

Of course, there was a long period between the beginning of farming around 10,000 years ago and the emergence of the first hydraulic empires around 7,500 years later. Much of this time involved the diffusion of agricultural species and knowledge through the movement of people and their crops and animals to regions far from their centres of origin. Often this diffusion was a mutual process either through exchange, or later, trade. Sometimes it was through coercion, but by controlling access to water, societies could establish physical centres of power that usually ensured acquiescence.

Akkadian Empire (Revolution Number 2)

Early civilizations were founded both on the biological potential of agriculture to produce food and the physical and economic power of those who controlled resources to maximize this potential. Mesopotamia was the site of the Akkadian Empire (from around 2,300 BCE), which is considered to be the first established human civilization and the first hydraulic empire.[3] The Akkadian Empire united the Akkadian speakers of what is now northern Iraq and the Sumerian speakers of what is now southern Iraq under one rule. Physically, the empire was divided into two complementary ecological domains that differed in the availability of water. The fertile alluvial plain of the Tigris marked a line between the rainfed agriculture of northern Iraq and the irrigated farmlands to the south. The agriculture of upper or northern Mesopotamia, which would later become Assyria, received sufficient rainfall to support rainfed agriculture throughout most of the year. In these circumstances, centralized control of irrigation through large institutional estates was not critical. In contrast, to be productive, the rainfed areas of lower or southern Mesopotamia, the lands of Sumer and Akkad, which later become Babylonia, needed large irrigation systems that were supervised by temple estates. Here, control and management of water provided the justification for a social hierarchy based on centralized government, bureaucratic oversight and fealty.

The Tigris and Euphrates, their tributaries, and Mesopotamia ("the land between the rivers") spanned all that lay between the foothills of the Iranian mountains to the northeast, and the deserts of the Arabian Plateau to the southwest. Over thousands of years, the Tigris deposited alluvial sediments that built up an immense fertile plain in southern Iraq. When irrigated, this rich soil allowed enormous agricultural surpluses to be generated. Cities in ancient Sumer were located in these fertile areas that were favourable for irrigated crops and sedentary agriculture, separated by landscapes that were suitable for herding and pastoralism. Without irrigation, agriculture in southern Iraq was restricted to the areas immediately alongside the banks of the rivers. However, an integrated system of water distribution allowed agricultural areas to expand and at the same time divert excess water that would otherwise cause flooding. Large canals were cut directly from the rivers to divert water through a network of smaller canals and when their water levels were high enough they were navigable for trade and communication.

The irrigation system also included raised canals and occasional aqueducts. Regulating mechanisms controlled the flow and level of water for irrigation, for example, into bunded basins where crops could be grown. Because of sediments carried by the rivers, their beds were raised above the surrounding floodplain, and water could be delivered to prepared fields using just gravity. However, methods for raising water, such as the *shadoof* (a hand-operated device) or the *noria* (a water wheel) also allowed water to be lifted directly from a river or well and distributed via gravity to prepared plots.[4] Because the agricultural potential was so high, irrigation was adopted not just to ensure crop survival but also to maximise yields. Authorities prioritized the maintenance, repair and dredging of irrigation infrastructure to ensure that water was continuously available. Their main crops were barley and wheat, but irrigation allowed Sumerians to grow date palm, pea, bean, lentil, cucumber, leek, lettuce, and garlic, as well as fruit such as grape, apple, melon, and fig, even where the centres of diversity of some of these crops were not from the same region (for example, apple from central Asia).[5]

The power of water for irrigation allowed the Akkadian Empire to flourish — lack of it caused its demise. From around 2,200 BCE, the terminal decline of the Akkadian Empire coincided with the failure of rains and the increasing aridity and salinity of the soil. River water from the Tigris carried dissolved mineral salts that compromised the once-fertile soil. Over-irrigation caused salts in the soil to rise to the surface and affect the growth of salt-sensitive crops. For this reason, cultivation gradually shifted from the preferred wheat to the more salt-tolerant barley. Leaving land *fallow* in alternating years was a common strategy for coping with the accumulation of mineral deposits in the soil. However, population growth and economic pressures caused farmers to reduce the frequency of fallows and ultimately to abandon the once *fertile crescent* for agriculture.

After four hundred years of urban life, the sudden disruption caused by climate change and human activity meant the collapse of agriculture and the loss of subsidies that had kept the Akkadian Empire viable. The water levels of the Tigris and Euphrates fell, disputes between pastoralists and farmers increased, and enemies invaded. The empire finally collapsed through invasion by the Gutians from the Zagros mountains of what is now southwestern Iran. Despite their newly captured region being a rich source of agricultural biodiversity, the Gutian administration showed little interest in agriculture, the need for written records, or responsibility for public safety. The result was famine and economic collapse. After the Akkadian Empire collapsed, the people of Mesopotamia split into two nations: Assyria to the north and Babylonia to the south. The bridge that the power of agriculture and water had to build and sustain an Empire between differing cultures and languages was broken.

Roman Empire (Revolution Number 3)

At its height, the Akkadian Empire exerted its influence across Mesopotamia, the Levant, and Anatolia.[6] At the time, this empire was comparable in scope and extent to those of Egypt, Greece, and Carthage in what is now Tunisia. It was the Roman Empire, however, that extended further and for longer than any of its predecessors. For almost 1,500 years between 27 BCE and 1403, the Roman Empire expanded from a city state to a contiguous area of 5 million square kilometres and a population accounting for between one-sixth and one-fourth of humanity at the time.[7] Not only did the Romans occupy these far-flung lands, they absorbed and adapted agricultural methods and systems from those they conquered. Earlier civilizations had been centred on rivers and their floodplains, but the Romans used their engineering skills to distribute water, especially through networks of aqueducts, to locations far from the original water source. Aqueducts were not unique to the Romans. Throughout history, many agrarian societies have built aqueducts to distribute water to irrigate crops and supply drinking water to major cities. In fact, many aqueducts were constructed in Greece and the Near East and Indian subcontinent, where the Harrapan people built sophisticated irrigation systems. However, it was the Romans who created the vast networks of aqueducts through which much of its economic power was expressed both from the area that was irrigated and the productive power of its irrigated agricultural systems.

Rome was still quite small when it became a republic in 509 BCE. In 396 BCE it captured the Etruscan city of Veii and surrounding areas just to the north of Rome. Over the next 250 years, Rome occupied the whole of the Italian Peninsula either by conquering territories, making them independent allies or absorbing them and their culture by awarding Roman citizenship. It was from the Etruscans that the Romans inherited the alphabet and many of their own cultural and artistic traditions. From others, they adopted new agricultural practices, species, and systems of land ownership.

Until the 5th century BCE, farms in Rome had mostly been family-owned. Through contact with Carthage and Greece, Rome adopted new agricultural methods that required consolidation of farmland into larger estates. Wealthier Romans purchased land from peasants who were no longer able to make a living from farming. The three Punic Wars between Rome and Carthage from 264 BCE to 146 BCE meant that Roman peasant farmers had to sell their lands that had lain idle for long periods while they were at war. As Rome started its conquest of overseas territories, it switched from a strategy of absorption and consolidation to one in which captured territories became provinces with their own governors appointed by Rome to oversee them. During the Second Punic War, Rome captured the capital city of Carthage and enslaved its inhabitants. It also absorbed all of Carthage's territory in North Africa and designated it as a Roman province that could provide agricultural supplies to Rome. Over the next century, Rome conquered all the coastal territories that surrounded the Mediterranean Sea. From here, Rome expanded into the Middle East and captured Jerusalem. It largely left the complex political systems of these territories intact while it absorbed suitable agricultural crops and management practices. During the following decades, Roman soldiers also occupied northwest Europe. Here, the Roman approach was different because these territories didn't have well established political systems. Instead, Rome introduced its own systems and agricultural methods while devolving power to local leaders.

At the time, the Roman Empire was the largest in history with contiguous territories throughout North Africa, the Middle East and Europe. As well as annexing large territories, the Romans altered their physical landscapes. Whole forests were cleared to provide enough wood for the expanding empire and new cities built to manage large agricultural estates. To control their expanding and increasingly diverse agricultural land bank, the Romans established four management systems — direct labour by the farmer and his family, leasing to a tenant, division of agricultural produce between owner and tenant, and slave labour supervised by slave managers or overseers. Whilst a few farms remained in the hands of poorer citizens and returning soldiers, most of the land was controlled by Rome's nobility which coveted land to distinguish it from the lower classes. Even though they owned most of the land, Roman aristocrats were rarely seen on their farms. Instead, these were managed by slaves and freed men who they paid to oversee them. Slaves were cheap and considered as property and farms were largely manned by slaves and managed by overseers whilst more and more Romans moved to the cities.

As the number of rural landowners decreased, power gravitated to the city of Rome from where an urban political elite ruled the empire. Political power resided in Rome, and agriculture rather than commerce was seen as the proper business of the senatorial class. As well as owning as much land as possible it was a point of pride to grow the best produce. This pride is not surprising because Italy has a range of natural conditions to produce a plentiful and varied food supply, and somewhere within the agroecology of Italy conditions exist that are suitable for almost all the grains and fruits of the temperate and sub-tropical zones that the Roman Empire occupied. The earliest inhabitants of the peninsula had already identified indigenous wild fruits and nuts to which the Romans added imported crops and animals that were suited to its benign climates.[8] In a very real sense, all crop roads led to Rome.

The main staple crops of early Rome were adopted crops such as millet and ancient species of wheat such as emmer and spelt.[9] Hard durum wheat became the preference of urban Romans who could use it to bake leavened bread and it was also more productive in the local region than soft wheat. Barley was also grown extensively on impoverished soils where it produced higher yields than wheat and was used both for human food and animal feed. Legumes included pea, bean, lentil, lupin, cowpea, and chickpea, none of which were indigenous to Italy but all of which were suited to local climates and seasons. The Romans also grew olive in areas of poor and rocky soils with little rainfall — conditions that were similar to the likely centre of diversification 6,000 years ago in Assyria. Viticulture was probably brought to southern Italy and Sicily by Greek colonists, but the Romans developed their knowledge of wine making from the Phoenicians of Carthage. The Romans also grew a huge range of vegetables including artichoke, mustard, parsnip, gourd, leek, celery, asparagus, radish, cucumber, onion, fennel, cabbage, lettuce, herbs such as coriander, rocket, chive, basil, mint, thyme, dill, caper, parsley, marjoram, cumin, garlic, and fruits such as fig, apricot, plum, mulberry, and peach.

Many of the crops grown by Roman farmers are still associated with modern Italian cuisine; their local cultivation was due to the suitable and varied climates of Italy. That they were adopted by the Romans was due to the commerce and movement between provinces of the empire. Some regions specialized in cereal production, including wheat, emmer, spelt, barley, and millet whilst others prioritized wine or olive oil, where their soil types were suitable. The main sources of grain to feed a Roman population of around one million at its peak were Egypt, northern Africa, and Sicily. With its incorporation into the Roman Empire, Egypt became the main supplier of grain to Rome. To meet this demand over 6,000 colonists were settled near Carthage and each given about 25 hectares to grow cereals. From their northern and western territories such as Gaul and

Britannia, the Romans brought turnip which was used as a food and also as winter fodder for cattle. In return, they introduced over 50 new food plants, many of which they were familiar with from the south and east of their empire.

The rule of Emperor Constantine helped accelerate the fall of a divided and unmanageable Roman Empire.[10] Again, the failure of agriculture was complicit. In 330 BCE, Constantine divided the empire into two — the western half was centred in Rome and the eastern half in Constantinople, (now Istanbul) which he named after himself. The western half of the Empire spoke Latin and was Roman Catholic and the eastern half spoke Greek and was Eastern Orthodox. Whilst the east continued to thrive, the west declined. After the fall of the western part of the Roman Empire, the eastern half continued for centuries as the Byzantine Empire. As the west declined in economic power, its agricultural production also fell resulting in higher food prices and leaving it with a large trade deficit with the eastern half. Whilst the west continued to purchase high value products from the east it could offer little in exchange. To make up for the deficit, the government reduced the silver content of its coins which caused inflation. Incursions from Germanic tribes disrupted trade routes in the west and the Roman Empire was ravaged by barbarian attacks from Visigoths, Vandals, Angles, Saxons, Franks, Ostrogoths, and Lombards, who eventually carved out their own territories in which to settle. The Angles and Saxons populated the British Isles, and the Franks occupied much of what is now France.

In 476 BCE, Romulus, the last Roman emperor of the western Empire, was overthrown by Odoacer, who became the first Barbarian ruler of Rome. More than a millennium of stability that the Roman Empire had brought to western Europe was over. Like the Gutians who destroyed the Akkadian Empire, the Visigoth invaders had little interest in agriculture and they established regimes in which rulers, nobility, and church owned most of the land. The burghers, who were responsible for municipal affairs, owned less than 10 hectares each, while serfs who cultivated the land were sold with it. Essentially, there was little reason to farm. The huge and productive agricultural estates that had fed Rome and its empire were no more.

Muslim Empire (Revolution Number 4)

During the Middle Ages (c. 5th–15th century) there was demographic, cultural, and economic decline in Western Europe that followed the collapse of the Roman Empire. Meanwhile, from the East, a new force was emerging — Islam. Less than 100 years after Islam was established, an area spanning the foothills of the Pyrenees to the frontiers of China had become one of robust economic and cultural development under Muslim influence. In this period, Muslims underpinned their swift expansion with huge advances in agriculture, medicine, and science. The development of agriculture was largely due to investment in irrigation. The Bahr Yussef canal was cut to connect the Nile with the Fayyum depression 25 km away, where irrigated cultivation was developed using animal, water, and wind power. Wind pumps were also used to pump water in what is now Iran, Afghanistan and Pakistan. In Islamic Spain (*al-Andalus*), the sakia irrigation wheel which used animal power was introduced in the 7th century and then spread further around Spain and Morocco. Under the Visigoths, the agriculture of the Iberian Peninsula had barely been at subsistence level and roles were defined by race. The Visigoths closely guarded their livestock interests while their subjects grew wheat, barley, grape, olive, and the vegetables that they had inherited from their earlier Roman masters. The only connection between the two systems were through tribute or taxes.

After Muslims took control of al-Andalus, they introduced new crops and expanded irrigation. Crops were introduced from Persia and India, including rice, sugar cane orange, lemon, banana, saffron, carrot, apricot, and aubergine, as well as the renewed cultivation of olive and pomegranate from Greco-Roman times.[11] Many of the high value crops such as sugar cane, banana, and cotton required irrigation through which the Muslims transferred their experience to new locations. For example, irrigation required devices to raise water by several metres to guarantee gravitational distribution across the whole system. Various forms of water wheels, or *noria*, already in use in the Middle East, were introduced as part of complex irrigation systems with centralized control of access and distribution. Such technologies were not novel, but the correct calculation of water levels was essential to ensure uniform distribution of water across the system. Muslims had the advantage of their advances in mathematics that made triangulation possible to accurately measure the height of water levels for irrigation.

The experience gained by Muslim traders enabled the spread of different farming techniques and crops across the Islamic world, the adaptation of crops to local environments and the modification of techniques from and to regions outside it.[12] In their travels, the Muslims had come across crops that were previously unknown to them and their merchants brought back new plants, spices and herbs from their many voyages. Crops from Africa (sorghum), China (citrus fruits), and India (mango, rice, cotton, and sugar cane), were distributed across the Muslim world. These introductions, along with irrigation, led to major changes in the economy, agricultural production, labour distribution, cooking, and clothing in the Islamic world. In addition to irrigation scheduling, the Muslims established agronomic trials based on soil and crop types, supported by written records and local knowledge. From their history as travellers, they linked knowledge gained from other locations with local experiments. A new generation of scholars conducted their own experiments and transferred their results through mosques and weekly markets in various languages. For example, Ibn Baytar's work was translated into Arabic, Berber, Greek, and Latin, while Al Biruni's *Pharmacopoeia* provides the names of drugs in Syriac, Persian, Greek, Baluchi, Afghan, Kurdish, and Indian dialects.[13] The Books by Ibn Bassal and Abū l-Khayr al-Ishbīlī, enabled scientific knowledge of useful plants to be widely disseminated and contributed to the growing Islamic scientific knowledge of agriculture and horticulture.[14]

As well as advances in the production and management of new and more diverse crops, perhaps the biggest impact of the Muslims was the social transformation that they introduced through changed land ownership. They recognized that real incentives were needed if increases in productivity could significantly enhance wealth and thereby generate tax revenues. Rather than the systems of ownership, tenancy, sharecropping, and slave labour practiced by earlier civilizations that had been enforced by penalties, the Muslims brought revolutionary incentives by introducing the rights of any individual to buy, sell, mortgage, inherit, and farm land according to their own preferences. Further, transactions that related to agriculture, industry, business, and even the employment of a servant required a signed contract with copies that were kept by each party. Another incentive was that those who farmed the land should receive a fair proportion of their outputs based on agreed contractual records between landlords and cultivators.

Underpinned by these incentives, two major developments in Muslim agriculture became possible. First, was a political decision to develop under-exploited lands to expand the land bank for agriculture. Second, was the introduction of new high-value crops and animals supported by advice and education — effectively the first agricultural extension service. Crops and livestock were initially introduced to ensure the subsistence and economic security of farmers and their families. The quality of diet and commercial opportunities were then enhanced through new crops such as artichoke, spinach, aubergine, carrot, and sugar cane. Vegetables became available throughout the year, reducing the need for drying and storage over winter. Citrus and olive plantations, market gardens, and orchards appeared near cities. These changes involved management practices that reduced soil fertility, but the techniques of soil, water, and nutrient management had now been mastered, and animal husbandry was incorporated not just for meat but also manure. Thus, in less than a century of the Muslim conquest, the lands that they absorbed had changed so radically that this transformation could be described as the Muslim Agricultural Revolution.

The successful elements of this revolution included an expansion in the land area under irrigation, improved agronomic management based on experimental evidence, changes in land ownership to reward cultivators with a harvest share that reflected their efforts, and the introduction of new crops. The advances in scientific techniques allowed agriculturalists to introduce and adapt crops and livestock breeds into new areas, supported by irrigation and improved soil management. Advances in astronomy also provided farmers with information about the time and onset of the seasons from solar movement through each zodiacal sign. This allowed for compilation of calendars on when to plant each kind of crop, when and how to fertilize crops and harvest their products.

It now became possible for a farmer to grow crops for a specific market at a particular time of the year. The dramatic spread of Islam to three continents in the 7th and 8th centuries led to the diffusion of an equally remarkable but less recognized agricultural revolution. A new form of agriculture, starting mainly in India where temperature, rainfall, and available crops all favoured its development and where it had been practised for some centuries before Islam, was carried by the Arabs or those they conquered to colder and drier lands through experiment and observation. By the end of the 11th century it had been transferred across the Islamic world and transformed the economies of much of central Asia, Persia, Mesopotamia, the Levant, Egypt, the Magreb, Spain, Sicily, the lands below and above the Sahara, parts of West Africa, and coastal East Africa.

In 1206, Genghis Khan built a powerful Mongol dynasty in central Asia and this Empire occupied most of the land mass of Eurasia, including China in the east and the Islamic caliphate in the west. In 1258, Baghdad and the House of Wisdom were destroyed by Hulagu Khan, grandson of Genghis Khan, and this marked the end of the Islamic Golden Age. The Grand Library of Baghdad with its repository of documents and literature on subjects ranging from medicine to astronomy and agriculture was destroyed. Survivors reported that the waters of the Tigris ran black with ink from the sheer number of books flung into the river.[15] Muslim scholarship has never recovered and nor, sadly, has the prestige of its agricultural sciences or scientists.

British Empire (Revolution Number 5)

The British Empire included dominions, colonies, protectorates, mandates and other territories ruled or administered, first by England and after its unification with Scotland in 1707, the Kingdom of Great Britain. In its time, it became the largest empire in history, and for over a century, was a global superpower. By 1913, the British Empire controlled over 412 million people, almost a quarter of humanity and by 1920 covered 35,500,000 square kilometres, a quarter of the Earth's land area.[16] At its maximum extent, the British Empire was more than seven times larger than the Roman Empire and over 700 times larger than the ancient hydraulic empire of Sumer and Akkad. The phrase "the empire on which the sun never sets" was used to describe the British Empire because its extent around the globe meant that the sun was always shining on at least one of its territories.[17] The start of the British Empire was in the early 1600s. The transfer of Hong Kong to China in 1997 probably marked its end.[18]

Before its expansion to distant territories, Ireland provided a test case for Britain's colonial ambitions. In the 16th and 17th centuries, *the plantation of Ireland* involved confiscation of land from Gaelic clans and its transfer by the English crown to settlers or *planters* from England and Scotland. In the 1650s the British government established the English Commonwealth and Cromwell's Protectorate through which thousands of Parliamentarian soldiers were settled in Ireland. The plantations that they established changed the demography of Ireland by creating large British and Protestant communities within the Irish Catholic population. Their gentry replaced the older Catholic aristocracy which had shared a common heritage with the local population. The new ruling class also brought with them different concepts of ownership, trade, and credit and during the 17th century created a Protestant Ascendancy, which secured the authority of British rule of Ireland from Dublin. The British also transformed much of the Irish countryside into extended grazing land to raise cattle for their own expanding consumer market. This demand for beef was devastating for the impoverished people of Ireland who were evicted from the best pasture land and forced onto marginal land where they turned to the potato as their staple means of survival.[19]

The Caribbean was the home of England's most profitable colonies when settlements were first established in St. Kitts (1624), Barbados (1627), and Nevis (1628). These new colonies quickly adopted the system of sugar plantations that had been developed by the Portuguese in Brazil, based on slave labour. In 1655, England captured the island of Jamaica from the Spanish, and in 1666 annexed the Bahamas. In 1660 the newly inaugurated Royal African Company was granted a monopoly by King Charles to supply and trade slaves to the British colonies of the Caribbean. From the outset, slavery provided the basis for the British Empire in the West Indies. Until 1807 it is estimated that Britain transported over 3.5 million African slaves to the Americas — a third of all the slaves that were shipped across the Atlantic.[20] To enable this trade, they built forts on the West African coast from which slaves were transported to the Caribbean on slaving ships in harsh and unhygienic conditions where one in seven died. Between 1680 and 1786, over 600,000 slaves were transported to Jamaica alone and in Antigua the proportion of slaves in the total population jumped from 48 per cent in 1678 to 84 per cent in 1720.[21] The extremely profitable trade in slaves became a major part of the economy of English port cities such as Bristol and Liverpool, which formed the third element of the triangular trade with Africa and the Americas.

By the late 18th century, goods produced through slavery had decreased in value to the British economy by more than the costs of suppressing regular slave rebellions. In 1807, the British Parliament approved the Slave Trade Act which prohibited but did not abolish the slave trade in the British Empire.[22] In 1808, Sierra Leone was established as British colony for freed slaves. In 1833, the Slavery Abolition Act abolished slavery in the British Empire and granted full emancipation to slaves after four to six years of *apprenticeship*. After further challenges from abolitionists, the apprenticeship system was abolished in 1838.[23] However, rather than the slaves themselves, it was the slave owners who received compensation from the British government.

The East India Company was the driving force behind British expansion into Asia.[24] Its private army, which had previously worked with the Royal Navy during the Seven Years' War, now extended this co-operation beyond India to include the acquisition of Penang Island in what is now Malaysia (1786), the removal of the French from Egypt (1799), and the Dutch from Java (1811), the occupation of Singapore (1819), Malacca (1824), and the defeat of Burma (1826). Since the 1730s the East India Company had already used its base in India to engage in the highly lucrative opium trade to China. This illegal trade helped balance the trade deficits from British tea imports from China. In 1839, the confiscation of 20,000 chests of opium by Chinese authorities at Canton was used as the justification for Britain to attack China in the First Opium War. This led to Britain's seizure of Hong Kong Island, which at that time was only a minor settlement.

By the early 19th century the British Crown assumed a greater role in the business of the East India Company. The company's affairs were regulated through a series of Parliamentary acts which established British sovereignty over the territories that it had captured. The Indian Rebellion in 1857 precipitated the demise of the company and resulted in casualties on both sides. In 1858, the East India Company was dissolved by the British government which then took direct control over India through the Government of India Act which established the British Raj, and appointed a governor-general to administer India. Queen Victoria was crowned Empress of India which became the empire's most treasured possession and largest single source of Britain's wealth. William Dalrymple's '*The Anarchy*' vividly, forensically and devastatingly describes how one of the world's most magnificent empires was replaced by a dangerously unregulated private company that was answerable only to its shareholders.[25]

English Agricultural Revolution

How did a small country detached from the north-western fringes of Europe colonize territories covering one quarter of the planet? Through the technical power of agriculture and industrialization and the economic power of empire. Advances in agricultural production through the English Agricultural Revolution released a workforce from farm to factory that could deliver the new technologies of the first Industrial Revolution. Together, they provided the catalyst for colonial expansion and a demand for raw materials that followed. It is estimated that between 1700 and 1870, total agricultural output and average output per worker in Britain each grew by about 270 per cent while the agricultural labour force decreased from 35 per cent in 1800 to 22 per cent by 1850, the smallest proportion for any country in the world.[26]

In the 1700s, English agriculture was transformed as landowners reclaimed and enclosed land, shifted to arable systems, reduced years of fallow, introduced fodder crops, diversified the range of food crops, managed soil nitrogen and moisture content, introduced mechanization and made profits. By 1800, England's unique rural structure allowed landlords to lease land to tenants who employed landless labour to grow crops and keep livestock under new systems of agricultural production. As a consequence of these transformational changes, agricultural surpluses provided the impetus for feudal landowners to become capitalist farmers. This agrarian capitalism led to better farm management and a more efficient workforce. Rather than any single factor, several contributed to the transformation that became known as the English Agricultural Revolution.

A critical stage in the transformation of English agriculture was the relationship between landowners and land. A series of acts were passed by Parliament that established *enclosures* through which the peasantry was excluded from open fields and *wastes*.[27] Open fields had been areas of agricultural land to which villagers had certain rights of access. The wastes were less productive areas such as fens, marshes, stony soils or moors to which the peasantry had traditional and collective rights of access to pasture animals, harvest meadow grass, fish, and collect firewood. By enclosing them, many of these wastes were reclaimed for arable agriculture. For example, draining the fenlands of eastern England allowed a simple system of agriculture based around fishing and fowling to be replaced by an intensive system of arable crops. Elsewhere, woodland was cleared and upland pastures enclosed. By demarcating certain areas from common access, enclosures allowed a number of small landholdings to be consolidated into a single unit of land. Once they had been enclosed, any use of the land became the prerogative of the owner rather than available for communal use.

Enclosures ended the ancient system of crop production in open fields. Prior to the enclosures, land was organized into many narrow strips where a tenant might farm a number of disconnected strips. In this *open-field* system, a single field was divided between the lord of the manor and his tenants. Poorer landless peasants had been allowed to live on the strips owned by the lord in return for working on his land. This arrangement allowed common grazing and small-scale cropping by tenants on several strips of land. To increase their financial returns, landowners looked to consolidate a number of strips into larger units so that they could utilize innovations in crop production, increase profits through higher productivity, and increase rents for tenants. This left many farmers with plots that were too small, too poor, or too distant to be viable. They became landless with no choice but to migrate to the city in search of work. This flight to the cities occurred at a time when technological advances in manufacturing needed large numbers of workers. Instead of farming, rural communities and their descendants became factory workers in cities. In many cases, their efforts in manufacturing new farming implements and processes further reduced the demand for rural labour in their own home towns and villages. Between 1800 and 1900, the English urban population increased from 17 per cent to 72 per cent, many of whom had abandoned farming for factory work.[28]

Through the Enclosures Acts and the redesignation of land use that they enabled, English agriculture changed from low-intensity systems based on small-scale cropping and pasture to high-intensity systems based on the production of arable crops. Output per unit of land increased through new farming systems that included the rotation of crops between seasons on the same land. The Norfolk four-course rotation, introduced a four-year cropping cycle of wheat, turnip, barley, and clover or grass. In the 18th century, Charles Townshend (Turnip Townshend) popularized this system that had first been established in Waasland in what is now Belgium. A sequence of four crops that included a crop for fodder and another for grazing meant that livestock could be fed throughout the year. The balance between arable and pasture also changed so that more productive arable land replaced grazing.

The loss of pasture for grazing was compensated for by turnips and clover for fodder. These new fodder crops were also involved in the reclamation of lowland heathland from rough pasture to arable production. The types of crop also changed with wheat and barley replacing lower yielding types such as rye. Cereal yields also increased. From about 1830, wheat yields increased dramatically once soil nitrogen was identified as a limiting factor for crop growth. Existing stocks of soil nitrogen were exploited by ploughing up permanent pasture for cereals. The soil nitrogen content was replenished through the use of leguminous species which could convert atmospheric nitrogen into nitrates in the soil for use by crops in the succeeding years. Legumes such as pea, bean and vetch had been used since the Middle Ages but from the mid-17th century farmers began to also grow red and white clover to increase soil nitrogen, and by the 19th century, these species had significantly increased the amount of nitrogen in the soil available for cereal crops.

Industrial Revolution

Industrialization represents perhaps the most significant event in human history since the start of farming.[29] Before the Industrial Revolution, there was little rural-urban migration because cities simply lacked the living conditions, employment opportunities, and food supplies for urban growth. It took decades before the impacts of industrial development triggered mass migration to cities. Nevertheless, as food supplies increased, supply chains stabilized, and with industrialization, cities began to absorb larger populations, sparking rural migration to urban centres. For this to happen, British agriculture had to adopt new technologies that could increase productivity and reduce the need for farm labour through mechanization. In 1701 (long before his musical success!) the English lawyer Jethro Tull developed a seed drill that could distribute seeds evenly and at the correct depth. In 1730, Joseph Foljambe's Rotherham plough was the first iron plough to be commercially produced.

Although the existing *ard* or scratch-plough was suitable for the soils and climates of the Mediterranean, it was not suited to the heavier soils of northern Europe. As a consequence, much of northern European agriculture was limited to lighter soils where the ard could be applied. The heavy plough helped overcome this limitation. In 1784, the threshing machine developed by the Scottish engineer Andrew Meikle reduced the need for labour to hand thresh grain. By the 19th century new machine tools and metalworking resulted in precision manufacturing techniques to mass produce agricultural equipment such as reapers, binders, and combine harvesters. For agriculture, this transition meant moving from manual production to machines, the manufacturing of chemicals, processes using iron, greater efficiencies in the use of water power, the introduction of steam power, development of machine tools, and the start of the industrial factory system for food processing and products. It also provided the resources to develop markets.

Markets

Larger consolidated units of land, facilitated by enclosures and new mechanical technologies, meant that a class of landowners could now become capitalist farmers with agricultural products that could become part of a regulated, private market. By the 19th century, rather than support the subsistence of the farmer, his family, and labourers, most agricultural production was destined for markets, initially locally, then nationally and ultimately internationally. This new agricultural trade required merchants, the use of credit and forward sales, market intelligence, pricing, and control of supply and demand for target markets. This trading led to a national market driven by demands of cities that were growing both by the influx of new labour from the countryside and rapid population growth supported by increasing affluence. Commerce benefitted from the expansion of roads and inland waterways so that goods could be transported rapidly beyond local markets. This freed producers from price fluctuations from over — and undersupply locally and offered an opportunity to sell surpluses elsewhere. Farming was now a business and not solely a means of subsistence.

Supply chains were an essential component of the new private markets. Even with expanding urban demand, better transport and marketing were only viable if surpluses were large enough, costs were low enough, and supply chains were sustainable. One way was to reduce transport costs through improvements in river navigation and roads to transport grain. By 1700, there was a national market for wheat. Legislation was introduced that required middlemen to be registered and enabled standard weights and measures, pricing, and tax collection by the government. In the late 18th century Britain was the largest coherent market in Europe. By the 19th century it was the largest economic force in the world.[30]

Capitalist Farmers

Changes in land use and adoption of new technologies were outward manifestations of the English Agricultural Revolution, but its foundations were built on a new means of measuring success — the pursuit of profit. From the mid-17th century, a new mentality drove feudal landowners to become capitalist farmers. This represented a broader national transition from feudalism to capitalism demonstrated by Britain's more rapid increase in labour productivity compared to continental Europe. Rather than inputs and outputs, success was measured by the surplus value to be extracted from the use of waged labour to produce commodities or services. In turn, this surplus allowed investment in private ventures, exploration, production, and trade through which capital from the land flowed through trade, and capital from trade flowed back into land owning and assets from it. Both of these flows required the systematic organization of labour. They also facilitated the conditions for colonial expansion in search of raw materials and, where necessary, the use of force to secure access to them and, where possible, new rules to control their trade. This agricultural transformation allied with its expanding technological and economic power provided the conditions and incentives to establish the British Empire.

Trade and Empire

During empire, Britain engaged in *triangular trade* with its colonies — natural resources, goods, and slaves were shipped across the Atlantic to benefit the colonial power. Mercantilism, designed to maximize exports and minimize imports, was implemented as an economic policy to promote colonialism and secure resources, tariffs to restrict imports, and subsidies to promote exports and achieve a positive balance of trade. By establishing colonies, the empire promoted its mercantilist goals through securing a cheap and ready supply of raw materials to English factories and a captive market for their finished goods. The government made its profit through taxes and customs duties on trade.

In principle, a ship could travel to Africa laden with cargo, buy slaves, sail to Barbados where it would sell slaves and buy sugar, return to England to sell sugar and buy guns, and then return to Africa to sell guns and buy more slaves. Before their independence, the British colonies of North America sent raw commodities such as rice, tobacco and timber to Europe for which they exchanged manufactured goods and luxury items. Europe also transported guns, cloth, iron, and beer to Africa in exchange for which it secured gold, ivory, spices, and hardwood. The main export from Africa to North America and the Caribbean was slaves who were forced to work on colonial plantations. In this way, goods and people flowed back and forth between Europe, Africa, and North America in the system of transatlantic trade. This Atlantic economy was driven through the unpaid toil, misery and bondage of enslaved labourers who worked on sugar plantations in islands such as Barbados and Jamaica. Some of those who survived the sugar plantations would be sold north to Virginia or the Carolinas to grow tobacco or rice. Although relatively few slaves worked in the colonies of New England, these colonies helped to maintain the system of colonial slavery by sailing slave ships and selling provisions and goods to the slave traders.

In 1651, the British Parliament approved the first of the Navigation Acts, whose purpose was to control the terms of trade between Britain and its colonies.[31] The Navigation Acts specified that *enumerated* goods (high value commodities, such as sugar, tobacco, and cotton) could only be moved on British ships and sold in British ports. This allowed the British government to collect customs duties on imported goods, British merchants and shipbuilders to make profits from the business, and Britain to rule the waves — literally. The history of the cotton textile industry provides an example of this process.

Cotton: Nexus of Agricultural Revolution, Industrial Revolution, and British Empire

The word *cotton* is derived from the Arabic *qutn* or *qutun*.[32] A long-fibred ancestor of cotton, *Gossypium barbadense*, was first domesticated in Peru from where the oldest cotton fabric can be traced to around 6,000 BCE. By around 3,000 BCE, cotton was being cultivated in the Indus Valley in what is now Pakistan and by the 6th century was being processed into fabric in India using cotton gins.[33] A cotton gin (*cotton engine*) is a machine that separates cotton fibres from their seeds after which the fibres are processed into various cotton goods such as calico. The remaining material is used to make textiles for clothing. The spinning wheel, a device for spinning cotton fibres into thread or yarn, was introduced from Iran to India in the 11th century where it was used by the local cotton textile industry until the Industrial Revolution. Cotton textiles were first brought to Europe when the Muslims conquered the Iberian Peninsula and Sicily. Expertise in cotton weaving diffused to northern Italy during the 12th century when the Normans conquered Sicily, and from there to the rest of Europe. The spinning wheel was first introduced to Europe in the mid-14th century and by the 15th century, Venice, Antwerp in Belgium, and Haarlem in the Netherlands were important ports for the cotton trade.

India, China, Central America, South America, the Middle East, and Muslim Spain all have long histories of hand manufacturing cotton textiles. In India, high-quality cotton textiles of exceptional quality were manufactured for distant markets. Cotton textile manufacturing of calicos and muslins was the largest industry of Mughal India, accounting for 25 per cent of the global textile trade in the early 18th century.[34] Bengal produced over 50 per cent of the cotton textiles exported to the Netherlands and substantial quantities to the rest of Europe, Indonesia, and Japan. By the late 17th and early 18th century, India supplied 95 per cent of the cotton that was transported to Britain through the East India Company mainly as cheap cotton calico fabrics that served a growing mass market for the urban poor. By 1721, to remove the threat that these imports represented to British manufacturers, Parliament passed the Calico Act.[35] The act banned the use of calicoes for clothing or domestic purposes. To compete with India, Britain invested in mechanization and introduced bans and tariffs to restrict imports. In 1774, the Calico Act was repealed by which time the development of industrial technologies and cheap labour had allowed British manufacturers to compete with imported fabrics. The East India Company's occupation of India also opened up a new export market for British goods. These could be sold in India without tariffs or duties and raw cotton could be imported from India to British factories. This gave Britain control over both India's significant export market and cotton imports from India for its own factories. By the 19th century, Britain had replaced India as the world's leading manufacturer of cotton textiles.

The expansion of the English textile industry in the 18th century illustrates the links between labour released through the English Agricultural Revolution, new technologies developed through the Industrial Revolution, and British colonial power to secure access to raw materials, ensure their supply, and set rules for trade of their finished products. By the late 18th century, textiles had become the predominant activity of the Industrial Revolution in relation to employment, business, and investment. The textile industry introduced modern methods such as mechanized spinning to the textile mills of Lancashire. However, mill workers, including children, had miserable working conditions, low wages, and long working hours. Richard Arkwright built his textile empire through factories powered by water and cheap labour and in the 1790s, steam power was introduced for textile production. Now, with these advanced industrial technologies, a cheap and plentiful workforce, and a captive export market, all the British textile industry needed was raw cotton.

Cotton, however, was not a viable crop for temperate European climates. In tropical and subtropical areas it was mostly grown by small farmers alongside their food crops, which meant that it was difficult for Europe to obtain cotton in sufficient quantities for its textile industries. An alternative system was needed in which single commodity crops could be grown in many of Britain's colonies under a uniform system of management, with the investment and economies of scale to make raw materials viable for the needs of the British Empire. That alternative system was the plantation.

The Plantation

The plantation system fuelled a huge increase in the supply chains of various commodity crops as plantation owners could make their products affordable for distant markets. For this they needed to maximize yields, achieve economies of scale, and reduce input costs. They also needed access to capital to establish the plantation, and secure land and planting materials. The concentrated effort on a single commodity made it much easier to acquire the expertise needed for any particular crop. Much of this knowledge was acquired through trial-and-error until the most suitable crop was found that matched the local climate and soil. It also required a mass migration of labour to work on plantations and a hierarchy to manage the operation. The challenge of securing enough labour for the plantations was solved by importing slaves and the use of indentured labourers.

Even though slavery was outlawed in 1833, working conditions remained largely unchanged as many plantation owners revived the system of indentured labour. Through this system, labourers had to work for several years to earn enough for the cost of their passage and a small wage. It was the system of indentured labour that created the massive migration within the empire as hundreds of thousands of workers from India and China were moved from colony to colony. In reality, working conditions remained as unrelenting, degrading, and brutal as those during the days of slavery. The work was similarly demanding, unpleasant and poorly paid. Plantations were often essential to the economic development of many colonies, which in many cases became one-crop economies. As well as moving crops and people to locations far from home, plantations shaped the cultures and economies of many colonies long after independence.

The campaigns that Britain waged in Europe, Asia, and Africa to secure the allied victory of World War Two virtually bankrupted the UK. Its subsequent debts severely compromised its economic and military power and led to the emancipation of its colonies. At its maximum extent, the British Empire comprised 57 colonies, dominions, territories, or protectorates from Australia, Canada, and India to Fiji, Western Samoa, and Tonga.[36] Agricultural supply chains from many of these territories, often through monoculture plantations of specific commodity crops, remain one of the most enduring and intractable legacies of its empire.

Revolution Numbers 2, 3, and 4 Revisited

Akkadian, Roman, and Muslim Empires: Fractured Legacies

The Akkadian Empire linked two cultures, Assyrian and Sumerian, around a single resource — water. When the water ran out because of insufficient rainfall and poor agricultural management, the empire became too weak and collapsed. In its time, however, the Akkadian Empire demonstrated that agricultural products could be beneficially traded within an empire and their volume supplemented through irrigation. For much of the empire's history rainfall in northern Mesopotamia was sufficient for crops without irrigation and in southern Mesopotamia soils were fertile enough to produce yield surpluses with irrigation. This allowed an internal *water market* to be developed that was beneficial to both communities despite their cultural and linguistic differences. However, after less than 200 years, the Akkadian Empire was over, a consequence of its reliance on a single natural resource that is variable in space and time but on which all other elements of civilization depended, and which others coveted. What is now the Middle East has been devasted once more, not least because of its dependence on another single natural resource — fossil fuel — on which its economies depend, and others covet. It almost goes without saying that much of the physical and cultural legacy of the agricultural systems on which the Akkadian Empire was built have been similarly destroyed.

The Romans had much greater impact on human history than the Akkadians both because of the *extent* of their empire, in its time the largest ever, and its *duration*. They also had the engineering skills and military prowess to expand their territories and enforce change both in the cultures and the geographies of those whom they colonized. The Romans had one more critical asset — slaves. Slavery was not particular to the Romans, in fact, the use of forced labour had also been central to the production and distribution systems of Mesopotamia. Both the Akkadians and Assyrians enslaved their prisoners of war, however the Romans moved large numbers of captured combatants and inhabitants to build and feed their empire. Water remained a critical resource, as evidenced by the vast irrigation systems some of which are still operating 2,000 years later but, in addition, the Romans created an internal *people market* to serve their needs and service their technologies. The enduring legacy of the Roman Empire is evident both in the resilience of its architecture and the diversity of its diet. For much of the area that it conquered, many of the crops and foods introduced and exchanged by the Roman Empire have remained part of the local diets and food cultures. Without slaves, however, the Romans could not have built their empire nor grown the crops and tended the animals that fed it.

The Muslim Empire built on what had gone before, but it also brought ideas that had never been previously applied to agriculture. For over four centuries, it was able to diffuse innovations, both through their initial transfer and then through secondary diffusion across the Empire. Changes in attitudes, social organization, institutions, infrastructure, technical progress, and economic development all played a part but fundamental to this diffusion was scientific enquiry in which hypotheses, experimentation, and conclusions were drawn not just for the local situation but for translation and communication across the empire. This golden period of Muslim agriculture was part of a wider blossoming of science and arts guided by enquiry, criticism, and conventions that could be challenged by evidence.

Perhaps for the first time, research on the use of plants for food, medicine, and aesthetics was conveyed to the public through written communication via libraries, verbally communicated in markets and mosques and through practical advice to farmers. The destruction of the Grand Library of Baghdad was such that the Tigris ran black with ink, conjuring an image not just of the lost scholarship that was contained within but also the barbarity of those who destroyed it. In its time, the Muslim Empire built a corpus of scholarship that provided a *knowledge market* across the Muslim world. Although much remains unsurpassed internationally and relevant locally, its descendants remain largely unaware or disinterested in the value of Muslim science or the science of Muslim agriculture.

In each of the above cases, agriculture developed within the lands of each empire and new crops, systems, and technologies were adopted and adapted to local needs. The adoption of new crops and livestock meant broadening their geographical distribution and the knowledge associated with them. Their adaptation meant new methods of local cultivation, selection of suitable ecotypes for different environments, and absorbing of information and beliefs associated with their properties and uses. The British Empire was different.

British Empire: An Enduring Legacy

The Muslim, Roman, Akkadian, and many other empires not mentioned here have had lasting legacies, but none has had as profound or far-reaching consequences on humanity and the planet as the British Empire. At its core, the development of agrarian capitalism in England divided agriculture into landowners, capitalist tenant farmers, and labourers. This societal structure, built on the physical rights to land enshrined in the Enclosures Acts, led to better farm management practices, more efficient labour use, and increased productivity and profits. This meant that people and capital could be directed to places across the empire for activities related to agriculture, including farming, exploration, transport, and technologies spanning the whole agricultural value chain for selected commodity crops. Technologies and the power that they generated both physically and economically provided an almost insatiable demand for raw materials to support industries such as textiles, food, and stimulants.

Unlike its predecessors, the commodities that the British Empire cherished could not be grown in the climates of north-western Europe. To meet this demand, the empire expanded its colonial ambitions and secured commodity supply chains across Africa, Asia, and the Americas.

There is a long history of relationships between *commodities* and *empires*. Imperial expansion has driven *commodity prospecting* and global trade in commodities to fuel industrial economies — often literally. Commodities that became part of imperial supply chains include food crops (e.g., wheat, rice, banana); industrial crops (e.g. cotton, rubber, linseed, palm oil); stimulants (e.g. sugar, tea, coffee, cocoa, tobacco, opium); ores (e.g. tin, copper, gold, diamonds) and fossil fuels (e.g. coal, oil). Their expanded production and global movement brought vast spatial, social, economic, and cultural changes with huge profits for some and incalculable costs for many. The first wave of British plantations were those in the New World to supply sugar, tobacco, and cotton. Others followed, however, and long after the end of the Empire that established it, the plantation system and the thinking behind it lives on.

Coffee in East Africa, oil palm in Malaysia, and tea in India are all examples of crops that have been farmed beyond independence. These commodities, however, no longer have a guaranteed market to buy them. The plantation system was built around a vast empire in which each colony was required to specialize in certain products and then trade these with other colonies. This mechanism could only operate when all the colonies were part of a common system, so once they were independent, each country was left to the mercy of international commodity markets. The inability of plantation economies to respond to world market changes remains an intractable problem for many now independent host nations. By focusing entirely on a single commodity for a distinct and largely foreign market, plantation economies were unable to switch when demand or prices fell. Agriculture is a long-term activity and plantations are often based on perennial crops such as rubber, tea, coffee, and oil palm. Planning and investment decisions have to be approved many years in advance and particular countries were frequently beholden to single crops — *one-crop economies*. Sugar in the West Indies was one of the first victims of this change when sugar beet grown in Europe was found to be economically more feasible than sugar cane from the tropics. Plantations across the Caribbean suddenly became economic liabilities, thousands of workers were no longer required to serve the needs of their former colonial masters and were left with few alternative livelihoods from their own agricultural crops and farming systems.

The British Empire has left a legacy of people, plants, and animals that have been moved to different parts of an empire that no longer exists. In some cases, such as New Zealand, Australia, and Canada, the movement of settlers led to independent nations that retain the status of Dominions of the British Crown. As well as people and their cultures, the British Empire moved plants and animals far from their centres of diversity and production. For example, sheep and cattle were unknown in New Zealand and Australia until the late 18th century but now New Zealand's economy depends on products grown on the country's pastures, in particular wool, meat, and dairy products. By 1982, the country had the highest density of sheep per unit area in the world — about 70 million sheep in a country of 3.2 million people. Its main crops are now cereals (wheat, oat, barley, maize), pulses (pea, lentil), brassicas (oilseed rape, herbage, forage) as well as ryegrass and clover.[37] Similarly, in addition to dairy and sheep, the main grain crops grown in Australia include wheat, oilseed rape, oat, pulses, and barley and in dry regions sorghum, cotton and sunflower.[38] Not one indigenous plant or animal forms a significant part of the agricultural sector of New Zealand or Australia.

Final Thoughts

It is for the citizens of former colonies to decide whether any benefits of empire justify the costs of colonialism on their assets, geographies, languages, cultures, societies and agriculture. It is clear, however, that the effects of colonialism on local agricultural biodiversity and the knowledge systems associated with it have been devastating — not just in what has been lost but also in what has not been used. We cannot easily reverse these losses, but we have a final opportunity to harness the existing knowledge of local plant, animal, and insect species and systems as a complementary asset to the mainstream agriculture that is now the norm. For example, the island of Borneo that is shared by Malaysia, Indonesia, and Brunei, hosts over 15,000 species of flowering plants — almost one in 20 of the plant species on the planet.[39] Yet, not one indigenous species of Borneo is part of the mainstream agriculture of its constituent nations, nor indeed elsewhere. In contrast, oil palm, a crop introduced by the British and Dutch a century ago from their other colonies in West Africa, now occupies 12 million hectares in Malaysia and Indonesia. This is not to say that oil palm should not remain a part of Southeast Asian economies but, rather, it is to ask why the indigenous biodiversity of Borneo and indeed the whole region, with a history of over 140 million years and in which people have lived for over 50,000 years, is not accorded the same status or considered to be worthy of serious investigation.

As an example of this myopia, we can take Australia. The continent hosts between 200,000 and 300,000 species that include 250 species of native mammals, 550 species of land and aquatic birds, 680 species of reptiles, 190 species of frogs, and more than 2,000 species of marine and freshwater fish.[40] There are also an astounding 24,000 species of native plants in Australia of which around 18,000 are flowering species. The continent has more than 1,200 species of Acacia, 2,800 species of eucalypts, and around 170 species in the Myrtle family that are notable for essential oils, including tea tree oil. Examples of Australian native fruits including little known species such as quandong, kutjera, muntries, riberry, Davidson's plum, and finger lime, various native yams and leafy vegetables and bunya nutis; the macadamia nut is perhaps one exception. This repository of Australian biological diversity exists, as with Borneo, but the popular narrative remains that aboriginal populations never adopted agriculture or domestication of animals, preferring a hunter-gatherer lifestyle. Early settlers viewed Australia as a blank canvas on which they could introduce agriculture and agricultural systems that they were familiar with from the other side of the world, and could impose these on the often-alien climates and soils of Australia without recourse to local knowledge. However, Aboriginal populations have been present in Australia for probably more than 60,000 years. During this time, Aboriginal people discovered and consumed a wide variety of foods, never being reliant on a few food types. Is it plausible that *none* of these crops, foods and knowledge systems are really of *no value* to future food systems and societies in crisis?

Studies of Aboriginal populations show that when one food is restricted, for example during a drought, they shifted foraging to other plants and animals, and often moved to a different area, or even into a neighbouring group's range, where they were allowed to hunt until times improved in their own area. By traversing their territory, they allowed each area to recover before returning, unlike modern farming practices that require inputs such as fertilizers and irrigation to maintain productivity. While not advocating a hunter-gathering lifestyle or abandoning modern agriculture, Australia as the oldest and driest continent and its original inhabitants, with 60,000 years of human history and over 200,000 species, has much to teach us about living on the edge of human existence. If we are to find agricultural and societal solutions for a planet in crisis, we need access to all the diversity of evidence, knowledge and wisdom available. To ignore that which has been gathered over millennia by those who have survived without, indeed in spite of, the power of empires would be unconscionable.

3

THE GREEN REVOLUTION

Revolution Number 6

Context

Over the last 50 years, a few 'staple' crops grown as intensive monocultures in some countries have come to dominate the global agrifood system.[1] In fact, this model of *modern* agriculture is a departure from all the production systems that preceded it, and possibly from those that will succeed it. Never in human history have so few crop and animal species fed so many people. After our decision to start farming 10,000 years ago, the transformation of agriculture now known as the *Green Revolution* represents perhaps the greatest experiment in human history. It is over seven decades since the start of the Green Revolution, but our modern agrifood, health, and economic systems are largely a consequence of it. That billions of people are alive who would otherwise have starved is a testament to its success. That there are so many unhealthy people and damaged ecosystems are a consequence of its failure.

This chapter describes how the Green Revolution (Revolution Number 6) became the template for modern agriculture. It then evaluates its legacy and considers whether it provides a suitable model for the future of the agrifood system for more people in ever more hostile and changing climates.

Agricultural Biodiversity

Over the tiny fraction of human history represented by agriculture, there has been an unprecedented standardization in the way we farm, a reduced range of crops and animals that we grow, and homogenization of the foods that we eat. This *one-size-fits-all* model gravitates towards a common human diet and a global agrifood system.[2] For the first time, irrespective of location, nationality, or culture, human beings are eating similar foods from limited ingredients derived from a few calorific crops grown far from where their products are consumed.

The reduction in the diversity of crop species and agroecosystems has inevitable, sometimes unintended, consequences for the range of natural biological diversity that exists within and beyond them. The human population has grown from between 0.001 to 0.1 billion people at the start of agriculture to about 7.8 billion people today.[3] Over this time, there has been an inexorable decline in agricultural biodiversity — the number of crop (or animal) species grown and the range of systems within which they are cultivated.

At a global scale, there are between 300,000 and 500,000 higher plant species.[4] Of these, about 30,000 have been identified as edible and around 7,000 have been cultivated or collected by humans for food.[5] According to the United Nations Food and Agriculture Organization (FAO), 95 per cent of the world's calories now come from just 30 species and over three-quarters of the world's total food supply comes from just 12 plant and 5 animal species.[6] In fact, almost half of the calories consumed by nearly 8 billion people are produced by just three staple grain crops (wheat, rice, and maize).[7] Just imagine if an unforeseen global disaster was to affect our access to any one of them.

Not only has the number of crop species declined from many thousands to at most a few dozen, so too have the *cropping systems* in which we grow them. High input industrial monoculture — one crop grown by itself in a field — is now the basis for the world's commercial agriculture. Oddly, despite its reliance on sophisticated technologies such as irrigation, fertilizers, and agrochemicals along with machines to deliver them, modern agriculture represents the simplest form of a food production system in our history. The transformation of agriculture to what has become known as the Green Revolution is forever associated with one man, Norman Ernest Borlaug.

Norman Borlaug — A Breeding Approach to Agricultural Biodiversity

Norman Borlaug was born in 1909, the great-grandson of Norwegian immigrants who settled in Saude, near Cresco, Iowa. He spent his childhood in rural Iowa and could have retired there in tranquil obscurity after a life on the family farm. But on the advice of his grandfather, "you're wiser to fill your head now if you want to fill your belly later on," Borlaug left home to pursue his education. He failed the entrance exam to the University of Minnesota in 1933, but was admitted into the school's General College and then joined the College of Agriculture's forestry programme.[8]

During his undergraduate degree, Borlaug attended a lecture by Elvin Charles Stakman in which Stakman discussed rust, a fungal parasite that feeds on cereal crops, and described new breeding methods that produced rust-resistant plants. This marked a turning point in Borlaug's development and he registered at the University of Minnesota where he studied plant pathology under Stakman. He completed a Master's degree in 1940 and a PhD in plant pathology and genetics in 1942. He then joined the DuPont Chemicals company at its headquarters in Wilmington, Delaware where he was responsible for research on industrial and agricultural bactericides, fungicides, and preservatives. In 1944, Borlaug was appointed to lead the Cooperative Wheat Research and Production Programme in Mexico, a collaboration between the Mexican government and the Rockefeller Foundation. The programme was a spectacular success and Borlaug's team developed high-yielding, disease-resistant wheat varieties that transformed the yields of Mexican farmers.

CIMMYT (Centro Internacional de Mejoramiento de Maiz y Trigo)

In 1966, the Mexican government established the International Maize and Wheat Improvement Center, CIMMYT (Centro Internacional de Mejoramiento de Maíz y Trigo), which later became part of the global network of research centres known as the Consultative Group on International Agricultural Research (CGIAR).[9] The Mexican government also established the Mexican Agricultural Programme (MAP) to raise national crop yields. Borlaug's wheat breeding activities, supported by the Ford and Rockefeller Foundations, helped Mexico become the showcase for the new High Yielding Varieties (HYVs) of wheat. The technology was quickly adopted in other areas of Latin America and Asia. In 1964, Borlaug was appointed director of CIMMYT.

Nobel Prize

In the 1960s, Borlaug travelled to India and Pakistan and demonstrated similar impacts of the new HYVs of wheat. There was a risk of mass starvation in South Asia, and both countries quickly adopted HYVs that were being produced through Borlaug's breeding programme in Mexico. Within five years, both India and Pakistan went from food shortages to self-sufficiency. In 1970, Norman Borlaug became the first agriculturalist to be awarded the Nobel Peace Prize. The title of his acceptance speech was *The Green Revolution: Peace and Humanity*.[10] There are only six Nobel Prize categories, with 935 Laureates since the first prizes were awarded in 1901. To date, Norman Borlaug remains the only agricultural scientist to become a Nobel Laureate. He is also one of only three Americans to win the Nobel Peace Prize, the Congressional Gold Medal (America's highest civilian honour), and the Presidential Medal of Freedom. Through his own efforts and the *army of hunger fighters* that he enlisted, the Green Revolution is credited with saving over a billion people from starvation.[11]

The Green Revolution

It can be argued that no single development in agriculture, and possibly in human history, has had more impact on humanity than the Green Revolution. Norman Borlaug is widely acknowledged as its father, but he didn't coin the term. It was first used in a speech on 8 March 1968 by the administrator of the U.S. Agency for International Development (USAID), William S. Gaud, who noted;

> *"These and other developments in the field of agriculture contain the makings of a new revolution. It is not a violent Red Revolution like that of the Soviets, nor is it a White Revolution like that of the Shah of Iran. I call it the Green Revolution."*[12]

Nowhere was the impact of the Green Revolution more evident than in India. In 1961, India was on the verge of mass famine. Borlaug was invited by Dr M S Swaminathan, the adviser to the Indian minister of agriculture, to visit India and introduce the Green Revolution approach. The Ford Foundation and the Indian government collaborated to import HYV wheat seed from Mexico. Punjab was selected to host the first trials of the new varieties because of its reliable irrigation system and skilled farmers. In the late 1960s India began its own Green Revolution programme of plant breeding, irrigation, and agrochemicals. By the 1970s it was self-sufficient in wheat and by 2013 it was the seventh largest agricultural exporter in the world.[13]

In 1960, the government of the Philippines, again with support from the Ford and Rockefeller Foundations, established the International Rice Research Institute (IRRI) near Los Banõs in the Philippines. In 1962, the first widely used HYV rice variety, IR8, was developed at IRRI. Like its HYV wheat counterparts, IR8 could produce more grains per plant and higher yields when grown with fertilizers and irrigation than traditional cultivars. Average productivity of rice in the Philippines increased from 1.23 tonnes per hectare in 1961 to 3.59 tonnes per hectare in 2009.[14] Plant breeding at IRRI was also geared towards incorporating insect, pest and disease resistance as components of the emerging approach to crop protection known as *Integrated Pest Management* (IPM), with the double objectives of reducing pest damage and pesticide applications. The switch to IR8 made the Philippines a rice exporter for the first time. India soon adopted the variety, which under optimum conditions achieved yields 10 times those of traditional rice, was a success throughout Asia, and became known as *Miracle Rice*. By the mid-1990s India was a major rice producer and exporter.[15]

Evolution involves the repeated interaction of many generations of a species with its environment. By contrast, *revolution* implies a transformative and often traumatic change between or even within a generation. In *natural* environments, we understand evolution as a process of *natural selection*, in which the environment applies selection pressures on successive generations of a species. Individual genetic traits may be positively or negatively associated with the selection pressure of the environment. When the natural environment remains fairly uniform, *stabilizing selection* ensures that genetic traits represent the optimum for a species. When the environment changes, *directional selection* moves the average of the population to a new optimum for the new conditions. In other words, species gradually *adapt* to their new environment. Darwin used the term '*survival of the fittest*' to describe how organisms that can best adjust to their changed environment are the most successful in surviving and reproducing through natural selection.

The Green Revolution involved transformative changes in *both* the genetics of particular crops (through plant breeding) *and* their environments (through management practices). These combined effects are called

GxE (Genotype x Environment) interactions. For the G (Genotype) component we can conventionally breed cultivars (*cultivated varieties*) with more desirable genetic traits, or more recently, by introducing specific traits through genetic manipulation (Genetically Modified Organisms, or GMOs). For the E (Environment) component, we can supplement rainfall with irrigation, soil nutrients with fertilizers, and use enclosures (glasshouses and polytunnels) to change factors such as temperature and light levels. We can also move crops to more favourable environments in space (better locations or soils) or in time (date of sowing) and can protect the crop against competitors using herbicides to kill weeds and pesticides to kill insect predators or diseases.

Before the Green Revolution, agriculture *evolved* relatively slowly because there were no intensive programmes of plant breeding, limited scientific understanding of genetic disciplines of physiology and agronomy, and limited transfer of innovations to farmers at a scale and speed that could significantly affect a large number of people. By creating a complete package of GxE technologies, the Green Revolution changed all of that. The new HYVs (Genotype) that were being bred by Borlaug and others required a package of supporting technologies (Environment) to achieve their potential. In the 1950s and 1960s, Green Revolution technologies increased agricultural production worldwide through high-yielding varieties of two cereal crops, wheat and rice. These *miracle* crops involved intensive plant breeding for specific traits, that in addition to high grain yield, made the new crop types suitable for irrigation, fertilizers, and agrochemicals. These new methods of cultivation required mechanization throughout the life of the crop, from soil cultivation, sowing of seeds, applications of irrigation, herbicides and fertilizers, and crop harvesting. Without each of these technologies delivered as a package by farmers with inputs synchronized in space and time, the HYVs couldn't achieve their potential — the Green Revolution couldn't revolve.

In essence, the Green Revolution combined a series of five novel technologies;

- High-yielding varieties (plant breeding)
- Crop nutrition (synthetic fertilizers)
- Crop health (herbicides and pesticides)

- Crop water use (irrigation), and
- Crop management (mechanization).

By building these technologies around the monoculture of a few important crops, the production system allowed for efficiency gains, economies of scale, input standardization, and targeted advice to farmers so that they could achieve uniform outputs in space and time to meet market demands. Of course, for all these technologies to be scaled up, farmers needed access to sufficient areas of uniform and suitable land either through consolidation of existing units or clearing new land for agriculture.

Plant Breeding

High-Yielding Varieties (HYVs)

The first stage of what became the package of Green Revolution technologies was breeding HYVs of wheat and later rice and maize. These new HYVs could absorb more nitrogen than their traditional counterparts. Nitrogen is essential for green plants to expand their foliage to capture sunlight for the process of photosynthesis. However, by absorbing extra nitrogen, taller, bigger crops typically *lodge*, or topple over before harvest. To reduce lodging, *semi-dwarfing* genes were bred into these crops to reduce their height and produce sturdier stems.

From these early successes, researchers identified several traits for grain yield and related genes that determine plant height and the number of branches or *tillers*. These traits were closely linked, for example, shorter, thicker stems are also less likely to lodge. With reduced lodging, photosynthetic products are directed to seed production, and shorter, sturdier crops are more receptive to mechanical application of chemical fertilizers, herbicides, and pesticides.

Fertilizers

Soils naturally contain chemical elements, some of which are essential for plant growth. Those considered as *micronutrients* are required in such small quantities that they are either already available in the soil or can be added occasionally. Others (*macronutrients*) are needed in such large quantities that they must be added as fertilizers. When a crop is harvested, proteins, carbohydrates, and lipids in leaves, stems, or seeds are removed from the field. This offtake of nutrients must be replaced through fertilizers. Nutrients can be provided through various sources — organic matter, chemical fertilizers, and certain plants such as legumes (pea, beans, and pulses such as groundnut) that can *fix* nitrogen from the atmosphere.

John Bennet Lawes

The management of soil fertility has challenged farmers for millennia.[16] Egyptians, Romans, Babylonians, and early Germans all used minerals or manure to increase the yields of their crops. In 1837, the English entrepreneur John Bennet Lawes began to test how various manures influenced the growth of plants growing in pots, and then extended his experiments to field crops at his estate at Rothamsted Manor, near Harpenden in the English county of Hertfordshire. In 1842, Lawes patented a manure formed by treating phosphates with sulphuric acid, which as *superphosphate* marked the beginnings of the chemical fertilizer industry. In 1843, he appointed the chemist Joseph Henry Gilbert as his collaborator and together they established the classical long-term studies on the Broadbalk field at Rothamsted that are still going on today. The scientific partnership between Lawes and Gilbert lasted 57 years, and together they set the foundations for agricultural science and established the principles of crop nutrition. In 1889, Lawes donated 100,000 pounds sterling to establish the Lawes Agricultural Trust. The Rothamsted Experimental Station remains the world's oldest operating agricultural research institution.[17]

Haber-Bosch

In 1840, around the same time as Lawes was working on plant nutrition, the German chemist Justus von Liebig presented a new vision for "a rational system of agriculture" based on scientific principles.[18] Rather than biology (the study of life and living organisms) von Leibig put chemistry (the study of the composition of matter) at the centre of his scientific principles. It wasn't until 70 years later, however, that two German chemists, Fritz Haber and Carl Bosch, provided a scalable application of Liebig's vision of chemical farming.[19] In 1908, Fritz Haber approached the German chemical company BASF to propose that ammonia (the basis of nitrogen fertilizers) could be synthesised from hydrogen and nitrogen under high pressure. In 1909, Haber achieved the first laboratory demonstration of the production of liquid ammonia. Carl Bosch extended Haber's *high-pressure synthesis* of ammonia from an initial proof-of-concept prototype to an industrial scale in what became known as the Haber-Bosch process.[20]

It was already known that ammonia synthesis could be achieved organically by microscopic organisms in the natural environment, but this natural mechanism could not compete with the scale and speed of the Haber-Bosch process. Nitrogen, which constitutes 78 per cent of the Earth's atmosphere, is an essential component of plants and is needed in much greater quantities than that which is freely available in soils. Plants cannot capture or *fix* nitrogen directly from the air, but must rely on the nitrogen fixing characteristics of microorganisms, often through a symbiotic relationship with plants through nodules that attach to the roots of living plants. Not all crop species have a symbiotic nitrogen fixing relationship with microbes. Although legumes such as soybean, groundnut, and field bean have this symbiotic relationship, most cereals do not, thus nitrogen must be applied to them through external sources — fertilizers.

Nitrogen is essential for plants, but it can also provide a feedstock for ammunition. This dual purpose was quickly recognized and commercialized in Germany for both agricultural and military use. The nitrogen industrial complex at Oppau, employing the new Haber-Bosch process, opened in 1913. Haber developed a new chemical weapon — poison gas, which as chlorine gas, was deployed under his supervision on the Western Front at Ypres, Belgium in 1915.[21] Through chlorine gas, the Haber-Bosch process contributed to many of the estimated 10 million lives lost in the First World War. So, we can see that the stimulus for industrial-scale fertilizer production for agriculture through synthetic nitrogen compounds from the atmosphere was the war demand for explosives and chemical weapons. Not surprisingly, the 1918 Nobel Prize award for chemistry to Fritz Haber was both controversial and unpopular. For many, he was considered to be a war criminal.[22] In fact, his name was on the first list of war criminals sought for extradition by the Allies and he had to hide for a short time in Switzerland. His process of fixing nitrogen still underpins modern agriculture by providing a cheap means of manufacturing fertilizer at large scale. Fifty-two years after Fritz Haber received the Nobel Prize for chemistry, in part for the development of poison gas from industrial synthesis of ammonia, Norman Borlaug received the Nobel Peace Prize, in part for the use of fertilizers derived from ammonia to feed humanity.

Crop Health

As with fertilizers, the discovery and use of chemical herbicides and pesticides also has a military connection, but their benefits for farmers are clear. The final yield of a crop can be reduced (indeed destroyed) by weeds that capture some of the light, water, and nutrients intended for the crop. The purpose of herbicides is to reduce that competition, usually by chemically defoliating the light-capturing green surfaces of the weeds. As well as having its resources *stolen* by weeds, insect pests and diseases can *steal* resources already captured by the crop. For example, insects may consume plant parts already produced by the crop or damage them through the effects of fungi and other organisms. The purpose of pesticides is to remove these competitors either before or while they are attacking the crop.

Herbicides

Until the widespread use of chemical defoliants, the only way to remove weeds was manually through the use of hoes and other implements whose history stretches back to the beginnings of agriculture. Manual weeding was laborious and tedious and led to the search for chemical methods of removing weeds. Chemical weed control for example using sea salt, industrial by-products, and oils has been around for a very long time — selective weed control in cereal crops started in Europe in the late 1800s. Sulphates and nitrates of copper and iron were used and even sulphuric acid was applied as a spray. Sodium arsenite was used both as a herbicide and as a soil sterilant. This highly toxic compound was widely used on tropical sugar cane and rubber plantations, often poisoning animals and humans in the quest for maximum yields.

Chemical methods were already in use, but the 1940s saw a new generation of herbicides developed as a result of research during World War Two.[23] In 1945, selective chemical weed control was introduced through the use of 2,4-D (2,4-dichlorophenoxyacetic acid) and 2,4,5-T (2,4,5-trichlorophenoxyacetic acid) as foliar sprays and IPC (isopropyl-N-phenylcarbamate) as a soil treatment. The new herbicides were transformative since their high toxicity allowed for effective weed control at very low dosage rates of a few kilogrammes per hectare. This contrasted with traditional applications of carbon bisulphide, borax, and arsenic trioxide, which were required at rates of over two tonnes per hectare, and with sodium chlorate which had to be applied at over 100 kilograms per hectare.

Herbicides are an essential component of Green Revolution technologies. HYVs selected for their yield potential must compromise on other traits. This leaves them vulnerable to aggressive competitors such as weeds that capture resources and ultimately suppress the crop completely. Rather than increasing the competitive ability of the crop (by trade-offs with yield potential), the preference is to remove the competitor species by chemical means. This can be done before sowing the crop, before its emergence, or when the established crop is photosynthesizing. Of course, this last option brings challenges in that the herbicide intended for the competitor species can damage the crop itself. In such cases selective herbicides might be deployed, however, these are also problematic in their accuracy against particular species and collateral damage to beneficial species such as pollinators.

In the mid-1980s, Herbicide-Resistant Crops (HRCs) were genetically engineered to withstand the effects of specific chemical herbicides, most notably glyphosate or *Roundup*. These GMOs enable non-selective chemical weed control, since only the HRC plants can survive in fields that are treated with a particular herbicide.[24] This may seem like an elegant solution, but HRCs remain controversial because of their environmental impact, health concerns associated with increased use of herbicides and ethical concerns about corporate monopolies.

Pesticides

Humans have been using various methods to protect their crops from pests for millennia. By the 15th century, chemicals such as arsenic, mercury, and lead had been introduced to kill insect pests. In the 17th century, tobacco leaves were used to extract nicotine sulphate which was used as an insecticide. In the 19th century, pyrethrum, extracted from chrysanthemums, and rotenone, extracted from vegetable roots were introduced for pest control.[25]

Once more, warfare accelerated the commercial development of pesticides. During the First World War, by-products of research into nerve gas and explosives were developed into insecticides. Others, such as cyanide and arsenic that were already in use to fumigate orchards, were switched to military purposes, causing shortages for agriculture. In the 1930s, the Nazis who controlled Germany's chemical industry, developed lethal nerve gases. Zyklon B, which was originally an insecticide, was used to kill people in concentration camps.[26] Dichlorodiphenyltrichloroethane, commonly known as DDT, was first synthesized in 1874 by the Austrian chemist Othmar Zeidler. In 1939 its insecticidal properties were discovered by a Swiss chemist, Paul Hermann Müller, who found it to be highly toxic to a wide variety of insects by disrupting their nervous systems.[27]

Irrigation

Of the total water on the planet, about 96.54 per cent is stored in oceans, seas, and bays, 1.74 per cent is locked in ice caps, glaciers, and permanent snow, and 1.69 per cent is stored as groundwater.[28] This leaves a tiny fraction as fresh water on which over 8 million animal and plant species, not to mention 7.8 billion people, must depend. Of this, most is used to irrigate crops.

Life originated in the oceans, and while many species have migrated to land, their organisms remain thirsty. This remains the case for human beings, crops and the animals on which we depend. Their thirst for water means that crops must extract it every day from the soil through their roots. For agricultural crops, water is not only essential for survival but must be available in sufficient quantities if they are to achieve the yields set by their genetic potential. Where rainfall is insufficient, water must be applied through various forms of irrigation.

Irrigation systems, and the massive investments that they require in the extraction, storage, and delivery of water to crops, form a major part of the cost of Green Revolution technologies. The semi-dwarf rigid stems of HYVs lend themselves to systems that can irrigate crops by spraying water above them. Such sprinkler systems, either spaced across the crop field, fixed on overhead gantries that traverse above the crop, or as centre pivot systems, individually or in tandem, form part of the armoury of irrigation techniques available to Green Revolution farmers. Other systems are designed to deliver water to furrows between crop ridges or to flood a section of the cropped field from which water permeates through the soil. In part to reduce the profligate loss of water that does not reach crop roots, drip irrigation and sub-surface irrigation systems have also been developed. However, none of these systems can reduce losses in water in its delivery to the field, ensure its quality for irrigation, or change the efficiency with which each crop uses water.

Mechanization

The development of HYVs was not just about maximizing crop yield, but was also about breeding semi-dwarf varieties of particular crops that could lend themselves to mechanization. By reducing the height of a wheat plant compared to traditional cultivars, plant breeders gave farmers the option to mechanize the entire operation, including application of fertilizer, crop protection chemicals, and especially irrigation. The design of machinery specifically for Green Revolution crops allowed mechanization to underpin all other technologies. It also allowed closely targeted and timely operations to be completed by a small number of trained operators, instead of large numbers of field workers, reducing costs and increasing management efficiencies. Again, economies of scale required that sufficient areas of suitable land had to be available to justify the capital investment in technologies for cultivation, inputs, and harvesting of crops grown as high input monocultures.

The Green Revolution in Perspective

A Success Story

Based on yield, food production, food calories *per capita,* and declining malnutrition, the achievements of the Green Revolution are impressive on a global scale. Those countries that adopted the new technology achieved record yields and between 1960 and 1985 the total food production more than doubled in emerging economies.[29] Between 1966 and 2000, there was a 275 per cent increase in Indonesian rice production.[30] By the 1990s, nearly three quarters of the rice produced in Asia and half the wheat grown in Asia, Latin America, and Africa was from the new varieties.[31] In less than 25 years, Asia's food production doubled with only a 4 per cent increase in the area under agriculture.[32]

Since the start of the Green Revolution, food production has regularly outstripped population growth. Between 1950 and 1990, global population increased by 110 per cent but over the same period global cereal production increased by 174 per cent.[33] Between 1961 and 2000, *per capita* food supply increased by 20 per cent and between 1970[34] and 1990 the number of hungry people decreased from 942 to 786 million.[35] Based on the evidence of its successes, re-evaluating the Green Revolution might seem to be cosmetic — even churlish. However, given the burgeoning human population, catastrophic loss of biodiversity, the growing area of deserts, agriculture-induced environmental pollution, and climate change, we need to reconsider the fundamental basis of the Green Revolution, not because it was wrong for its time but to decide whether it is right for the future. The next section reviews the legacy of the Green Revolution through four lenses — political, people, plants, and planet. A final section considers how an understanding of the Green Revolution can guide our thinking about the future of the agrifood system. First, we review the legacy of its architect and the *peace and humanity* that he espoused.

Borlaug's Legacy

Norman Borlaug died in 2009, but the Green Revolution that is forever associated with his name remains alive. Every day we buy its products in supermarkets; consume them at home; become satiated, obese, malnourished, or stunted by eating them; are employed to grow, process, or sell them; or remain unemployed because of them. For many, Borlaug will be remembered as the man who fed the world. Indeed, it is estimated that a billion people are now alive who would otherwise have starved without the Green Revolution.[36] This is a powerful and visible testament to its success, but it isn't the entire story. For that, we need to look deeper at the context of the Green Revolution and its legacy. Norman Borlaug himself warned that the Green Revolution could only provide a transient solution to global hunger. In his Nobel Prize lecture in 1970, he said;

> *"The green revolution has won a temporary success in man's war against hunger and deprivation; it has given man a breathing space. If fully implemented, the revolution can provide sufficient food for sustenance during the next three decades. But the frightening power of human reproduction must also be curbed; otherwise the success of the green revolution will be ephemeral only."*[37]

Borlaug actually underestimated the breathing space that the Green Revolution would provide against the *population monster* that he predicted would reach 6.5 billion people by 2000. In fact, by then the global population was around 6 billion — and is now around 7.8 billion people. Technically, there is more than enough food for the world, suggesting that crop yields have increased faster and population more slowly than Borlaug anticipated. Indeed, yield increases of the major cereal crops have been spectacular. We now have wheat crops that can achieve yields of nearly 18 tonnes per hectare when the global average in 1970 was 4.2 tonnes per hectare.[38] Since Borlaug's crop breeding programmes there has been a generation of novel genetic technologies that have the potential to break the yield ceiling set by the photosynthetic capacities of existing crops, increase their fraction allocated to grain, or protect them from competitors. Norman Borlaug believed that the scientific power of plant breeding and modern technologies could break the cycle of food insecurity that has afflicted much of the world for generations. At its core, the Green Revolution was about applying Western technological solutions to decrease poverty and hunger in the global South. However, in 2021 we can now ask whether a top-down package of technologies is enough for our future, and if so, at what cost?

Politics

At the start of the Green Revolution in about 1943, much of the world was on the brink of famine and starvation, and many of its poor and disaffected were looking for radical solutions to their parlous condition. What was initially a technical approach to increasing food production became a political battle between conflicting ideologies. The term *Green Revolution* was a deliberate attempt to provide an alternative to the *red revolution* of communism. The choice presented was that *developing* countries could achieve more through a peaceful agricultural revolution by keeping the present system in place, rather than through a radical political transformation that would overturn the existing order. It was essential that investment, subsidies, technology transfer, and results, literally on the ground, could stall insurrection before it became an unstoppable force. The two countries most closely associated with the Green Revolution and Borlaug's legacy are Mexico, where it was born, and India where its impact was greatest. This section considers the historical context of the Green Revolution in each country and then in an entire continent, Africa.

Mexico

The US government had always been interested in the affairs of its neighbour to the south. In 1933, discussions began at the Rockefeller Foundation to establish an agricultural research programme in Mexico, but these plans were frozen by the election of Lázaro Cárdenas as Mexican president in 1934. As president, Cárdenas redistributed land, provided loans to peasants, supported confederations of workers and peasants, and nationalized foreign-owned industries.[39] Under the agrarian reform programme, the amount of redistributed land to peasants was nearly double that of all of his predecessors combined, and by the end of his administration around half of Mexico's agricultural area was in the hands of previously landless farmers. Cárdenas also allowed publicly owned banks to lend money to the peasants who had received land under the government reforms. Beyond Mexico, Cárdenas was recognized for his efforts to nationalize foreign-owned industries. In 1937, his government nationalized the country's railways, and in 1938, the country's oil industry. These included the refineries of the Standard Oil company which was the main source of funding of the Rockefeller Foundation. Relations with the US deteriorated.

In 1940, Ávila Camacho was elected as president of Mexico. His administration established a new relationship with the US, and conflicts existing before his term were resolved. In response, the United States provided Mexico with financial aid to improve its railway system, construct the Pan American Highway and reduce its foreign debt. Camacho instituted electoral reforms to ban communists from standing in elections, reversed the Cárdenas policy of socialist education, and repealed the constitutional amendments that mandated it. The Mexican government decided that agriculture would underpin national economic growth. US Vice President-Elect Henry Wallace helped persuade the Rockefeller Foundation to provide support for the Mexican government in agricultural development, which he saw as advantageous to US economic and military interests.

The Rockefeller Foundation proposed establishing a new organization, the Office of Special Studies, as part of the Mexican government, but to be under its direction. Although nominally a collaboration, the Mexican Agricultural Plan (MAP) was also driven by the Rockefeller Foundation which saw it as an experiment in international technical cooperation and Mexico as an ideal subject. Mexicans would be beneficiaries of this experiment, but the programme would be aligned with US foreign policy and economic interests. The cooperation envisaged by the foundation involved resources to transform local production with American technologies supported by plant breeders, agronomists, and scientists.

Wheat-growing highland areas were chosen as the focus of the programme. For most Mexicans, maize was their staple crop with almost 10 times more area than wheat, but wheat was grown mainly by commercial farmers with agricultural systems and resources that were similar to their US counterparts. In fact, the MAP suited a particular group of farmers that were mostly based in northern Mexico. Between 1941 and 1952, 90 per cent of the agricultural budget (18 per cent of the federal budget) was allocated to large-scale irrigation projects in the northwest.[40] As well as being the president, Camacho was also a farmer in northern Mexico. The model for the MAP was that of capital-intensive US agriculture supported by expert training. Since technology transfer would be North to South — from the US to Mexico — local technologies, culture, and agricultural knowledge were seen as counter-productive and even as backward. The MAP required approval by the government of Mexico, but it was removed from the mechanisms of democratic oversight.

India

By 1968, American foreign policy was increasingly in conflict with American public opinion. At the time, there were over 485,000 US troops in Vietnam and more than 20,000 had already died. Over 500 unarmed villagers had been killed by US troops in the village of My Lai and across the USA millions of people took to the streets to protest against US involvement in Vietnam as part of the largest protests in US history. With China already Communist, and US forces mired in Southeast Asia, the US switched its interest to South Asia, principally India and what was then West Pakistan. The US government priorities were focussed on managing domestic as well as international challenges.

In 1954, the challenge of excess US grain production was partially resolved through the US Agricultural Trade and Development Assistance Act (PL 480), which enabled the US to export surplus grain as aid on US carriers to the Global South. India became the largest recipient of the programme. By the late 1960s, US President Lyndon Johnson believed that India viewed the programme as an entitlement based on the disparities between its continuing poverty and America's wealth and agricultural surpluses. Johnson personally intervened in each food aid decision that related to India and sought to extract policy concessions, more commitment from India to agricultural reform, and more gratitude for America's largesse. As a result, food aid to India was only provided on a month-to-month *short tether* basis and only with presidential approval. Acceptance of Green Revolution technologies was linked with that approval.

During the British Raj, India's agriculture had been geared towards Britain's colonial interests and not its own food security. Consequently, when India gained independence, its fragile economy was subject to frequent famines, economic instability, and low agricultural yields. These factors provided a rationale for India to implement the Green Revolution as a development strategy and for the US as a foreign policy objective. The urgent need for immediate and tangible actions was reinforced by two successive years of severe droughts in 1964 and 1965, which led to food shortages and famines in India that threatened insurrection. Modern agricultural technologies, recently tested in Mexico, offered quick wins to increase food production and reduce political instability. For rapid impact, HYVs of wheat and rice developed in Mexico were made available to India. The stage was set for an Indian Green Revolution.

In its early years, the Green Revolution in India produced great economic prosperity. First introduced in Punjab, it transformed the state's food production and bolstered India's national economy. By 1970, Punjab was providing 70 per cent of the nation's total food grains, and farmer incomes were also increasing by over 70 per cent.[41] Punjab's prosperity became a model to which other states could only aspire. Farmers in states like Punjab and Haryana with good irrigation and infrastructure were able to secure credit, benefit from the new technologies, and achieve faster economic development than elsewhere in the country. However, many peasant farmers couldn't afford the new HYV seeds and the cost of fertilizers, irrigation systems, pesticides, and machinery. In addition, not all farmers were able to maintain these costly technologies, especially when their harvest was poor. This meant that to cover the high cost of cultivation, they had to take out loans — usually at high interest rates from money lenders. The result for many farmers was not economic empowerment but an endless cycle of debt. The Green Revolution in India was not simply a technological response to food insecurity but a part of American domestic and foreign policy and Indian national policy. In Borlaug's words, "*food is the moral right of all who are born into this world*". That right was conferred, however, not to the poorest but primarily to those already best equipped to gain from it.

Africa

Mexico and India were active participants in the Green Revolution, Africa was not. Although many of its countries were on the fault lines of the Cold War, Africa received less interest from the West than did Asia or Latin America. The 1981 Berg Report[42] implied that Africa's problems lay not in its transition from European colonialism, nor the ongoing European, US, Chinese, and Russian interference, but because African governments were failing to manage their own economies. Since African states were among the most indebted, multilateral agencies could influence and shape the domestic policies of these countries. Later, the *structural adjustment policies* that were imposed on African states had profound impacts on their agriculture.[43]

Despite the issues of governance, corruption, and mismanagement, the starting point for any analysis of the Green Revolution and Africa is that the crops favoured by the Green Revolution were not widely grown in Africa. The main food crops of Africa included root crops such as sweet potato and cassava, or tropical white maize and not Green Revolution cereals such as wheat, rice, and yellow maize. Furthermore, the HYVs brought in from Latin America and Asia were not suited to African conditions and their early failure deterred local farmers from adopting them. In the 1960s and 1970s efforts were made to introduce HYVs into Africa but with limited success because the international development agencies that were introducing the varieties had tried to circumvent the time-consuming and expensive process of identifying and testing locally adapted plants. This failure was later addressed through more location-specific breeding programmes but by then international assistance programmes had begun to decline because donor governments had to a large extent lost interest in agriculture for development.

Another reason why African farmers failed to embrace the Green Revolution was the lack of farmland that was suitable for intensive mono-cultures. Instead, farming involved a complex mix of cropping systems that were less suited to conventional irrigation. Since only 4 per cent of African farmland is irrigated farmers must depend on unpredictable rainfall, further reducing the incentive to plant Green Revolution varieties. Another factor is that most farmers in Africa are women, without the political or economic power to demand investments in rural education, infrastructure, and electrical power that were prerequisites for the uptake of Green Revolution technology in Asia. Even now, more than half a century after its successes in Latin America and Asia, the Green Revolution has still had little impact on much of sub-Saharan Africa. Between 1970 and 1998, while the area planted to Green Revolution varieties increased to 82 per cent in the developing countries of Asia and up to 52 per cent in Latin America, only 27 per cent of farmed area was sown with new varieties in sub-Saharan Africa.[44] As a result, average cereal yields in Africa were only 1.1 tonnes per hectare against 2.8 tonnes per hectare in Latin America and 3.7 tonnes per hectare in Asia. In sub-Saharan Africa, per capita food production actually declined between 1980 and 2000, and today one-third of all Africans remain undernourished.[45]

People

Whilst it has had huge global impacts, the Green Revolution has boosted the yields and profitability of a few crops in a few targeted areas in some countries. Its rationale was that bumper yields in focus regions would encourage subsistence farmers to move towards commercial production. For this to happen, however, the package of technologies would have to be *scale-neutral* so that subsistence farmers had equivalent access to inputs, sales, and know-how as their richer brethren. In fact, the technology was not scale-neutral and subsistence farmers had to bear the risks of switching to commercial production. A further barrier for poorer farmers was that Green Revolution technology was not divisible — all of its components had to be in place for the HYVs to achieve their yield potential. In many cases, the whole package was combined into fertilizer-and-seed deals which again favoured farmers with access to cash or credit. It is clear that the political ecosystem favoured commercial farmers. Extension agents focus on big farmers; credit agencies only accept low risk borrowers; those who sell chemical inputs such as fertilizers and pesticides prefer clients who buy large quantities; state agencies provide services to those from whom the government solicits approval and support, such as large landowners. Since they have preferential access to knowledge, finance, and inputs, it is inevitable that innovation favours the wealthy and secure at the expense of the impoverished and weak.

In a review of published studies, over 80 per cent of the sampled papers concluded that the Green Revolution resulted in greater social inequality.[46] Even if not its intention, that is its product. The narrative of state and development agencies has remained that Green Revolution technologies offer the best solution for agricultural development. Nevertheless, in response to widespread criticism of this narrative, international donors have begun to recognize its limitations and are beginning to direct more resources towards smallholders. Their needs now feature on international research agendas and where funding is available might attract interest from researchers. That said, the mindset remains around transitioning subsistence farmers towards commercial production by replacing their traditional crops with HYVs of staple crops. Often, plant breeding, technologies, and support systems are applied to subsistence farmers with little consideration for their own crops, agricultural knowledge, or belief systems. At best, this represents reparations by plant breeders and scientists seeking to make major crops more relevant to the needs of smallholders. At worst, it threatens to replace the generational and vernacular knowledge of subsistence farmers and communities about their own crops, conditions, and values.

Plants

The Green Revolution did not initiate a decline in the number of crops that are grown, but it certainly accelerated it. It is no coincidence that wheat, rice and maize were the poster crops of the Green Revolution. Indeed, we might as well call it the *Grain Revolution* or the *Calorie Revolution*, for that is what it was. In fact, it was not Borlaug's intention that only staple grain crops would benefit from the new plant breeding technologies. In his Nobel Prize lecture, he said;

> *"the only crops which have been appreciably affected up to the present time
> are wheat, rice, and maize. Yields of other important cereals, such as sor-
> ghums, millets, and barley, have been only slightly affected; nor has there*

been any appreciable increase in yield or production of the pulse or legume
crops, which are essential in the diets of cereal-consuming populations".[47]

So, does it matter that so few crops now feed so many people? If it does, there is a logical argument that we should simply give more mainstream crops the 'Green Revolution treatment'. This is now happening for some important crops, including non-cereals through the international CGIAR system as well as through national and regional programmes. Whilst there has been some investment in other crops, these efforts are swamped by those on the main staples. The argument remains that a dollar spent on a major staple will have greater return on investment and impact than that spent on other cereals, let alone crops that are not mandates of the international research system. The counter argument is that if the funding dedicated to mainstream crops was matched by that directed to other less favoured crops we could diversify the global food basket. Such a shift in funding emphasis is not that simple without a new paradigm.

The HYV breeding strategy of the Green Revolution was geared towards crops requiring fertilizers, irrigation, and other material inputs. We have already seen that poorer farmers do not have ready access to such technologies. In fact, the Green Revolution varieties were tested and bred in far better conditions than those of most smallholder farmers. This leads to a persistent yield gap between commercial farmers with access to capital and good soils and poorer farmers cultivating marginal soils. This disparity cannot be addressed by plant breeding alone. Marginal farmers without infrastructure, cash, or credit were already disadvantaged in areas where wheat and rice are viable, but this disparity is even greater where they cultivate less-favoured crops. In India, this still results in the seasonal migration of labour from disadvantaged regions to states that are equipped to benefit from Green Revolution technologies.[48] Not only does this migration reduce the capacity to farm in disadvantaged regions, it further marginalizes their traditional crops that have been stigmatized as *poor people's crops* of little value.

Breeding HYVs was based on yield performance in monocultures under high input conditions. An important feature of less-favoured crops is that they are often grown in multiple cropping systems on marginal soils by the poorest communities. In such circumstances, as well as identifying yield targets for such crops grown alone and with inputs, there is a need for research and technologies that can be applied to more complex cropping systems. For this, we need to look at the diversity of the whole agroecosystem.

Nutritional Diversity

The basis of the Green Revolution was to increase yields of major staple crops through HYVs that were responsive to inputs. The measure of success was the yield advantage of the crop in tonnes per hectare of agricultural land. Not only was this comparison made between HYVs and traditional varieties of the same species, but also against that of other crops under similar conditions. When we compare crop yields only in terms of weight per unit area of ground, we skew that advantage to those crops that produce the most calories. In simple terms it takes much more energy to produce one gram of protein than it does to produce the same amount of carbohydrate. Until very recently the preoccupation with the harvested weight of the crop has led to farming calorie-dense crops rather than nutrition-dense crops. The consequences of this strategy are considered in Chapter 4.

Knowledge Diversity

The Green Revolution focused exclusively on crops for which there is an established body of research literature and published evidence. The management practices for these crops were already familiar both for commercial farmers in developed countries and those in developing economies. Varieties and management practices needed improvement and application across geographies supported through extensive scientific research. In the case of less-favoured crops, there is a huge reservoir of local knowledge, management practices, and even belief systems that govern farmer behaviour. Technologies based on experience gained on the major crops cannot easily be transferred to less-favoured crops unless there is a recognition and attempt to include, codify, and test local knowledge and expertise in their improvement and management.

Planet

The Green Revolution has had profound impacts on the integrity and resilience of global ecosystems. Claims for its merits rest on the achievement
or aspiration for ever greater crop yields, in which issues of sustainability
are secondary or ignored. Environmental degradation is justified through
narratives of the urgent need for food security, development, modernization,
and progress — *what else could we have done?* However, dramatic
increases in the yields of *miracle* crops in single fields have significant
ecological consequences at a global scale in which other people and the
planet pay the price. These consequences are presented to consumers as
cost-effective necessities for cheap food and the need to consolidate major
players in the global food system to provide the best value for money
through ever greater efficiencies and economies of scale. For example, the
ownership of global seed supplies is now so concentrated that only a few
companies control most of the global proprietary supply of seeds for the
world's major crops. This is not simply a case of privatizing global seed
systems but also controlling the agricultural biodiversity that is associated
with them.

By trading the *agronomic resilience* of traditional varieties in exchange
for higher yielding but less robust HYVs, the balance between crop and
environment is further compromised. Agricultural biodiversity and multiple cropping systems provide a buffer in adverse seasons and options to
deliver improved nutrition and better diets. Agricultural resilience is most
useful in exactly the kinds of impoverished and fragile environments in
which the poorest and most disadvantaged communities live and farm.

A systemic feature of the Green Revolution is not in the technologies
themselves but in their application without regard to the ecological consequences of their use. The rationale of HYV breeding programmes in
wheat and rice was to incorporate pest and disease resistance into new
varieties to reduce herbicide and pesticide inputs. Indeed, work at IRRI
showed that pesticides did not need to be part of the Green Revolution
package, but overwhelming pressure from the agrochemical industry to
apply pesticides, usually prophylactically, often prevailed. Similarly,
applying luxury levels of nitrogen fertilizers (*just in case*) was encouraged
by their relatively cheap economic costs with little regard to environmental

consequences. It is only by separating these costs that some agricultural operations can be presented as a success story. A recognized failure of economic models is that by off-loading indirect costs and forcing negative effects to a third party, a business can be seen as profitable. In this case the third party of these negative externalities is the planet and those who share it with us.

The misapplication of pesticides can be hazardous for farmers, damage local ecology, and endanger the health not only of workers but ultimately consumers. However, these costs, both economic and moral, are not factored into the Green Revolution balance sheet. Nor are the unintended consequences of pesticide use. For example, insect pest outbreaks are less controlled when their natural predators have been eliminated by the pesticides. Some of those early herbicides, including 2,4,5-T and DDT, were later recognized as unsafe for humans and the environment and have been discontinued in many countries, but their damage has not.[49]

Between 1961 and 1998, global fertilizer consumption increased by 4.1 per cent each year.[50] By incentivizing the application only of chemical fertilizers, soil organic matter and health are compromised. Soils become compacted and impenetrable through the impacts of heavy machinery, decrease in stored organic matter, and destruction of microbes. The consequences are soils whose erosion rates eclipse the rates of soil formation, increase runoff from cropped fields, and leach nutrients and pesticides into waterways and reservoirs. Where fertilizer has ended up in lakes and streams, eutrophication can kill aquatic plants and animals.

Green Revolution crops are thirstier than their traditional counterparts, leading to lower water tables and increased salinization.[51] The allocation of most of humanity's available water to the irrigation of HYVs is simply an untenable prospect in a world that is increasingly short of water. That so much irrigation water leaks before it reaches the field or is lost through non-productive evaporation is a moral as well as an economic issue. The very crops that are most efficient at using water to produce nutritious products are those that have been marginalized by decades of Green Revolution thinking. We return to this in Chapter 7.

Final Thoughts

"We need both productivity and sustainability — and there is no reason we can't have both. Many environmental voices have rightly highlighted the excesses of the original Green Revolution. They warn against the dangers of too much irrigation, fertilizer, or chemicals. They caution against a con-solidation of farms that could crowd out small-holder farmers. These are important points, and they underscore a crucial fact — the next Green Revolution has to be greener than the first. It must be guided by small-holder farmers, adapted to local circumstances, and sustainable for the economy and the environment. Let me repeat — the next Green Revolution must be guided by small-holder farmers, adapted to local circumstances, and sustainable for the economy and the environment. The last thing any-one should do is create short-term gains for poor farmers that have long-term costs for their children". (Gates, 2009).[52]

Key criticisms of the Green Revolution have been so eloquently articu-lated by none other than Bill Gates. But there is little evidence that much has been done to change the dynamic amongst donors and government agencies towards new approaches to achieving both productivity and sus-tainability. In fact, questions related to sustainability and the environment also appear in the original Green Revolution narrative. One of Norman Borlaug's biggest claims, the "Borlaug hypothesis", is that the Green Revolution *"prevented land from falling under the chainsaw and the plough"*. Borlaug argued that because of improved Green Revolution tech-nologies, between 1950 and 1998 over 1,200 million hectares of land were saved from deforestation.[53] The narrative remains that *there is no alterna-tive* to Green Revolution agriculture and anything else would have been worse. This analysis is flawed for several reasons.

First, without subsidies the Green Revolution could not have suc-ceeded. In 1966, the Government of India with the assistance of the Rockefeller Foundation purchased 18,000 tonnes of Mexican wheat seeds under its HighYielding Varieties Programme — the largest purchase of seeds ever. In the Philippines, within a year of the launch of its food self-sufficiency programme, subsidies for rice increased by 50 per cent.[54] The Mexican government purchased domestically grown wheat at above world market prices.[55] The governments of India and Pakistan paid 100 per cent more for their own wheat.

Second, the assumption is that over the same period of massive investment in HYVs, the yields of non-Green Revolution crops did not increase. In fact, between 1967 and 1970, Indian production of barley, tobacco, jute, chickpea, tea, and cotton all increased by 20–30 per cent without the benefit of Green Revolution technologies nor the subsidies afforded to it.[56]

Third, the ecological consequences of the Green Revolution have been substantially underestimated. Rather than the principle of `producer or polluter pays', the costs of environmental degradation, biodiversity, habitat loss, and damage to human and planetary health have largely been passed on to the state, the planet, and ultimately our children. We cannot claim that we were not warned. As early as 1962 Rachel Carson wrote *Silent Spring,*[57] the seminal work which led to the banning of DDT and the raft of legislation that was introduced to protect the planet's air, land and water. Whilst she is now rightly recognized as one of the founders of Environmentalism it is almost forgotten how much fury Silent Spring and its author attracted from the Chemical Industry and their political allies. It was not until 1970 that Congress passed the National Environmental Policy Act[58] which established the Environmental Protection Agency. Interestingly, whilst the domestic production of DDT in the USA was banned, its export wasn't and its traces are still found everywhere; in every bird and fish on every island and in the breast milk of every mother. Silent Spring serves as a lesson for us all nearly six decades later.

Any true evaluation of the Green Revolution would need an environmental audit of negative externalities that need to be included as part of the analysis, and as a consequence, a change in practices. In 1997, Sir Gordon Conway, a British agricultural ecologist and former president of the Rockefeller Foundation and the Royal Geographical Society, proposed *The Doubly Green Revolution: Food for all in the twenty-first century* in which he articulated the shortcomings of the Green Revolution in terms of its pursuit of productivity at the cost of sustainability and the environment, and how these could be addressed without losing its benefits.[59]

Fourth, the suggestion that the only alternative to the Green Revolution was to do nothing is simply wrong. Rather than ask what would have happened had it never occurred, we should ask what might have happened had the substantial resources, marketing, and research effort devoted exclusively to the Green Revolution been deployed differently. What could have been achieved had these resources been used to improve other crops and cropping systems that were both productive and sustainable and have now been shown to be resilient to extremes of climate?

We cannot with any confidence provide a comparative analysis of alternatives to the Green Revolution over the last decades because many of them were never given the chance to happen. However, we can analyze and invest in credible options for the next decades that are fit for our future and that of the planet. The rest of this book attempts to show how.

4

THE TITANIC AGRIFOOD SYSTEM

Revolution Number 7

Context

It is sometimes said that the global agrifood system is broken. It is not. It is, however, heading in the wrong direction, and given its global nature, its future path has consequences for us all. This chapter looks at how our current agrifood system evolved, where it is now, and what challenges it must overcome in the next decade if we are to avoid a health and environmental catastrophe for humanity and the planet.

> *The English word titanic, meaning "enormous, exceptionally strong, massive," comes from the name of the Greek gods called Titans, the children of Gaia (Earth) and Uranus (Sky)*

As a metaphor, we can think of the global agrifood system as Titanic. A sleek, modern ocean liner — in fact, the largest ship ever built. Its tanks are full of fuel, it has a payload of multinational passengers, and enough food and entertainment to keep them happy throughout the journey. The ship is part of a fleet owned by an international consortium registered in a faraway tax haven. The crew all wear identical smart uniforms donated by a corporate sponsor. A team of professional technicians and engineers ensures that the ship's state-of-the art technologies and entertainment systems are in perfect working order and that most passengers are happy with the food and level of service. Captain Horace de Piffle Reckless, who is in charge of the ship has earned a reputation for being popular with his passengers. He wears a number of medals on his chest, earned from military campaigns before he joined commercial shipping.

The ship is already full, but at its last port of call it took on a number of new passengers who, because all the cabins were occupied, must sleep on the deck. The ship is moving at maximum speed when it strikes a few small submerged icebergs causing it to lurch alarmingly and take on water. The captain reassures the passengers that a 'few bits of ice' cannot possibly cause his huge ship to capsize. However, the ship's helmsman reports that the sophisticated navigation equipment has located more obstacles ahead and if the ship doesn't change course immediately it will eventually hit a large iceberg that could sink the ship. Nevertheless, the captain continues on the same route insisting that all will be well because the ship is more than capable of surviving an iceberg, and anyway, a safe port is already in sight. Some of the wealthier passengers have already seen the danger and occupied all the lifeboats, leaving none for the other passengers. The ship's owners are not on board, their company has full personal indemnity against disasters, and they have plenty of other ships in their fleet. Meanwhile, the band plays on.

This chapter uses the Titanic narrative as a metaphor for the current state of the global agrifood system. The passengers are all of us apart from those who disembarked at early destinations and others who couldn't afford the passage. The submerged icebergs ahead are called Environment, Society, and Health, and, beyond them, the looming iceberg is called Climate Crisis. It is for the passengers to decide whether the ship can survive the iceberg or that its captain must immediately change course to avoid sinking. We do not yet know if they will decide to take charge of the ship's direction or leave the decision to Captain Reckless and his crew.

Cheap Food Paradigm

We have never had it so good. Since the Green Revolution, food production has consistently outpaced population growth and yield increases of the major staple crops have ensured global food security with little change in land use. The global agrifood system, with its complex supply chains from production, transport, processing, packaging, storage, retail, consumption, loss, and waste of foods and their ingredients still successfully feeds most of the world's population. It also provides livelihoods for more than 1 billion people. Since 1961, food supply *per capita* has increased by more than 30 per cent despite population growth of 4.8 billion over the same period.[1] A highly efficient agrifood sector has also been underpinned by national and international investments in research and development, liberalized trade and regulatory frameworks, global marketing campaigns, and huge subsidies.

Through the competitive nature of the free market, the global agrifood system has driven higher yields and economies of scale in the production and processing of the major commodity crops and new records in food production and efficiency that have delivered enough food to feed everyone on earth.[2] Competitive forces have stimulated the adoption of intensive and large-scale agriculture, especially in food-exporting nations, and the development of sophisticated supply chains from producer nations to ensure just-in-time delivery of products to consumers across the globe. This intensive and productive agrifood system has made food items more available to more consumers, and thanks to a globalized food system and free trade agreements, the relative price of food in most of the major markets is at an historical low. On a global scale, food has never been more plentiful, more accessible or more affordable to billions of consumers. Indeed, the proportion of the family budget spent on food in many countries has been declining for decades — the typical US household now spends only 10 per cent of its income on food.[3]

The global agrifood sector has achieved a *Cheap Food Paradigm* for its investors, consumers, and advocates, and products that are increasingly available, attractive, and affordable across societies and geographies. In real terms, food has never been cheaper nor its preparation easier. This has left many consumers with more disposable income and spare time to spend on non-food purchases. To achieve the Cheap Food Paradigm, the global agrifood system has consolidated its businesses, industrialized its processes, and delivered massive economies of scale through technologies, standardized ingredients, and slick marketing — the public wants what the public gets. The global agrifood system is truly titanic: it is Revolution Number 7.

Industrial Agrifood System

The basis of the modern agrifood system is the use of preservation and processing techniques to transform raw materials, primarily carbohydrates, into ready-to-eat food products and deliver them along vulnerable global supply chains to distant consumers. Food technologists, engineers, and factory operations have enabled thousands of products to be manufactured for global markets, and feedstock for the agrifood system has been derived from the main staple crops of the Green Revolution (Revolution Number 6) using technologies that are the descendants of the empires of power (Revolution Numbers 2, 3, 4 and 5) that made it possible.

Food processing involves transforming agricultural ingredients into edible products or converting one form of food into others. These transformations can be done at the primary, secondary, or tertiary level. Primary food processing makes agricultural ingredients such as grain or meat edible. To achieve this, it employs ancient methods such as drying, threshing, winnowing and milling grain, shelling nuts, butchering animals, and more recent methods of freezing fish, meat, and vegetables, extracting and filtering oils, canning food, and homogenizing and pasteurizing milk. Food processing initially involved the conversion of seeds (grain) into flours. This would have been done in-house until the advent of mechanical mills at the village or community level, where individuals were allocated milling time to process their own grain. Meat and fish were air-dried or smoked, spices dried, and vegetables pickled. Secondary food processing involves making foods from already prepared ingredients, for example baking bread, fermenting fish, and making wine and beer. By converting grain into flour and then baking it, communities started developing supply chains, initially these were simple and local because of the perishability of products. Packaging allowed goods to be transported and protected. Some secondary food processing methods are commonly known as *cooking*. More recently, tertiary food processing has involved the industrial scale production of processed foods from primary or secondary processed ingredients into large amounts of identical products destined for mass markets.

Food processing is as ancient as food itself; humans have been cooking food using fire for at least 250,000 years.[4] Ancient processing techniques include fermenting, sun drying, salt preservation, and different kinds of cooking such as roasting, smoking, steaming, and baking. Such methods cause chemical changes that restructure the food from its natural form and help serve as a barrier to microbial contamination. Techniques such as salt preservation were evident in ancient Greek, Egyptian, and Roman civilizations and remained largely unchanged until the Industrial Revolution. Convenience and processed foods also have an ancient history. The Aztecs of central Mexico developed several convenience foods, for example *pinolli* from ground and dried maize meal, which could be taken on long journeys and only needed water for preparation. In 2005, a bowl of 4,000 year-old millet noodles was discovered at the Lajia archaeological site in China.[5] In 2018, cheese found in the tomb of Ptahmes the Major of Memphis in Egypt dates back 3,200 years,[6] and in the same year in

Jordan's Black Desert, archaeologists unearthed 14,400-year-old bread made from wild cereals in a stone fireplace.[7] Even popcorn is not new — Peruvians were eating it and other maize-based foods up to 6,700 years ago.[8] In fact, processed foods have made up a significant part of the human diet whenever it was not possible to eat fresh foods because of seasonal limitations or as a result of crop failures or wars.

In England, mass-scale food processing was introduced in the 18th and 19th centuries to feed manual labourers arriving from rural areas to manufacturing centres during the Industrial Revolution. Many modern food processing technologies were also developed in the 19th and 20th centuries to meet military needs.[9] For example, in 1809 the French chef Nicolas Appert developed a technique for bottling that could hermetically preserve food to supply to French troops in Napoleon's War of the Fifth Coalition. In 1810, Peter Durand developed a technique for canning foods again to support military needs.[10] Pasteurization, invented by Louis Pasteur in 1864, improved the quality and safety of preserved foods. Pasteur demonstrated that contamination by microorganisms caused the spoiling of beverages, such as wine, milk and beer.[11] The process of pasteurization involves the heating of liquids to between 60 and 100°C which kills most of the bacteria and moulds within them.

Pasteur patented the pasteurization process which was then applied to wine, beer, and milk. Processes such as canning and pasteurization have allowed longer storage life of countless food products through an understanding of the microbiology of spoilage and food pathogens. In the 20th century, technologies developed for warfare and also now for the space race, were applied to food processing techniques. These include spray and freeze drying, evaporation, juice concentrates, and the incorporation of artificial sweeteners, colouring agents, and preservatives into foods. In the late 20th century, processed products such as dried instant soups and ready meals became the mainstay of American urban households, supported by new appliances such as the microwave oven, electric blender, and automatic bread maker.

An important aspect of food processing was improvements in food safety, including the removal of toxins, extension of shelf life, and maintenance of food consistency and quality. These processes allowed many foods to be made available throughout the year, enabled perishable foods to be transported along supply chains, and for them to remain safe to eat by deactivating spoilage and pathogenic microorganisms such as salmonella that can cause serious illnesses.

Not all food processing was designed with safety and quality in mind. To make cheap ingredients acceptable to the consumer, many heavily processed foods are high in fat, with added sugar and salt, preservatives, artificial colours, and flavourings. Food processing can also decrease the nutritional content of processed foods. Nutrient losses depend on the type of food and method of processing. For example, heat treatment destroys vitamin C and canned fruits contain less of it than the fresh equivalent. Damage to food processing machinery can also introduce contaminants during production processes. Such contaminants can be transferred from a preceding operation, through animal or human bodily fluids, microorganisms, and metallic and non-metallic fragments. The incorporation of these contaminants into the production process can lead to equipment failure and ultimately to human consumption.

Hygiene and prevention or elimination of pathogens have now become an essential requirement for the food industry, and a modern food factory for any consumable product now resembles a hospital environment rather than the craft-based industry from which it derives. Factory operations require protective clothing, separation of raw and processed materials, air flow control of the environment, and packaging. Food packaging, based on petroleum industry products, has allowed the development of highly processed foods encased in breathable or gas-impermeable materials, resealable packets, and edible coatings. These incredibly useful technologies to package the food that we eat are also a principle source of the plastic that permeates the global environment and the carbon dioxide that permeates the atmosphere. There really is no such thing as a free lunch.

Cost of Cheap Food

According to FAO, to feed 2.3 billion more people by 2050, farmers will have to produce 70 per cent more food.[12] The first challenge is whether our current business-as-usual model can deliver this extra food from the same cohort of staple crops that already feed nearly 7.8 billion people. This is not simply a question of intensifying food production, but tightening the strait jacket of the food industry on the global food supply chain. These huge increases in crop yields will need even greater control of raw materials through farming contracts that stipulate not just the species but also the varieties used, sowing times, application of pesticides and herbicides, and harvesting times that are determined by factory needs. The second challenge is whether higher yields, further economies of scale, greater consolidation, and new technologies can deliver food at the same relative price. Is the Cheaper Food Paradigm sustainable?

At what point does further industrialization of the whole food system risk its collapse? The third challenge is whether more cheap food can be delivered to more people without using more land — can the planet cope with *sustainable intensification* of industrial agriculture? If we cannot say yes with certainty to all of the above questions, there is a global problem. Note that all the above challenges assume that food production simply has to feed more people without any other considerations — no pandemics, no global warming, no disturbance of weather, transport or retail systems, no plagues of insect pests and diseases, no loss of soils or decreases in water resources, no changes in human behaviour, no consequences on human health or social justice, no food or nutritional insecurity.

Global food production has now been reduced to a few staple crops grown as industrial monocultures. The term *staple* implies that these crops are primarily grown to feed us. However, having such a tiny number of such crops has led some of the best scientific minds to focus on remarkable innovations in their use. Perhaps we should call them *ubiquitous* crops since their products permeate almost every aspect of human societies. Take, for example, maize. A ubiquitous component of processed foods is high fructose corn syrup (HFCS) derived from maize. HFCS is a sweetener made from maize starch. It has a similar chemical composition and effect on the body as sugar, but it is much cheaper. HFCS is sweeter and more soluble than glucose so can be used as a cheap sweetener in a variety of products such as carbonated soft drinks.

Of course, we can always claim that since we never consume fizzy drinks it is easy to avoid HFCS, so how then is it ubiquitous? Here are some of the other products that contain HFCS — candy bars, yoghurt salad dressing, breads, canned fruit and fruit juices, breakfast cereals, doughnuts, cookies, sauces, condiments, crisps, cookies, crackers, energy bars or health bars, coffee creamer, sports drinks, jams, jellies, and ice cream. Any of these products can masquerade as healthy by simply including *low fat* or *fat free* on the label. It is hard to imagine a modern consumer who can avoid all of them. Similarly, we can always claim that all of these products are technically foods and therefore maize is still a staple crop. However, the industrial production of HFCS produced on a large scale using high-tech methods involves several enzyme processes that require specific varieties of maize that produce high levels of starch that is easily hydrolyzed. So, the HCFS producers have control of large amounts of maize that could have been be used as a staple crop to feed people, but is now used as a sweetener and therefore displaced from its use as a staple food. Similarly, the use of maize-derived alcohol for gasohol, a mixture of gasoline and ethyl alcohol, has diverted a large amount of maize from feeding people to feeding machines.

It is increasingly clear that the Cheap Food Paradigm comes with huge costs and increasing risks for people and planet with environmental, social justice, and health consequences. When we include these negative externalities, our titanic food system looks less cost-effective, more damaging to humans and ecosystems, and riskier. The reality is that we can no longer ignore the negative externalities on environment, society, and health caused by our current food system if it is to be fit for the future. And then there's climate change.

Returning to our maritime metaphor, the future of the Titanic food system will depend on how our ship negotiates the submerged icebergs of Environment, Society, and Health and avoids the calamitous iceberg of Climate Crisis.

Environment

Industrial agriculture is characterized by large-scale monocultures, heavy use of chemical fertilizers, herbicides and pesticides, and industrial meat production in restricted spaces used for intensive animal farming. As well as processed foods, staple crops are increasingly required to provide ingredients for animal feeds and biofuels accompanied by greater use of nitrogenous fertilizers and water for irrigation. These feedstocks for food, feed, and fuel must be transported along carbon-heavy supply chains to locations where industrial processes reaggregate them into foods. Mainly calorie-rich processed foods are then distributed around the world to distant consumers, again along carbon-heavy distribution systems to consumers who are often in different countries or continents far from where the original crops and animals were farmed or their products constructed.

Like all extractive industries, industrial farming uses the soil as a resource to be tapped and if the supplies of, say, nutrients or water are insufficient, these can be replenished by fertilizers and irrigation. In monocultures, the assumption has been that such replenishment, plus the removal of weeds and plant pathogens, provides a continuous cropping cycle without the need for years of fallow for the soil to recover or leguminous cover crops to restore soil nitrogen. The increasing evidence is that continuous monocultures exhaust soil fertility, need higher and higher doses of fertilizers, and damage soil structure.[13] A degraded soil is more vulnerable to erosion, especially top soil, leaving the cropped field susceptible to nutrient and water runoff to adjacent fields and water courses. Water pollution through fertilizer runoff contaminates downstream drinking water supplies and natural water courses. Irrigation accelerates this process where runoff and drainage beyond the crop root zone add to water movement from rainfall.

By their very nature, monocultures do not support the rich range of biodiversity in natural systems or even diverse cropping systems. When we then include the impacts of pesticides and herbicides, agricultural lands suffer from lack of ecosystem services such as pollination that more diverse landscapes offer.

Society

The Green Revolution has helped food production outpace the needs of a rapidly growing global population. At the same time, it was assumed that greater efficiencies and economies of scale would enable farming to remain a profitable business. However, the pressure to consolidate has meant a *get big or get out* culture that has taken its toll on communities. There are now fewer and fewer ageing farmers growing fewer and fewer mainstream crops to feed more and more increasingly malnourished people. The impacts are not just on the farmers themselves but on the economies of rural communities and farm states. In the early 1900s around 40 per cent of the total US population lived on farms and 60 per cent lived in rural areas. Today, only about 1 per cent live on farms and about 20 per cent live in rural areas.[14] With the industrialization of agriculture, farms were consolidated into ever larger units and food is produced in ever greater quantities from the same area of land.

Studies around the world have shown that farming is a stressful occupation with higher rates of suicide than that amongst the general population.[15] Suicides are particularly high amongst small-scale farmers experiencing economic distress. In the United Kingdom, Europe, Australia, Canada, and the United States, farmers have the highest suicide rates of any industry and are more likely to develop mental health problems.[16] As well as deteriorating mental health, the causes of farmer suicides include, the physical environment, family problems and economic stress and uncertainty. In India, farmer suicides have been a national catastrophe since the 1990s. The National Crime Records Bureau of India reported that since 1995, nearly 300,000 Indian farmers have committed suicide.[17] Of these, 60,750 were in Maharashtra or in other states with weak financial regulations. For India, researchers have identified various reasons for farmer suicides, including high debts, poor government policies and corruption, crop failure and personal and family problems.[18] Whilst there is no single cause, greater debt burden and the inability to repay loans to landlords, banks, and money lenders are common factors that will only increase where societies suffer further disruption and instability. The current farmers protests in India further illustrate the depth of the crisis in agriculture — in a country lauded as a success story of the Green Revolution.

Food Insecurity

Concepts of food security have evolved in the last 30 years to reflect changes in official thinking and societal attitudes. The term first appeared in 1974 when the first World Food Conference defined food security in terms of food supply;

"Availability at all times of adequate world food supplies of basic foodstuffs to sustain a steady expansion of food consumption and to offset fluctuations in production and prices".[19]

This concept of food security was clearly intended to ensure the supply, availability, and price stability of food at the international and national level. It reflects the Green Revolution zeitgeist of the time — expanding the supply of staple crops (basic foodstuffs), encouraging their greater consumption (making them widely available), and stabilizing their costs to the consumer (policy interventions to reduce price volatility). The 1974 World Food Conference was held in Rome under the auspices of the FAO, but the impetus for it was the devastating famine in Bangladesh that occurred in the preceding two years.[20] This preventable human tragedy further justified the global priority to ensure basic food supplies and their availability for consumption.

In 1983, FAO expanded the concept of food security to include access to food and provided a new definition that reflected the balance between the demand and supply side of food security;

"Ensuring that all people at all times have both physical and economic access to the basic food that they need".[21]

The analysis was revised to include the individual and household level and national and regional scales. In 1986, the World Bank Report on Poverty and Hunger focused on temporal and economic dynamics of food insecurity.[22] It made a distinction between *chronic* food insecurity, associated with continuing or structural poverty and low incomes, and *transitory* food insecurity, caused by natural disasters, economic collapse, or conflict.

In 1996, the World Food Summit supported a revised FAO definition of food security;

"Food security exists when all people, at all times, have physical, social and economic access to sufficient, safe and nutritious food which meets their dietary needs and food preferences for an active and healthy life".[23]

From this comprehensive and now widely accepted definition, two phrases are especially striking. *All people, at all times* means the reliable and timely distribution of food across geographies and generations: food must be globally available for all. *Sufficient, safe and nutritious food which meets their dietary needs and food preferences* says that having enough food is not enough: it must be safe to eat, nutritious, and part of a desirable diet. We might consider food security at the individual, household, national, or even regional level, but to meet the FAO definition it must be global. If not, we have *food insecurity* that occurs wherever and whenever food is unavailable, unsafe, unhealthy, or undesirable.

The crowning achievement of the Green Revolution was that it generated food surpluses which outpaced population growth. It is sometimes argued that there is plenty of food on the planet, and to achieve global food security we simply need to distribute it better and waste less of it along the supply chain and in the household. Both better distribution and less waste are critically important factors, but they are not enough to ensure food security. From the 1996 FAO definition, we can see that irrespective of its quantity, availability, or distribution, food security must include *nutrition*, and therefore, food insecurity causes *malnutrition* or malnourishment. Malnutrition can be either because individuals lack access to enough healthy food (undernourishment) or may choose to eat too much unhealthy food (overnourishment). So, the definition unequivocally puts nutrition at the heart of the food security agenda.

Undernourishment

After decades of steady decline, since 2015 the prevalence of undernourishment has remained stubbornly around 11 per cent of the global population, but with population growth the absolute number of people who are undernourished has slowly increased.[24] In 2019, more than 820 million people in the world were still undernourished, a figure that is equivalent to the combined populations of the United States of America, Indonesia, and Pakistan (respectively the third, fourth, and fifth most populous countries on the planet).[25] The number of undernourished people is increasing across almost all of Africa with a regional level now at almost 20 per cent. Undernourishment is also slowly rising in Latin America and the

Caribbean, although its prevalence is still below 7 per cent.[26] Western Asia shows a continuous increase since 2010, with more than 12 per cent of its population currently undernourished.[27]

Overnourishment

In 2019, 2 billion adults, 207 million adolescents, and 155 million children were overweight or obese — a total approaching 2.4 billion people.[28] Again, for comparison, the combined population of China and India is 2.8 billion people. Of those under-fives who were overweight, half were in Asia and a quarter in Africa.[29] Between 2000 and 2018, the number of overweight children under five increased by an astonishing 44 per cent.[30] We can see that obesity is not simply a problem for rich countries, it is a global pandemic that affects every continent. Middle-income countries such as South Africa, Brazil, and China have increasing rates of overweight and obesity across all age groups and economic levels. The 10 most obese countries are all developing small island states in the Pacific, ranging from Nauru, with an obesity level of 61 per cent, to the Federated States of Micronesia with an obesity level of 45 per cent.[31] Of the next 10 most obese countries, nine are in the Middle East and the Gulf region and the tenth is the USA.[32] Obesity is not only a question of geography or wealth. We can see dramatic differences in obesity level between developed countries, say the USA (36.2 per cent) and Japan (3.6 per cent).[33] In addition to its costs to economies and health systems, obesity is increasingly killing us. It is now one of the leading risk factors for premature death, accounting for 4.7 million deaths globally in 2017 — around 8 per cent of the global total.[34] Again, there are large differences in death rates from obesity across the world. In the Pacific Island of Nauru, obesity rates are above 60 per cent whilst in Vietnam they are as low as 2.1 per cent.[35]

Triple Burden of Malnutrition

Over 2.7 billion people are either underfed or overfed — over 20 per cent of the global population is malnourished.[36] As with many global issues, it is tempting to look for simple solutions to complex problems. For example, undernourishment and overnourishment have historically been considered as separate challenges affecting distinct populations. Why not simply provide more food to those who are undernourished and persuade those who are overnourished to eat less? Looking for more nuanced policy decisions around risk factors, we might also link undernourishment with poverty, food insecurity, and infection, and overnourishment with affluence, junk foods, and sedentary behaviour. Such simplistic analyses and solutions are not helpful. Overnourishment and undernourishment are inextricably linked because they both occur within communities, families, and even individuals at different stages of their lives.

The popular media image of malnutrition is of an emaciated individual (usually an African child) who is *hungry* and therefore food insecure. Hunger, however, can also apply to an individual who consumes enough or even too much food but is *hungry* for nutrients that are lacking in the diet. An individual may not be physically hungry or underweight, but their poor-quality diet means that they are still food insecure through *hidden hunger*.[37] The phenomenon of hidden hunger occurs when the quality of food an individual eats is deficient in micronutrients such as vitamin A, iodine, and iron that are essential for growth, development, and ability to work. The prevalence of hidden hunger across societies, geographies, and economic groups provides a common link between the undernourished and overnourished. We can now see that there are three forms of malnutrition — undernourishment, overnourishment, and hidden hunger. Together, they represent the *triple burden* of malnutrition.

To address this triple burden, we must step outside our disciplinary, geographical, and economic silos, seek common approaches, identify priorities, and enforce actions based on evidence. It is difficult to agree on priorities because malnutrition is a global issue that affects people irrespective of their location, nationality, socio-economic status, sex or gender or whether they live alone, in large households or supportive communities. Anyone can experience malnutrition, but the most vulnerable groups are children, adolescents, women, and those who are immunocompromised. Of these, malnourished children are those most vulnerable with life-long consequences for themselves and humanity.

Malnourished Children

If we do not address the issues of malnutrition amongst the world's children even before they are born, we will ensure a global generation of physically, mentally, and socially malnourished adults. Children are affected by all forms of malnutrition either during childhood or subsequently throughout their lives. Worldwide, 5.6 million children die before their fifth birthday every year, with 80 per cent of these deaths occurring in sub-Saharan Africa and Asia.[38] Malnourished children are susceptible to infectious diseases because their immune systems are compromised. Childhood malnourishment is part of a malnutrition cycle in populations where pregnant women are themselves malnourished. As a consequence, infants born to undernourished mothers are of low birth weight, cannot achieve their full growth potential, and are at risk to infections, illness, and early mortality. This cycle is perpetuated when low birthweight female babies grow to become malnourished children and adults and subsequently produce low birth weight infants. The visible evidence of undernourishment is *stunting* (chronic malnutrition), or *wasting* or low weight for height (acute malnutrition). Overnourishment includes overweight and obesity, which are frequently linked to diet-related non-communicable diseases (NCDs) such as diabetes mellitus, heart disease, and some forms of cancer and stroke.

Stunting

Stunting is the impaired growth and development of children caused by poor nutrition, frequent infection, and decreased resistance to diseases such as diarrhoea and pneumonia.[39] Children are defined as stunted if their height-for-age is more than two standard deviations below the World Health Organization (WHO) Child Growth Standards median.[40] Although children are only classified as stunted from around two years of age, their condition is the result of poor nutrition of the pregnant mother and their own low birthweight. In 2015, one in seven new-borns, or 20.5 million babies globally, suffered from low birthweight.[41] The effects of stunting are largely irreversible — a stunted child will become a stunted adult who may never reach full height and whose brain may never achieve full cognitive potential. Globally, 144 million children under five suffer from stunting.[42] Their lives are blighted by learning difficulties in school, poorer income opportunities as adults, and barriers to full participation in their communities. Childhood stunting may also lead to increased mortality and a greater risk of adult obesity.

Wasting

Wasting involves a rapid decline in nutritional status in children under five years of age. Children are defined as wasted if their weight-for-height is more than two standard deviations from the median of the World Health Organization (WHO) Child Growth Standards.[43] Wasting in children is the outcome of poor nutrient intake and/or disease and is life threatening. As well as being more likely to die, children suffering from wasting have reduced immunity and are at risk from long-term developmental effects. In 2019, 47 million children under five were wasted, of which 14.3 million were severely wasted.[44] Wasting and stunting are often described as distinct forms of malnutrition that require different interventions for treatment or prevention, however both forms of malnutrition are closely linked and may occur concurrently in the same populations and in the same children. Wasting and stunting are both causes of increased mortality, especially when they are present in the same child.

Obesity

Children who are obese are above the normal weight for their age and height. Obesity in children often leads to health problems in adulthood. Resulting chronic diseases may include diabetes, stroke, high blood pressure, cancers, and heart disease. Body Mass Index (BMI) is a measure used to categorize overweight and obesity. Overweight is defined as a BMI at or above the 85th percentile and below the 95th percentile for children and teens of the same age and sex.[45] Obesity is defined as a BMI at or above the 95th percentile for children and teens of the same age and sex.[46] BMI is calculated by dividing a person's weight in kilograms by the square of their height in metres.[47] For children and teens, BMI is age- and sex-specific and is often referred to as BMI-for-age. Since their body composition changes, a child's weight status is calculated using an age- and sex-specific percentile for BMI rather than that used for adults.

Climate Crisis

Atmospheric carbon dioxide concentrations are at their highest levels in over 800,000 years.[48] Over this period, fluctuations in atmospheric CO_2 coincide with the onset of ice ages (low CO_2) and interglacial periods (high CO_2) associated with changes in the Earth's orbit around the sun. Since the Industrial Revolution, there has been a rapid rise in global CO_2 concentrations that cannot be explained by the habits of our planet. Over the previous 800,000 years the atmospheric concentration of CO_2 never exceeded 300 parts per million (ppm), it is now above 410 ppm.[49] The cause of these increased concentrations is us — mainly by our burning of fossil fuels. Coal, oil, and natural gas are all considered as fossil fuels because they were formed over millions of years from the carbon-rich, buried and fossilized bodies of plants and animals. When they are burned, fossil fuels emit CO_2 — usually into the atmosphere. In 1817, at the heart of the Industrial Revolution, the global total emission of CO_2 was 49.43 million tonnes.[50] In 1917, the figure was 3.53 billion tonnes, and in 2017 global CO_2 emissions were 36.15 billion tonnes — a ten-fold increase in a century.[51] Today, China is the world's largest CO_2 emitter — accounting for over 25 per cent of emissions. It is followed by the US (15 per cent), EU-28 (10 per cent), India (7 per cent), and Russia (5 per cent).[52] If these figures are expressed *per capita*, the emission per US citizen is 16.5 metric tonnes per year, while the comparable *per capita* emissions for each citizen of, say, Malawi, Chad, and Rwanda are 0.1 metric tonnes, or 0.006 per cent of the US value.

Increases in atmospheric CO_2 and other greenhouse gases such as methane and nitrous oxide have resulted in an increase in global average temperatures of more than 1°C since the Industrial Revolution. Together, increases in greenhouse gases and associated global warming caused by human activities are the main causes of climate change — natural variability has played a very minor role. Given their severity and consequences, global warming is increasingly described as *global heating* and climate change as the *climate crisis*. Most scientists, observers, and governments agree that the climate is changing and that its impacts are existential for humanity. In 2015 in Paris, UN member parties set a target of limiting global warming to 2°C above pre-industrial temperatures and agreed to *"pursue efforts"* to limit this to *1.5°C.*[53] More than five years later, the world is off track to meet the agreed target of 2°C let alone 1.5°C. Without drastic changes in our current actions, predicted warming will be between 3.1–3.7°C.[54] Eight of the 187 UN member signatories have still not ratified the Paris Accord. The contributions to greenhouse gas emissions of these countries amount to less than 4 per cent of the total — Iran (1.66 per cent), Turkey (1.04 per cent), Iraq (0.48 per cent), South Sudan (0.24 per cent with Sudan), Angola (0.16 per cent), Libya (0.14 per cent), Yemen (0.07 per cent), and Eritrea (0.01 per cent).[55] In April 2016, the United States signed the Paris Climate Agreement, and accepted it by executive order in September 2016. This was reflective of the critical role that the United States has in addressing climate change; a result of both its high emissions and its economic ability for climate adaptation. However, on June 1, 2017, in a televised announcement from the White House, former President Donald Trump said;

"In order to fulfil my solemn duty to protect the United States and its citizens, the United States will withdraw from the Paris climate accord."[56]

The US began the withdrawal procedure in 2019 and formally left on 4 November 2020. At the time, its carbon emissions exceeded the combined total of India, Japan, Russia and Germany.

At noon on 20 January, 2021 Eastern Standard Time, President Joe Biden was sworn in as the 46th President of the United States of America. Amongst a raft of 19 Executive Actions that he signed on the same day was the following statement;

"I, Joseph R. Biden Jr., President of the United States of America, having seen and considered the Paris Agreement, done at Paris on December 12, 2015, do hereby accept the said Agreement and every article and clause thereof on behalf of the United States of America."[57]

A changing climate is responsible for a range of potential ecological, physical, and health impacts, including extreme weather events (such as floods, droughts, storms, and heatwaves), sea-level rise, disturbed agricultural cycles, and disrupted water systems. Since large amounts of CO_2 are embedded in traded products, the carbon emissions of some countries increase while those of others decrease when national emissions are calculated on consumption rather than production.

Role of Agriculture in Climate Change

Over the past 250 years, the three main causes of increased greenhouse gas emissions have been burning of fossil fuels, changes in land use, and agriculture — which of course is linked to the two other causes.[58] In addition to the use of fossil fuels and changes in land use, agriculture contributes directly to emission of greenhouse gases through methane and nitrous oxide emissions from livestock, crop cultivation, use of synthetic fertilizers and manures, burning of crop residues, and biomass for land clearance. Today, agriculture is directly the cause of more than 25 per cent of total global greenhouse gas emissions, the entire global agri-food system contributed 37 per cent of total emissions, and is on course to increase these by 30–40 per cent by 2050.[59] Agricultural activities are responsible for 54 per cent of methane emissions, 80 per cent of nitrous oxide emissions, and virtually all CO_2 emissions linked with land use.[60] Livestock farming contributes in two ways — in addition to enteric fermentation in ruminants, crops such as maize and alfalfa are used as livestock feed. Worldwide, livestock occupies 70 per cent of the land area used for agriculture — around 30 per cent of the land surface of the Earth.[61]

Impacts of Climate Change on Agriculture

Climate and agriculture are interrelated, and many of the flows between them take place at a global scale that makes it difficult to predict local effects. Nevertheless, the climate crisis is already affecting agriculture through changes in atmospheric CO_2 and ozone level, temperatures, rainfall, evaporation, ground water and sea level, climate extremes (heatwaves and floods), and the incidence and distribution of insect pests, diseases, and their vectors. These changes affect the cultivation, yield, and distribution of crops and even their nutritional content, as well as the quality of food derived from them. The climate crisis is a global reality, and the challenge is to decide how much, how fast, where, and what we should do about it.

Plant and animal species (including humans and even policy makers) respond to the amplitude in climatic factors as well as their average. For example, a crop might grow quite comfortably at a constant temperature of perhaps 28°C, but not at the same average if the temperature varies between 20°C and 36°C. Planning for agriculture cannot simply rely on average values, but also on the degree and frequency of variations. Indeed, it is the magnitude and frequency of such *extreme events* that are more likely to persuade cynics, sceptics and governments of the need to address underlying trends. Another challenge is that we now have a reasonable understanding of the rate of *global heating*, but we are far less certain about its impacts on the pattern and distribution of rainfall and the sensitive feedback loops between temperature and rainfall, such as evaporation and atmospheric humidity. This makes it even more difficult to predict the impacts of climate change at any location. Even as our understanding of changes in *abiotic* factors such as atmospheric and soil conditions improve, we remain unclear as to how these changes will interact with *biotic* factors such as plants, animals, insects, and microbes.

Likely Impacts on Agricultural Production

We continue to argue about its culprits, causes, and consequences, but the climate crisis is already having major impacts on agriculture in much of the world. In 2014, the Intergovernmental Panel on Climate Change (IPCC) predicted that the planet may reach;

> "*a threshold of global warming beyond which current agricultural practices can no longer support large human civilizations*".[62]

Irrespective of average trends, our climates are increasingly volatile, making crop yields difficult to predict. Modelled scenarios show large positive and negative variations in the yields of major crops at specific locations — not exactly a sound basis for decision making. The best current evidence suggests that regions toward the poles, where agriculture is limited by short growing seasons, are more likely to gain from temperature increases, while subtropical and tropical regions (where most of us live) may suffer the effects of heat and drought on productivity. Droughts are likely to become more frequent and greater in their intensity in Africa, southern Europe, the Middle East, most of the Americas, Australia, and Southeast Asia.

These broad conclusions hardly provide the basis for a long-term strategy for agricultural adaptation in response to climate change. Rather, they tell us that the future for agriculture is uncertain. Our most urgent challenge is how to manage the uncertainties, risks, and shocks that changing climates are increasingly imposing on the global food system. Unlike viral pandemics, the consequences of climate change are not immediately evident at the individual or societal level. Nevertheless, we cannot self-isolate, wear masks, or find a vaccine for climate change — we must live with and manage its impacts. As with most human activities, the climate crisis first affects those who are least responsible for its causes and most vulnerable to its impacts. The poor in developing, tropical regions face the most negative short-term impacts, especially from the greater frequency and severity of adverse weather events such as heat and drought and the effects that even small changes in seasonal weather can have on their production systems.

Within a region, the impacts of climate change differ between communities and are associated with factors such as age, ethnicity, disability, gender, wealth, and class. Poverty, socioeconomic status, and political marginalization put women, children and the elderly in the weakest position to cope with the adverse impacts. In terms of the agrifood system, women are especially vulnerable because of their relative power status and responsibilities at the household and community level. Men are usually in charge of cash and staple crops and may have access to inputs such as fertilizers and irrigation, but women are often responsible for cultivation of subsistence crops (with minimal access to inputs) to feed the family. Through impacts on food prices, household nutritional security is particularly threatened as more nutritious crops may be displaced by calorie-dense but nutrient-poor diets from staples. It is not simply a question of changes in farm yields. In most developing regions, climate change will increase the vulnerability of those involved in agriculture who are already suffering from poor infrastructure and lack of technological inputs. As well as the uncertainties at any location, much remains unclear about how farmers and rural communities will respond to the impacts of climatic change on length of growing season, heat and drought stress, cropping calendar, yields of individual crops, and the viability of farming under routine practices.

Within the limitations of our current understanding of climate change and agriculture, a few snapshots illustrate likely trends. Maize accounts for almost half of the cultivated area of Southern Africa and average yields are predicted to decrease by 30 per cent.[63] Export crops such as coffee and cocoa are also highly susceptible to climate change and the fragility of poor transport and infrastructure systems that will also be vulnerable to erosion and flooding. Detailed analyses of rice yields by the International Rice Research Institute (IRRI) forecast a 20 per cent reduction in yields over the region per degree Celsius of temperature rise.[64] It is not only a question of whether plants can remain physically alive under high temperatures but whether they will produce economic yields. For example, rice becomes sterile at temperatures above 35°C for over an hour during flowering, and consequently produces no grain.[65] There are similar risks for the yields of 12 million hectares of oil palm plantations in Malaysia and Indonesia.[66] In Northern Europe, climate change is initially projected to increase crop yields but projections for Southern Europe predict that yields of non-irrigated crops of wheat, maize, and sugar beet will fall by up to 50 per cent by 2050. Knock-on effects include local decreases in land value as farms become abandoned and national losses in trade through lower crop yields. As with all evolutionary processes, rapid climate shocks will have more impacts than gradual trends where there may be time for crop selection and plant breeding efforts and the ability of farmers to adapt to the new conditions. Rapid climate change will be disproportionately harmful where impoverished soils and hostile conditions are already marginal for crop production.

Effect on Quality

As well as yields, there is evidence that climate change can affect the nutritional content of crops by reducing specific macronutrients such as carbohydrates and protein, and micronutrients such as iron and zinc. This would adversely affect poorer countries whose populations are less able to simply eat more food to replace nutrient loss or don't have access to varied and nutritious diets. Not only would lower concentrations of micro-nutrients have negative consequences for human nutrition, it may also mean that livestock will need to eat more feed to gain the same amount of nutrition. One of the most contentious and concerning meta-analyses of climate change effects on nutrient content of crops was authored in 2002 by the mathematician Irakli Loladze.[67] The website Politico offered this headline — *"the great nutrient collapse, the atmosphere is literally chang-ing the food we eat, for the worse"*.[68] The analysis showed that in temper-ate climates, and in mainstream crops such as wheat, elevated concentrations of atmospheric CO_2 resulted in plants with less protein, more carbohy-drates, and reduced levels of 25 important minerals. In tropical crops such as maize and millet, elevated concentrations of atmospheric CO_2 also resulted in lower micronutrient levels, but these differences were not sig-nificant. Another study estimated that at projected atmospheric CO_2 levels in 2050, food crops could see a reduction of protein, iron, and zinc content in staple food crops of 3 to 17 per cent.[69]

Insect Pests, Diseases, Weeds

Climate change doesn't only affect the behaviour of crops and livestock but also that of their insect pests and diseases. Increases in temperature affect the metabolic rate and frequency of insect breeding cycles and their likely expansion into areas that were previously too cold for part or all of the year. It is not just the movement of insects further north and south of the equator, but also to many higher latitudes that are already experiencing a dramatic change in the populations of insects. Climatic factors such as heat, rainfall, and humidity may alter the developmental stages of crop pathogens and their dispersal and that of their vectors to previously unin-fected areas. Greater rainfall in tropical areas would lead to more humid and longer wet seasons. Combined with increased temperatures, these conditions could favour development of fungal diseases, as well as increased populations of insect and disease vectors. As well as the crops themselves, associated weeds will also change their development and growth in response to changes in temperature and rainfall. Together, greater infestations of insect pests, diseases, and weeds will need higher levels and frequency of crop protection chemicals.

Food Prices

According to the UN IPCC special report on climate change, food prices will rise by 80 per cent by 2050 and food shortages are likely to occur.[70] Although 2050 seems a long way off, food shortages are already affecting poorer parts of the world much sooner and by more than in richer areas.

Weathering the Perfect Storm

In 2009, Sir John Beddington, England's chief scientific advisor, warned that;

> A *"perfect storm"* of food shortages, scarce water and insufficient energy resources threaten to unleash public unrest, cross-border conflicts and mass migration as people flee from the worst-affected region . . . It is predicted that by 2030 the world will need to produce around 50 per cent more food and energy, together with 30 per cent more fresh water, whilst mitigating and adapting to climate change".[71]

He went on to add;

> "If we don't address this, we can expect major destabilisation, an increase in rioting and potentially significant problems with international migration, as people move out to avoid food and water shortages".

Humanity is already facing much of the *perfect storm* predicted by Sir John Beddington for 2030. The earlier part of this chapter described the impacts of the *submerged icebergs* of Environment, Society, and Health on the titanic global agrifood system in the context of the Climate Crisis *iceberg*. Taken in isolation, we might hope that our *Titanic* could remain on course if we simply included better environmental and social safeguards and developed healthier products. Taken together, however, the submerged and approaching icebergs represent the impending perfect storm described by John Beddington. We cannot steer a course that completely avoids it — it is already too close. We need options for how societies across the globe can weather the perfect storm. For the global agrifood system, there really are no quick fixes, silver bullets, or simple solutions, no more tinkering with our current model. This means reimagining agriculture and rethinking our approaches to make it more resilient to the challenges that it faces. Rather than reducing the agrifood system to its bare essentials and all the risks that uniformity brings, we need to embrace diversity — of people and their cultures, species and systems, and ideas that broaden the base of the system. Rather than a *Cheap Food Paradigm* that is driven principally by private profit, we need to move towards a *Climate Resilient Paradigm* that is measured by the resilience of the agrifood system in the face of the climate crisis, serves the nutritional needs of humanity and does less harm to the planet.

So, what exactly is a Climate Resilient Paradigm? For the agrifood system it means connecting the three elements in which human beings have agency — the systems in which we grow crops and keep livestock, the foods that we derive from them, and the livelihoods that depend on them. Instead of reducing each component to its bare essentials, resilience requires that we integrate the various elements of the agrifood system, find ways to harness diversity (biodiversity, cultural diversity, and livelihood diversity) throughout the entire value chain from farm to plate and manage the complexity that diversity brings.

Climate Resilient Cropping Systems

Our current model has treated agriculture like any other extractive industry. The soil is there to provide a medium (preferably inert) in which to grow crops and extract goods. This means removing all competitors (weeds, pathogens, microbes, insect pests) and adding inputs (fertilizers and irrigation) that provide returns on their investment. Decades of extraction have left the world's soils impoverished, eroded, and lifeless. Without more and more inputs and greater and greater protection, our cosseted crop varieties cannot achieve the yields that we demand from them. The vicious cycle of higher inputs and diminishing returns means that monocultures of favoured crops are vulnerable to disturbances that are outside our control — heat stress, floods, pandemic diseases, and insect pests. For resilience, we need to move from the concept of soil as an *extractive* resource to being a *renewable* one. This means putting soil health at the core of the agricultural system, preserving its natural biodiversity, and increasing the biodiversity that we add to it. In a rush to uniformity and an obsession to master the biological environment, we have abandoned the ecological principles on which natural biological systems work and the agroecological principles in which managed biological systems such as agriculture can work. This means harnessing nature rather than taming it. There is an entire scientific discipline called *agroecology* that can show us how.

The retreat of agriculture to a handful of staples has coincided with its reduction to one production system — monoculture. Growing a single crop variety in the same field, often over several years, is now the model for modern agriculture. There are scant examples of where nature, when left to its own devices, favours monocultures. For high-input agriculture, of course, there were, and still are, very good reasons for monocultures — they allow synchronized planting and crop harvesting, timely application of fertilizers, insect pest and disease control chemicals, irrigation, agronomic expertise, and advisory services to all be targeted to the specific needs of a specific crop variety. Machines for soil preparation, crop management, and harvesting have been designed specifically for monocultures, especially *combinable* crops such as wheat, maize, and soybean that can be harvested using a combine harvester. This level of coordination allows for economies of scale and efficiency gains that would have been unthinkable before mechanized monocultures. For much of the world, monocultures will remain the *stock-in-trade* of modern agricultural economies. However, monocultures of major crops on the best land in grain-exporting countries with access to high inputs, advice, and credit will not be enough for all humanity's food needs. We need a new landscape for agriculture — literally.

Multicropping: A Model for Climate-Resilient Cropping Systems

Whether it is a tiny seedling growing on a windowsill in London or one of 3,000 plant species growing in the 100 million year old Daintree Rainforest in Australia, all green plants obey the same principles. They capture resources (principally sunlight, carbon dioxide, water and nutrients) and convert them into products (principally carbohydrates, lipids, proteins and minerals). This requires flows of resources to and from plants through two media (soil and air) and in two dimensions (space and time). Communities of plants must interact with other living organisms in search of these resources whether they are part of a complex natural ecosystem in the jungles of Borneo or in a soybean monoculture in the American midwest. The difference between a natural ecosystem and an agroecosystem is that human beings influence the resource flows to (fertilizers, agrochemicals, irrigation) and from (harvested yields) agroecosystems. Ecology is the study of how living organisms interact with the environment and agroecology is that branch of ecology that relates to the behaviour of living organisms in agricultural systems. Whether they are in a prepared field, natural landscape, urban space, rooftop or stacked on shelving in a vertical farm, all forms of agriculture are agroecosystems and *inter alia* a part of nature. At any particular location a complex natural ecosystem is usually more productive and biologically more efficient than a simple agroecosystem — that is to say that it makes more total products without external inputs. The retreat to monocultures is to maximize the fraction of outputs from an agroecosystem that are useful to humans and minimize the cost and efforts involved in managing it. This comes at a cost to biodiversity, including 'ecosystem services' such as pollinators and predators of crop pests, the environment, including pollution of soil and surrounding water bodies, and humans through exposure to harmful compounds.

For a climate-resilient future, agriculture needs to break out of the confines of prepared fields of single crops and start to include landscapes, urban spaces, common lands, and even our gardens as food sources. Put simply, diversity underpins resilience. As well as encouraging a wider array of climate resilient crops, diverse cropping systems provide a greater buffer to climatic variations and protection from shocks than regimented monocultures. There are plenty of examples of multiple and complex cropping systems that humans have developed and adopted throughout our agricultural history.[72] *Multicropping* is still widespread, particularly in tropical and low-input production systems. It involves cultivation of two or more species on the same piece of land where their growth cycles overlap for at least part of the growing season. Multicropping includes *intercrops* of different species grown over a single growing season, *relay cropping* where different crops are cultivated in staggered planting, and *agroforestry*, where the space between trees or shrubs is occupied by one or more companion crops. Agroforestry can include high-value plantation crops such as oil palm, coconut, cocoa, or rubber interplanted with annual food crops. Examples of more complex systems are multipurpose home gardens containing vegetables, staples, and fruits. Recently, there has also been a resurgence of interest in using confined urban spaces and rooftops to grow multiple crop species and even for aquaculture. Like the natural system that it mimics, multicropping is ecologically more stable than monocropping, especially where inputs such as fertilizers, insect pest and disease control, and irrigation are limited or unavailable. The lack of external inputs means that plants must capture and share resources. Whilst the maximum yield of any particular species might be compromised, the performance of the whole system improves.

There is evidence that multicrops can produce greater total yields and use less fertilizers than their component crops grown alone.[73] In China, a meta-analysis of 226 experiments showed that intercropping can achieve the same or significantly higher yields than intensive monocultures and at the same time use less fertilizer.[74] Reported yield advantages of multicrops vary but typically range between 25 per cent and 40 per cent. The preference of farmers for multicropping is not just because it may be more productive but because of its stability in variable environments.

There are several reasons for this greater stability.[75] Insect pest and disease effects on one species may be reduced because specialized pathogens may settle on non-host component species of the multicrop. This *fly-paper effect* increases the stability of the whole system where resistant species can occupy the space left by diseased individuals. For example, virus diseases may spread more rapidly through adjacent plants of a particular species than when such plants are *socially distanced* by dissimilar, non-susceptible plants of another species. Insect vectors are also discouraged from attacking plants that are intermingled with other species than when a single species is grown in a block. Multicrops of species with different growth cycles suppress weed establishment and can protect the soil against erosion and wind damage. By providing continuous soil cover, multicrops complement systems of regenerative agriculture that build topsoil, increase biodiversity, and improve the water cycle. Multicropping is often chosen as insurance against the risk of complete crop failure in hostile and unpredictable environments. However, rather than a last resort, in an uncertain future multicropping could become the preferred option in much of modern agriculture.

If multicrops are more productive, stable, and resilient, why aren't they already the model for agriculture? The main reason is that whilst complex multicrops are more biologically efficient, productive, and stable, their complexity and need for human labour often makes them less economically profitable than simple monocrops. This obvious limitation has deterred investors, extension advisers, researchers, and technologists from looking at improving such systems, advocating their use, or monetizing their biological advantage. However, the ambivalence of the agricultural industry to take multicropping seriously is also rooted in inertia. The industry often prefers to stick with familiar crops and systems that they have refined and improved, sometimes over decades in the case of plantation crops, rather than introduce new crops and the complex management that they require. Multicrops are usually seen as backward, traditional, and unworthy of attention from researchers looking for technological breakthroughs, development professionals and sponsors looking for immediate impact and politicians looking for quick wins.

At its simplest level, multicropping is, well, complex. Multicropping and agroecology more generally require an intimate knowledge of the relationship between agricultural species and their environments to guide decision making on what grows where, with and without what, when, and how. The first place to access such knowledge is not in the conventional disciplines of academia but in the communities that have maintained multiple cropping systems for millennia. Not only is this knowledge vernacular (unwritten), it is being lost faster than it is being generated through the race for modernity. Funding research and publishing high impact papers on complex and traditional biological systems may require several years of detailed measurements, local knowledge, and persistence. There are more alluring options for investors and the brightest and best researchers and technologists — such as finding more ubiquitous uses for high fructose corn syrup!

Of course, there are also many practical hurdles to overcome to modernize multicropping and reduce costs. How can we mechanize the management of multicrops, reduce their labour requirements, and optimize their productivity? In the past, the need for manual labour was clearly a deterrent. However, new agricultural machinery, the use of remote sensing, robotics, and artificial intelligence can all improve management operations, and mathematical modelling and big data can help researchers identify the most suitable species combinations, spatial configurations, and management techniques that suit local growing conditions. The missing ingredient in the revival of multicropping is the will of the agricultural industry and the commitment of researchers to seek innovative solutions rather than explanations for why they should be ignored. Why would they otherwise reject the potential of greater than 25 per cent increases in total yield that multicropping typically generates? Regardless of this antipathy, the climate crisis demonstrates the need to harness the great potential of multicropping to sustainably increase yields and lower environmental costs of both low- and high-input agriculture. Taking their lead from the natural environment, we now know that diverse and complex multicropping systems are more stable, resilient, and productive than monocultures, and can become the model cropping systems for a climate-resilient future. The stability of natural ecosystems provides the framework on which our agroecological systems are managed, and soil health provides a foundation on which these systems should be built.

Climate Resilient Crops

A handful of crops and animals now provide almost all of the food consumed by over 7.8 billion people through a globalized agrifood system.[76] Advocates will say that this narrow cohort of elite agricultural species has served us well. But to rely on so few species is already risky, to depend on them for more people on a hotter planet is sheer folly. The paradox is that we have a treasure trove of over 7,000 crops that have been grown throughout our agricultural history, not to mention the 30,000 edible species that we have collected from the wild or from forest gardens before the start of sedentary agriculture. Are they really *all* useless and undeserving of attention, not just as sources of food and feed, but for medicines, discovery chemicals, bioactive compounds, building materials, and renewable energy? Taken to its logical conclusion, if just three crops are now enough, why not choose two or indeed one crop to provide all the world's food needs? Surely, with modern technologies, industrial monocultures, economies of scale, free trade, and slick marketing we can persuade *the public to want what the public gets* from one supercrop. For this, we simply need a guaranteed supply of feedstock, mainly calories, from which we can design, test, manufacture, and transport processed products around the world.

There is a logical argument to focus all research, development, innovation and marketing on one crop and insert any missing flavours, additives, and even micronutrients into its perfectly formed food-like products. Of course, this suggestion may be rejected as absurd, but in reality, we have already reduced most of the world's food supply from thousands of crops to three. Why not adopt just one crop as the lodestar of our titanic food system? Think of the economies of scale and savings in duplication of supply chains — and profits. Before a smart consultancy company proposes a *one crop, one world* moniker and sponsors and researchers arrange international conferences to decide on whether the crop should be wheat, rice, or maize, there are good reasons why this option — indeed our current tiny selection of favoured crops — will not be enough for humanity to weather the perfect storm. Multicropping agricultural systems that are resilient to climate shocks also require diverse crop species that are resilient to climate change.

One of the first questions asked of a farmer is *what is your crop yield*? The price per tonne of a commodity crop determines profitability after accounting for inputs, and is the yardstick by which all farming is compared, as well as the main target of plant breeders and researchers. Without positive returns on investment from its crops (or livestock), a farm cannot be profitable. However, yield expressed in tonnes per hectare is a crude measure of value and no measure of values. In staple crops, the yield is heavily dependent on the carbohydrate or calorific content of the grain. At harvest, other non-grain components such as leaves and stems attract little interest and are often regarded as an inconvenience rather than an asset. Even more telling is that the weight of grain is a poor measure of its content of macro- and micronutrients, lipids, and bioactive compounds. When we include some or all of these other components the *value* of a crop begins to look very different in terms of its usefulness to society.

For example, the protein content of drought-tolerant and nitrogen fixing legumes, such as pigeonpea, is much higher than that for drought vulnerable nitrogen demanding cereals such as maize.[77] Whilst the *yield* of maize (primarily carbohydrate) will invariably be higher than that of pigeonpea under high-input growing conditions, its nutritional value, drought tolerance, and resilience to extreme climates will not. In drought-prone regions where maize is increasingly failing, why not bolster it with drought resistant and nutritious legumes as an intercrop that can also fix nitrogen and reduce the need for fertilizers?

As climates become more volatile and soils more degraded, many locations will become too hot, too dry, or too fragile to support the production of high-yielding varieties of major staple crops. In some cases, while production may still be possible, its costs in terms of irrigation, fertilizers, and crop protection chemicals will become increasingly prohibitive with lower or more unreliable yields. If we now consider value as *climate-resilient-nutrition* rather than simply *yield-for-profit*, pigeonpea becomes a very different prospect. There are dozens of examples of well-established but undervalued crops such as pigeonpea. There are many hundreds of climate-resilient and nutritious underutilized crops that have never had the investment, support, and subsidies afforded to the major staples. We ignore them at our children's peril.

Climate-Resilient Nutrition

Nutritionism

In one of his wonderful books, *In Defence of Food: The Myth of Nutrition and the Pleasures of Eating*, Michael Pollan introduces the term *nutritionism*.[78] The term is not his, but was first coined by the Australian sociologist Gyorgy Scrinis. However, it is Pollan who demonstrates that *nutritionism* — measuring the nutrient content of foods — is not the same as *nutrition* — the process of nourishing or being nourished. Nutritionism reduces food to the sum of its parts, and its guiding principle is that the whole point of eating is to maintain bodily health — food is medicine, nothing more, nothing less. To maintain optimal bodily health, consumers are introduced to types of ingredients and products (*low cholesterol, low fat, high fibre*) that describe what is in their food and what is not. Rather than eating a recognizable food, consumers can now eat a processed food-like product to which good things have been added and bad things removed. All the better if the food-like product requires little or no preparation and avoids unnecessary distractions such as cooking and eating around a table with other people or for pleasure.

The foundations of nutritionism can be traced to 1827 when an English doctor, William Prout, proposed the classification of food into *sugars and starches*, *oily bodies*, and *albumen*.[79] These later became known as the three principal components of food — carbohydrates, fats, and proteins. In 1840, the German chemist Justus von Liebig, who had already identified nitrogen, phosphorous, and potassium as essential components of plant nutrition, presented a vision for "a rational system of agriculture",[80] in which chemistry rather than biology was the foundational scientific discipline. Von Leibig proposed a theory of metabolism that explained life strictly in terms of a handful of inert chemical compounds rather than biological forces related to *vitalism*. Unfortunately for von Leibig, his list of chemical compounds was insufficient to explain the complex metabolism of human beings.

When sailors went on long voyages, many developed scurvy with symptoms such as gum disease, changes to hair, and bleeding. This condition could be rapidly reversed if they ate plant foods such as oranges and potatoes — plants and animals contain compounds other than those on von Liebig's list and some of these in very small amounts are essential to life. In 1911, the Polish biochemist Casimir Funk noted that individuals who consumed brown rice were less vulnerable to the neurological condition beri-beri than those who ate only the fully milled (white) product.[81] Funk isolated the substance responsible and called it *vitamine* (*vital* for life and *amine* a nitrogen base). He proposed the existence of at least four vitamines — one preventing beri-beri, one preventing scurvy, one preventing pellagra, and one preventing rickets. The "e" at the end of *vitamine* was later deleted when it was realized that vitamins need not be nitrogen-containing amines. Funk had added essential elements to the list proposed by von Leibig and the earlier identification of carbohydrates, fats, and proteins by Prout. This more complete list of chemical compounds strengthened the case for nutritionism — by extracting ingredients from certain foods, the consumer didn't have to eat the food itself. How apt that the founding fathers of our *global diet* and the foods that provide it were an English doctor, a German chemist, and a Polish biochemist!

Nutritionism also provided a perfect justification for the American food industry. By referring simply to *good* and *bad* ingredients rather than saying *don't eat too much meat or dairy products and eat more vegetables*, the industry could avoid being challenged by vested interests. Helpfully, in 1982, a US National Academy of Sciences report codified official new dietary advice in terms of compounds rather than foods.[82] A new language was introduced to the consumer with such terms as *polyunsaturated, cholesterol, fibre, polyphenols, amino acids, flavanols, carotenoids, antioxidants, probiotics, prebiotics,* and *phytochemicals*.

By eating these compounds whose properties and uses were only known to scientists, the public no longer had to eat what it previously knew as food. Fish (*high in omega 3*), chicken (*low in saturated fats*), and red meat (*a rich source of zinc*) were now simply delivery systems for nutrients that could be added to processed foods. Instead of eating boring old vegetables that needed cleaning, chopping, and cooking, individuals could now consume convenient processed products to which antioxidants derived from vegetables could be added. Similarly, instead of vegetables such as broccoli and citrus fruits such as oranges that are protective against cancer, the consumer could eat processed foods that have been fortified with beta carotene extracted from vegetables and vitamin C extracted from citrus. The food industry could now achieve the *win-win* formula of being politically expedient, scientifically credible, and hugely profitable — all it needed to do was to convince the public that processed foods are actually healthier than whole foods. This required marketing consumer-friendly brands supported by product labelling.

In his book, Michael Pollan describes how a major obstacle to consumer acceptance of processed foods was the label *imitation*. In 1938, the US Food, Drug, and Cosmetic Act imposed strict rules that stated whether a product was the real thing or an imitation of what it purported to resemble. In other words, when consumers buy, say, bread, milk, and cheese, they should get the foods that they expect. If it is not bread, milk, or cheese it should clearly state that it is an imitation. This meant that a discerning consumer could decide whether they wanted the real thing and not an imitation of it — not good if you are trying to sell the imitation product and a major obstacle to adoption. In 1973, the food industry succeeded in getting the Food and Drug Administration (FDA) to rescind the original imitation rule with one that said that as long as an imitation product was not *nutritionally inferior* to the food it was impersonating, the term *imitation* could be removed. So, the fats in yoghurt and sour cream could be replaced with hydrogenated oils or guar gum and still be called *yoghurt*, the cream in whipped cream could be replaced with corn starch and still be called *cream*. It also opened the door to a limitless list of other products developed by engineers to replace the foods whose names and cultural associations they were commandeering. Apart from the questionable nature of calling something that it clearly was not, the nutritional equivalence of the new product assumes that we know enough about nutrition to know what is equivalence. The history of baby formula and margarine suggest otherwise.

And then there's labelling. If there is an art that the food industry has mastered it is that of the label. For example, *low fat* yoghurt can be exactly that without the consumer being aware that they are actually buying *high fructose* yoghurt in which the almost universal ingredient of corn syrup has been added. On the back of the pack will be an exhaustive list of ingredients that only those with multiple advanced degrees could understand. Again, if we return to our example of bread, over the thousands of years since our ancestors discovered breadmaking, it has been composed of the same ingredients (flour, yeast, water, salt). They would be surprised to return to earth and visit a supermarket selling *bread* containing ingredients including fats, flour treatment agent (L-ascorbic acid; E300), bleach (chlorine gas), reducing agent (cysteine hydrochloride; E920), soy flour, emulsifiers and preservatives (calcium propionate). Of course, these may all be on the label, but transparency (list of ingredients) is not the same as trust (getting what you thought you were buying). As with labelling, there is the minefield of health claims. Any consumer buying *heart healthy* chips could be forgiven for thinking that deep fried potatoes could actually improve heart function and that eating chips fried in polyunsaturated fats was relatively less damaging to *heart health* than those fried in saturated fats.

Over recent decades, nutritionism has reduced the agrifood system to its minimum components — fewer farmers in fewer exporting countries growing fewer crops to be made into more products to be distributed to more people around the world. Mass production systems have made processed foods much cheaper and more convenient than cooking individual meals from raw ingredients. Advances in food processing technologies and delivery systems have produced a vast array of processed foods based on raw materials, mainly carbohydrates, from major staple crops. Essential micronutrients and vitamins that are missing from the feedstock have been added through fortification of the final product. Global uptake of convenience foods, ready meals, snacks, and confections have provided large potential profits for food manufacturers and the suppliers of processed food products. They provide sustenance and nutrition, but do they provide nourishment? The price that we have all paid for the benefits of nutritionism and the global diet that it delivers has been the loss of biological and culinary diversity and the joys of eating — should we live to eat or simply eat to live?

Diversifying Diets

Our increasingly globalized diets and the food production systems on which they depend have dramatically reduced diversity of the global food base. Agricultural research and crop breeding have targeted increases in crops yields to achieve food security rather than their nutrient content to achieve nutritional security. Researchers and their sponsors have focused on crop selection and management practices that increase the yields of staple crops, often cereals. This approach has ensured global food security — there are now more than enough calories to feed the world, however, nutritional insecurity still prevails across the globe. Nutritional security is about the quality of the diet — having access to sufficient nutrients to ensure a healthy and productive life. The prevalence of hidden hunger shows that we have not made sufficient progress on nutritional quality of the global food system.

Looking at global production, whilst calories available *per capita* have increased worldwide the nutritional diversity of the agrifood system has not.[83] Reduced dietary and agricultural diversity are inversely related to the increase in the incidence of diet-related diseases. Nutritional quality is correlated with a more diverse diet — one that contains ingredients from both plant and animal sources. Adherence to nutritionism is suspect on at least three counts;

- We don't fully understand the complex interactions that occur among the foods that we eat and our microbiome — health effects of diverse diets are much greater than the constituents alone
- The nutritional composition of raw ingredients, especially plant-based materials, varies immensely — from variety and cultivar through to geographical location, season, agronomic practice, use of fertilizer, soil type, rainfall, and production method. Further variations are caused by the effects of processing and storage on nutrient content. The *nutritional equivalence* that imitation foods claim is based on uncertain evidence
- It is becoming increasingly evident that dietary diversity — the number of foods that we consume — is more than the sum of its parts. Once again, it seems that the number of species on a plate of food plays a role in nutrition — and a more interesting meal!

The food industry can justifiably claim that it provides a complete inventory of uniform and safe processed food products with a long shelf life, with efficient transport to consumers around the world. If you walk down a supermarket food aisle in almost any country in the world, shelves are stacked with thousands of the products of nutritionism. Careful inspection will show, however, that these products contain the same ingredients, often blended into processed products that have been transported from elsewhere. These diverse products, but not their ingredients, mean that thousands of *forgotten ingredients* have been displaced by a modern, uniform, and processed diet. Forgotten ingredients include those from crops and animals that are no longer part of our diet, as well as forgotten foods that have been lost in the onslaught of the titanic agrifood system.

Nowhere is this onslaught more evident than in the Pacific region where, as well as having the world's 10 most obese countries, the region includes 10 of the 20 countries with the highest rates of diabetes. The International Diabetes Federation reports that about 87,000 adult Fijians — 15 per cent of the whole population — have diabetes and estimates that another 46,000 have the disease but are undiagnosed.[84] In 2019, amputations related to diabetes accounted for 40 per cent of all hospital operations in Fiji and the top three causes of deaths were all non-communicable diseases — diabetes, heart disease, and stroke.[85] How have Pacific islanders, whose culture is centred around outdoor activity and abundant tropical crops, become so unhealthy? The key is a loss of pride in their traditional, healthy cuisine. Pacific islanders export their own natural foods, but eat cheap processed imported foods and only occasionally have a traditional meal of fresh fish, seaweed or shellfish, green leafy vegetables, and coconut.

Across the world, diets have shifted from diverse locally-available traditional foods to a single modern Western diet. With this shift, nutritious, traditional, and local foods have become relatively more expensive than those made from rice, maize, and wheat. Increased profits and subsidies have encouraged farmers to grow these staples instead of their own crops. Traditional foods, methods of food production and preparation, use of local ingredients, and knowledge of their preparation are being lost across generations of each family. Not only are children unable to cook but nor can many parents, especially the foods of their ancestors. Kitchens are often not included in the design of apartment blocks and communal housing — fast food and take-aways based on universal brands are the only options.

Throughout our history humans have adapted and thrived on diverse diets — all are very different but each uses locally available and nutritious ingredients, is limited in processing, and based on local culture. There are many sustainable and healthy diets across the world. Unfortunately, the modern Western diet is not one of them. How can we rediscover our lost foods and their recipes? Knowledge about these forgotten foods, their ingredients, and methods of preparation is often held in the heads of our parents, grandparents, and great-grandparents. Before they are lost forever, we must rediscover these recipes and help our elders share their knowledge with younger generations. There is a renewed interest both in the foods that we eat and the sustainable sourcing of their ingredients. If we can rediscover recipes for these foods, we can diversify our diets. To achieve nutritional security and optimal health, the human body requires more than just calories and nutrients. Consuming a diverse diet has long been recognized as the best way to achieve the nourishment that is essential for health.

Climate Resilient Food Systems

Diversifying Value Chains

Forgotten foods and the crops from which they derive have the potential to feed the future but which and how require evidence. Local foods are often best suited to their natural environments, based on crops that are resilient to climate shocks, are nutrient dense, and address micronutrient deficiencies. They can diversify diets and at the same time localize supply chains, reduce the carbon footprint of agriculture, and support livelihoods through diverse cropping systems and diversified value chains. But the question is, *who has the knowledge and incentive to manage diversified cropping systems*? Over recent decades, agriculture has become increasingly separated from the rest of society. We now have a growing, predominantly young, urban population (more than half of us live in cities) that depends for its food on a declining, predominantly ageing, population of rural farmers who cultivate crops and manage livestock far from where their products are consumed.

By separating the bulk of society from the source of its food, we diminish the dignity and status of farmers on whom we depend for cheap, plentiful, and timely food supplies, and we divorce consumers from an understanding or interest in the source of the food products that line their supermarket shelves. The massive consolidation and mechanization of agriculture means that there has been a generational flight to the cities from once vibrant rural communities. Many would have continued and some would have started to farm if it was seen as a desirable vocation and attractive business. Agriculture is now one of the least attractive career options in most industrialized societies and is increasingly seen as a last resort for young, rural *digital natives* who are in touch with the more appealing world of the city, whether in an office or a factory. As a result, we have fewer, ageing farmers who cultivate larger fields of monocultures of fewer crops on bigger farms with only machinery for company. It doesn't have to be this way.

Final Thoughts

The RMS Titanic sank early on 15 April 1912 in the North Atlantic Ocean. At the time, the largest ocean liner ever, the Titanic had an estimated 2,224 passengers, of which over 1,500 died when she struck an iceberg. During 14 April, the ship's crew had received no fewer than six warnings of sea ice, but these were ignored and the Titanic was travelling at about 22 knots when her lookouts sighted the iceberg. The ship was unable to turn quickly enough and received a glancing blow that allowed water to fill 6 of her 16 compartments, in which most of the third-class passengers were located. Although Captain Smith and his crew knew that there was ice nearby, they didn't reduce speed because the owners of North Atlantic liners guaranteed the arrival of their ships on schedule. This meant that ships frequently sailed close to full speed, and their crews treated hazard warnings as advisories rather than instructions. There was a belief that ice posed little risk. In an interview the Titanic's captain, Edward Smith had proclaimed that he could not "*imagine any condition which would cause a ship to founder. Modern shipbuilding has gone beyond that*".[86] Most of the passengers who escaped were from first- and second-class cabins who could quickly reach the lifeboats. Most third-class passengers were trapped by gates that barred them from the first and second-class areas. Before the steerage passengers could escape, the Titanic's crew had locked the access routes from the third-class section.

Nature of Food

The story of the Titanic serves as a lesson for the global food industry. Over confidence, hubris, speed, and cutting corners to meet targets that generate profits brings risks for us all. As passengers, we have to decide if the Titanic agrifood system is going in the right direction or whether we must slow it down and change course so it can weather the perfect storm. This needs not a leap of faith, but imagination of opportunities.

Imagine a different agrifood system where;

- there is no idle land, and agricultural fields are filled with diverse crops selected to grow in changed climates with minimal use of pesticides, herbicides, fertilizers, and irrigation
- children do not starve or suffer from stunting or malnutrition
- people are conscious and aware of the health and environmental implications of their diet and choose foods that deliver balanced nutrition without distressing the planet
- people rediscover recipes and foods that were known to their great-grandparents
- people are rarely obese or overweight and diabetes rates are falling
- healthy food is consumed from diverse crops, wastage is negligible, and carbon emissions are minimized and;
- trees, crops and people co-exist beneficially.

There are thousands of forgotten foods that can make our diets more interesting, nutritious, sustainable, and better suited to weather the perfect storm. There is growing interest in the food that we eat and the sustainable sourcing of its ingredients. If we can rediscover forgotten foods, our diets can be diversified now and in future climates. By finding our own forgotten foods, we can identify which climate-resilient and nutritious crops could help feed the future. It is not only about replacing existing products but about adding diverse options that are attractive, affordable, nutritious, and desirable. For this, we need to help consumers to reconsider their food habits.

Promoting local produce and species can also limit the carbon footprint of our food and help us celebrate diverse cultures, cuisines, and ingredients while also reducing our impact on the environment. We often think of food, nutrition, environment, and climate separately but if we do not care for one, then the others will suffer. The current global pandemic forced behaviour change in a very short time. Individuals and food supply chains had to adapt. For many, food habits changed — more time was available to rediscover and enjoy food and its preparation, supply chains became shorter, and more foods were locally sourced. Was this change an aberration, or can we use the experience to prepare our food system for the future shocks that lie ahead?

This pandemic should serve as a lesson for how we can consume and enjoy many more local and diverse foods by reducing over-reliance on a small number of staple crops and global food systems that produce identical products transported along vulnerable and crowded supply chains across the globe. It also demonstrates how bad diets and health-related diseases contribute to our likelihood of surviving the virus. We have the opportunity to rediscover diverse local traditional diets that can be healthier for us and the planet. The question is whether we have the political and social will, access to scientific and historical evidence and partnerships with those, often indigenous and marginalized, communities that have cherished and protected these diets and their foods, despite the onslaught of the global food industry and the scepticism of modern science.

5

GLOBAL AGRICULTURAL RESEARCH AND EDUCATION

Revolution Number 8

Context

At the end of Second World War, the political and economic circumstances of many societies across the globe changed substantially. A simultaneous expansion in education at both the primary level and in universities sought to equip people in recently liberated and often newly established countries with the skills and qualifications needed for leadership, management of national assets, and nation building after decades or even centuries of subjugation.

A rapid transition to independence required countries to consider the role of agriculture in their future plans. They needed to decide if agriculture could contribute to economic development and national food security or whether it should be displaced by industrialization as the major engine of economic growth. Whilst agricultural research has been a feature of most societies, the speed and extent of societal transformation in the decades after the Second World War, allied with changes in educational and political systems, resulted in the biggest revolution in agricultural research in human history. This chapter is about that Global Agricultural Research Revolution (Revolution Number 8).

A Global Model for Higher Education

According to the Merriam-Webster dictionary, a university is *"an institution of higher learning providing facilities for teaching and research and authorized to grant academic degrees"*. Within this broad definition there appear to be around 30,000 institutions around the world with 'university' in their title.[1] For many that label is where their similarities end — not least in their wealth, power and influence. Take, for example, Harvard University in Cambridge, Massachusetts, USA. Harvard has an endowment worth 40.9 billion dollars,[2] which is greater than the GDP of 98-member states of the United Nations.[3] Over 150 Nobel Prize winners and eight US presidents have been associated with the university as alumni, researchers or faculty.[4]

Arguably all universities have some link with agriculture, if only because we all have to utilize its products for food, clothing, and medicine. In fact, many hundreds of universities have *agriculture* or *agricultural sciences* in their title, countless others have faculties, departments, and centres related to agriculture or specific elements of it such as agronomy, crop science, animal science, genetics, plant breeding, food science, marketing and policy. Not only are there a huge number of academic institutions currently involved in some aspect of agriculture, many have a long history. The University of Al-Karaouine in Fez, Morocco was founded by Fatima al-Fihri in 859 and later became one of the leading centres of education in the Muslim world.[5] The University of Bologna, Italy, is generally recognized as the oldest university in the world[6] (Al-Karaouine was not initially established as a *university*), was founded in 1088 and began the European tradition of universities as *communities of scholars*. In 1224, University of Naples Federico II was founded as the first public university by Frederick II, Emperor of the Holy Roman Empire.[7] These earlier universities did not specifically teach agriculture, but the University of Naples Federico II provides an unbroken continuum of teaching, research, and artifacts related to agriculture that are housed in the royal palace and botanical gardens of Portici near Naples, Italy.

Given their sheer number, global distribution, longevity, academic staff, researchers, and graduates, the world's academic institutions host a huge repository related to agriculture. When we add their combined physical assets in terms of facilities, research stations, and farms, the *land bank* occupied by agriculture is probably greater than that for any other subject in academia. Along with teaching, many academic institutions also have a long history of agricultural research. Initially, this research focused on the basic sciences that relate to agriculture: physics, biology and chemistry. Later, it encompassed an understanding of agricultural processes such as physiology and agronomy, along with crop and animal genetics which underpin agricultural systems and trade policies that govern their exploitation. More recently, agricultural research has moved from experiments in fields and controlled-environments to embrace the concept of using data to create *models* of agricultural systems. These models attempt to show the behaviour of particular crops and animals, their interactions with physical, biological and socioeconomic environments under current and future climates and the effects of these crops and animals on human demography when influenced by natural disasters, famines and political upheavals.

Until recently, and in almost all countries, the bulk of funding for the fabric of the university system came from the public purse. This was either directly from the government as block grants to institutions or through fees for students in undergraduate or postgraduate programmes. These public funds were complemented by those for specific activities, charitable donations, alumni, and philanthropy. In the United States, an additional model was adopted through which federal land was granted to individual states for them to "*sell, raise funds, and establish and endow educational institutions*". The mission of what became known as the *Land-Grant* universities was set out in the US Morrill Act of 1862,[8] "*an Act donating Public Lands to the several States and Territories which may provide Colleges for the Benefit of Agriculture and the Mechanic Arts*". In 1994, the act was extended to give Land-Grant status to several tribal colleges and universities. Eventually, most Land-Grant colleges developed into large comprehensive public universities. Some became prestigious private institutions such as Cornell University, which became a world-leader in agricultural sciences; Massachusetts Institute of Technology, which is amongst the top ten rated universities in the world; and Tuskegee University an historically Black college that was founded by Lewis Adams, a former slave.

By the 1970s, governments began to question the need to provide tax payer money to institutions whose graduates went on to earn above-average incomes. This reluctance was encouraged by commentators in mainstream media who described academic institutions as *ivory towers*, their teaching as *out of touch with the needs of the private sector* and their graduates *lacking the skills for employment* in the *real* world of industry and commerce. By inference, these opinions assume that training employable graduates for the 'real world' of the private sector is the sole or primary purpose of a university.

Until 1992, British higher education had been organized on what was called the *binary system*, which separated universities from other institutions such as polytechnics and colleges.[9] The universities received block grants from the government, which also gave them considerable autonomy and discretion on how to use these funds for periods of up to five years at a time. This allowed for long-term planning, setting strategic priorities and a distinct character for each university. Of course, critics would say that it also led to complacency and lack of accountability. Generations of former undergraduate students (including me) will attest to the frequently boring, repetitive, and often incoherent lectures delivered by academics who, like their students, appeared to wish that they were somewhere else. Everything changed in 1992 when the binary system was abolished and most polytechnics became universities. All universities (old and new) now had to compete for funds and follow the same rules and standards to justify their share of the national allocation.

A *modular system* was established in which students could choose a selection of *credit bearing* modules usually lasting only one semester. To provide quality control across the higher education system, a Teaching Quality Assurance (TQA) audit was periodically conducted in which external inspectors reviewed the syllabus, teaching materials, work samples, and examination papers for each course. The aggregate TQA score for each course was then published to provide a transparent comparison between undergraduate and master's degree courses across the British higher education system. The TQA for any subject allows students, and often their parents and sponsors, to make their choice of university using a simple index, rather than having to seek a wide range of opinions and evidence before taking such a life changing decision.

For research activities, individual university staff and departments needed to apply for funding from various sources including national Research Councils. These Research Councils were funded by government, but retained significant autonomy in how their funding allocation was distributed across academic disciplines under their jurisdiction. Again, a Research Assessment Exercise (RAE) was established to assess the research outputs of all university departments over several years. As well as a transparent mechanism to compare across institutions, the RAE provided a degree of competition among researchers within and among their departments. Through a complex formula, the performance of individual staff was aggregated and converted into a single grade for the department. This single figure was used to determine the department's level of government research funding over the three or four years until the next RAE.

Emergence of Global Rankings

As universities increasingly sought greater international recognition, their global ranking became a natural development from national league tables. In 2003, the Shanghai Jiao Tong University in China published its Academic Ranking of World Universities (ARWU).[10] The ranking started in 1999, initially as a benchmarking exercise for Chinese universities and a mechanism to help Shanghai Jiao Tong University establish its strategic priorities. China's aspiration was to establish world-class universities but to do this it needed to define what was a world-class university against which top Chinese universities could be compared and ranked. This benchmarking exercise resulted in the first academic ranking of world universities.

The ARWU Rankings gained worldwide interest from mainstream media and became the most widely recognized university ranking system at that time. Universities had long been comparing themselves with their international counterparts in terms of research performance based on peer-reviewed publications, career destinations for top researchers, and international awards. However, the ARWU provided an independent global assessment of academic status across all participating universities. A number of different global league tables quickly followed the launch of the ARWU. Of these, the Times Higher Education (THE) world university rankings and the Quacqarelli-Symonds (QS) rankings have become the most widely recognized global indices.[11] In 2012, both QS and THE started ranking universities that were less than 50 years old and QS also started ranking student cities based on a combination of debatable criteria that include *student mix*, *desirability*, *employer activity*, and *affordability*.

The university ranking industry has become good for business and has provided profitable products for scientific publishers and newspapers. External scrutiny by ranking organizations created an incentive for universities to reorganize their systems and compete for the best students and faculty, and also for research funding. Whilst the criteria and weightings used to calculate the university rankings have caused concern, these criticisms have led to continuous *improvement* of ranking methodology rather than abolition of the rankings.[12] Ranking critiques have given rise to a new set of rules and safeguards (the Berlin Principles) and a watchdog institution (International Ranking Expert Group [IREG] Observatory on Academic Ranking and Excellence) which promotes good practices within the ranking industry.[13] In 2006, members of IREG established a set of principles of quality and good practice for the continuous improvement and refinement of methodologies used in establishing university rankings.[14] The broad consensus amongst institutions of higher learning is that university rankings are here to stay and that they must continually devise communications strategies for when they move up or down any particular league table, reflecting a football cliché that 'the table doesn't lie'. Of course, unlike football, universities can be in more than one league at the same time and so with judicious marketing, can promote themselves in whichever table they are currently ranked highest.

Modularization

Implicit in the move towards teaching and research assessment is a need for metrics to evaluate the performance of staff, the courses they teach, and the research that they and their colleagues undertake within their own institutions and with collaborators. For teaching, implementation of a modular degree system altered UK higher education — along with the higher education of many other countries — to more closely resemble that in the USA. Effectively, it now forms the only model for the delivery of higher degree programmes across much of the global university system. By its very nature, modularization reduces the academic cycle into discrete components or *modules* each of which is composed of *credits*. Modules are independent of each other and typically short in duration. An entire degree programme might be composed of 36 modules of ten credits each. The final degree class is then calculated from the number of modules that a student completes and whichever of those modules are included in the final course assessment.

Modularization offers a practical way for universities to respond quickly to the needs of a changing marketplace and improve the relevance of degrees to employers through a more *corporate curriculum*. Modularization can also provide institutions with the flexibility to make changes to the curriculum or to design new qualifications in response to perceived demand both by students and potential employers — again reflecting an assumption that the main purpose of a university is to train students for employment. Of course, with the increasing *massification* of higher education, another primary objective of modularization is to increase access for more and a wider range of students whilst reducing the cost of delivery per student. As well as reduced cost, in principle modularized course structures allow better planning to use resources and centralized administrative functions (such as timetabling, admission, enrolment and graduation), which, in theory, should allow staff to focus on core academic activities and research.

There are advantages to the modular system. It is unambiguous in that marks awarded in each module are based on objective and verifiable criteria, details of which students are informed in advance. It is timely and allows students to follow progress throughout their course rather than wait for final marks at the end of a programme. It is transparent with the anonymous marking of scripts, internal and external evaluation procedures, and calculation of final marks leaving little room for bias against or preference for particular students. And it is fair in that students are rewarded for continuous effort throughout the course rather than for an intense effort at the end of the programme and/or luck in *question spotting* for final exams.

Agricultural Sciences and Modularization

Despite the advantages of modularization, there remain concerns about its impact on the future of higher education, in which learning is fragmented into *bite size chunks*, degrees commodified into products and students transformed into customers. For many, modularization is the academic equivalent of the supermarket, in which products (degrees) are exchanged for cash (fees). Similarly, the term *quality control* — with its Orwellian overtones — seems more at home on the factory floor or supermarket shelf than the lecture theatre and dispossesses academic staff of their professional agency over subject matter and delivery. For agriculture, the modular degree stands in contrast to the notion of the traditional university subject with its emphasis on the unity of subject knowledge that is developed sequentially and largely for a common cohort of students who start and complete their degree programmes together. Departure from this model to modularization sets two major challenges for teachers and students. The first is how to integrate learning across modules of an agriculture degree that might span crop sciences, animal sciences and management. The second is how to develop student capability in a subject that must marry the theoretical elements of taught modules and the practices of a profession.

As with other subjects, for students of agricultural sciences modularization offers a number of advantages over traditional *linear* degree programmes. Its flexibility allows for interdisciplinarity, curricula breadth and the possibility of attracting students to specific modules who would not otherwise consider agriculture as a subject. In particular, it provides flexibility because modules can be taken in a different order and times of the calendar year. This gives an opportunity for students to move between institutions nationally and internationally. In addition to the cultural aspects of visiting other countries, students can benefit from exposure to environments and agricultural systems that are different from their own and even participate in projects that relate to their own interests but which for climatic or resource reasons cannot be offered by their home institution. What better opportunity for a student in, say, northern Europe, than working for part of their degree in a tropical environment and on unfamiliar crops or agricultural systems? In being independent of a rigid degree programme, individual modules can also be offered as short in-service courses or conversion options for individuals who are otherwise employed. Perhaps the main advantage of modularization is the opportunity that it offers to establish cost-effective degree programmes in agricultural sciences, both in developed countries and for developing countries. In such cases, the costs can be shared across institutions and external expertise brought in through visiting professorships and guest lectures.

Research Assessment and Agricultural Sciences

Like its cousin, *teaching quality assurance*, in many countries universities and their staff have become familiar with periodic assessment of their research. Argument against either is difficult. Who could reasonably expect that the quality of teaching and significance of research in public institutions should *not* be measured or that their evaluations should *not* be independently derived and made available for public scrutiny? Transparent assessment should have the multiple benefits of driving up quality, allowing for self-reflection and improvement, providing metrics to compare performance across subjects and institutions and an objective mechanism to justify the allocation of limited resources.

Key phrases that guide assurance and assessment metrics are *value for money* and *accountability*. An example of procedures to ensure both is the Research Assessment Exercise (RAE) and its successor the Research Excellence Framework (REF), which are applied to UK higher education institutions.[15] Research assessment generates ratings for quality across all UK universities and disciplines. These metrics are used by government funding agencies to distribute public funds for research based on *quality rankings* and *value for money*. As well as the funding bodies, industry, commerce, and other users of university research contribute to the assessment of research quality. Inputs from beyond academia allow for expertise from industry and commerce to help guide decision making. Institutions that achieve the highest research ratings receive a greater proportion of available funds; in short, 'points mean prizes'. Not only do high-ranking institutions receive more public support, their score allows them to attract more funding from the private sector, charities, foundations, philanthropic organizations and other sponsors. In this way, the most successful institutions have a *compound interest* in research assessment; the perceived quality and value for money of their research attracts investment at home and suitors abroad.

Measures of performance and quality — under a common rubric of *accountability* — are an essential component of value for money since they provide a benchmark for *return on investment* and *value addition* from research. Accountability also can be used to evaluate whether research support falls within the remit of the institution and department. In that regard, it is used to assess the suitability of the institution to carry out research using its facilities, expertise and human resources. In principle, this should deter researchers, for example working in basic or fundamental research, from taking on projects that require application beyond their competencies. Terms like *quality* and *value for money* provide justifications, even imperatives, to assess research by individuals and across institutions and disciplines. Of course, the metrics of assessment are determined against the criteria on which they are based and influenced by those who decide them. Across the myriad of subjects and specialisms that exist in modern academia, simplifications and assumptions are inevitable if we are to *compare apples and pears* (say, theology and magnetic resonance) and still provide a universal formula. Winners — those who score highly in the exercise — usually insist that the assessment criteria are fair, whereas losers frequently question their validity. However, setting aside argument over implementation details, research assessment exercises are with us for the foreseeable future and pose several challenges for the evaluation of university research in agricultural sciences and related field-based disciplines. Each challenge has significant implications for the future of agricultural research, its practitioners, and its funding, and, since agriculture affects us all, our common future.

Not only must agricultural research secure funding from public sources, increasingly, it must also endeavour to attract support from the private sector. Even with public funds, there is now a greater emphasis on *wealth creation* than societal impact, which privileges research that has potential to generate income. In the case of private funding, industry can and often does impose additional confidentiality terms and restrictions on the publication and dissemination of research results, with obvious limitations on its social value and research impact. Rather than contributing directly to communal benefit, privately funded research outputs are often retained by companies for commercial advantage. In this way, agricultural research is increasingly driven by profit rather than enquiry, short-term outputs rather than long-term benefits, commercial advantage rather than social impact. It is also increasingly reactive to the needs of industry and government rather than proactive to the needs of society. As industry gains ever more control over the funding of research, it increasingly determines what is done, how, by whom, for whom, on which terms and at what cost. This is a marked departure from research done as a public good for the wider benefits of society or the environment. As industry increasingly funds university research, it expects to gain maximum commercial advantage over competitors through confidentiality agreements and control of innovation and new technologies. Even the distribution and dissemination of student theses may often be restricted for a period of time after their completion so that the sponsor can extract any commercial advantages first.

Whilst commercial funding might provide interim support, agriculture as a global industry cannot survive without long-term and strategic planning. This will need an approach which recognizes that investments may take many years to lead to impacts that themselves are for more than profit alone and relate to complex challenges beyond the immediate interests of the sponsor.

Quality

In the research assessment process, individual academics submit a record of their publications and the *citation index* or *impact factor* of the journals in which their work appears. The number of their publications provides a measure of their performance and that of their departments, and the quality of their publications is based on the citation index or impact factor of the related journals.[16] The citation index or impact factor of a journal is based on the frequency to which journals and papers are referred by other authors. This crude measure of quality clearly favours academics who frequently publish in high-impact journals, in areas of current academic interest and with potential for commercialization or *wealth creation*. Research that is equally important but of less topical or commercial interest will get lower scores and therefore be seen as of lower quality. Of course, academics are well aware of the rules and often choose to work on topics which appear in high-impact journals both to further their careers and secure research funding.

Quantity

By its very nature, research done in laboratories does not directly depend on the weather outside nor on the time of day or day of the week. This means that the workload for measurements, data analysis, and preparation of papers can be distributed throughout the calendar year. In contrast, field research depends on location, season and sometimes even on the time of day, and must often proceed throughout the growing cycle of a crop or animal. Because seasons vary, quality journals require replication of measurements in time as well as space. Consequently, the same experiment must be done over several seasons (often where there is only one season per year) to collect enough data for publication. Whilst this long period for experimentation might not restrict the final number of papers from a study, it does affect the time course from preparation of a research proposal, securing funding and final publication of results from the funded project.

Increasingly, funding agencies will only support short-term research. Again, agriculture loses out when it needs years to provide outputs and even longer to measure impacts. Funding agencies may insert milestones and clauses that allow them to terminate or redirect funding for research programmes that have already been started. The challenges of justifying longer-term funding and the nature of fixed-term and temporary contracts of non-tenured researchers adds further complexity to agricultural research projects. Short-term funding and temporary contracts have knock on effects on the motives and careers of researchers who must always be on the lookout for new appointments when and wherever they appear. Telling a young researcher on a short-term contact or their sponsor that it may be three or four years before results are published is unlikely to attract enthusiasm let alone loyalty. If the job security of researchers is tenuous, why would they invest time and effort in research that may continue after the end of their contracts?

Consequences

Agricultural sciences are caught between two ends of the research spectrum. Fundamental or basic research seeks breakthroughs (usually technical and in laboratories) in which application of research results is not a priority. For this, researchers need to demonstrate a track record of publications and evidence that they and their institutions can achieve outputs that maximize quality (high-impact publications) and accountability (cost-effective use of resources). Whilst applied research also requires accountability (*money well spent*), it must also provide agricultural solutions to practical challenges. When industry is the paymaster, the funder will insist on milestones, deliverables, targets and outputs that justify an investment, which usually means short-term commercial benefit to private companies from research trials done in universities. In fact, agricultural research entails more than commercial field trials since its outputs have multiple impacts on society and the planet. A constant stream of publications based on the response of particular crop varieties to, say, fertilizers or agrochemicals does little to enhance the reputation of agricultural scientists beyond being *trials officers* for commercial companies and research that contributes little to our understanding of how agriculture can support economic growth, livelihoods, social development, and environmental integrity.

Agricultural Research Systems

Human beings have conducted agricultural research since the start of agriculture. Much of this has been through trial-and-error observations by farmers on their own land. With the advent of civilizations and empires, agricultural research became more systematic. Dedicated experiments were used to decide which crops, management techniques and storage conditions were best suited to different practices across territories. Empires also adopted agricultural methods learned from conquered peoples and, in turn, left a legacy of knowledge for their successors. In many cases, good practices were transferred and methods translated from one part of the empire to another. This early *technology transfer* allowed ideas and innovations to be disseminated rapidly across the empire and the exchange of knowledge between communities allowed new crops and livestock to be introduced in disparate locations.

Whilst academic research in universities began to formalize local and empirical (observed) knowledge, it was *gentlemen farmers* who established the first institutes dedicated to agricultural research. Initially in Britain, and then in other European countries, the Agricultural and Industrial Revolutions created an increasing demand for agriculture to feed a burgeoning population. Wealthy landowners had the resources and education to undertake independent research to help meet this need. These early agricultural researchers were usually landed gentry with an interest in how to improve agricultural methods for themselves and their local communities. Even if they were not exactly *sons of the soil* they were certainly *lords of the manor* motivated to improve agricultural systems and conditions. For example, Jethro Tull, an Oxford scholar and rich landowner, used his experiences of travelling in Europe to develop the first seed drill that became a widely adopted innovation to uniformly sow seeds on prepared agricultural land. Charles Townshend, who first introduced the idea of crop rotations using turnip as a *break crop*, was a wealthy politician who as Secretary of State worked with his brother-in-law Prime Minister Robert Walpole to direct British foreign policy. Townshend was the eldest son of Viscount Sir Horatio Townshend and his wealth, contacts and farmland allowed him to develop and test new methods without having to worry about the consequences of crop failure, or what neighbours might think. The use of crop rotations is now a widely accepted model for arable farming. Sir John Bennet Lawes, who established the first agricultural research station on his estate at Rothamsted, in Hertfordshire, UK, also came from a wealthy family of landowners and studied at Eton and Oxford before becoming actively involved in agricultural experimentation. He was typical of the great pioneers of agricultural research of the 18th and 19th centuries. He financed his own research and that of a brilliant young chemist, Henry Gilbert, with whom he developed chemical fertilizers that earned additional income through patents and sales of widely adopted inorganic fertilizers. It helped that John Bennet Lawes was an important and influential figure who was well known in all the best academic, political, and social circles, each of which included similar wealthy landowners.

At least in England, scientific progress was driven by individuals with the education and privileges to devote their energies to subjects that intrigued them, including agricultural problems of the day, without the need to convince funding agencies of the commercial potential of their work. Many were either politicians or enjoyed positions of power and influence and were answerable to no one but themselves. They did publish their research and their papers were subject to review by their peers (in many cases literally!) and their principle objectives were to show how improvements in agriculture could enhance the livelihoods of the local population most of whom still worked on the land or were connected with farming. As well as scientific enquiry, they saw that communicating their findings was important and many agricultural societies were formed to freely disseminate information on novel techniques and practices. Eminent amongst these was the Royal Agricultural Society of England (RASE), which was founded in 1838 as a private society to disseminate agricultural innovations to both the farming and scientific communities. The stated objectives of RASE were *the acquisition of agricultural knowledge, the union of science with practice, and the communication of agricultural information.*[17] No mention in these objectives of wealth creation, return on investment, impact factor, employability or quality assurance mechanisms.

Colonial Agricultural Research Systems

At the same time that agricultural research was being developed in England by wealthy individuals to benefit local agricultural communities, its government and those of other colonial powers were scrambling to sequester the lands and natural resources of much of Asia and most of Africa. Whilst there were also many other motives for colonization, agriculture was seen as the *new frontier* to exploit resources solely for the benefit of the colonial power. Many governors who were sent to manage pockets of the British Empire were themselves landed gentry with an interest in agriculture. They were keen to apply the techniques, farming systems, and approaches that were being developed in England in their newly acquired territories. Their primary purpose was not to uplift the local population or improve traditional agricultural practices, but rather to accelerate large-scale production of high-value crops such as sugar, tea, and rubber that could not be grown in cool northern climates, and to transport these as commodities to the home country or elsewhere across the empire. Staple food crops of lower economic value for the British population were not the purpose of these plantations because such crops could either be grown in Britain or imported from elsewhere in Europe or North America. Indigenous crops that were grown primarily by local people for their own subsistence were largely ignored or discouraged.

The aim of agricultural research in England was to improve existing systems in which both landowners and tenants had a common interest, but its purpose in colonial territories was to impose completely new systems to benefit the colonial power. These new systems were not shaped or fashioned over generations by or with local farmers, thus their introduction and maintenance were traumatic for local populations, cultures and environments and their legacies continue long after independence. The natural environment and messiness of local agriculture were there to be tamed by the serried ranks of uniform monocultures on plantations stretching to the horizon. Before formal colonization, an informal British empire already existed in private companies with outposts and administrations that were protected by the naval and armed forces of the crown. In many cases, the purpose of these companies was to grow and trade products from high-value commodity crops such as sugar cane, tea, spices, oil palm, rubber, and tobacco. These crops were grown in plantations that were owned mainly by European companies or individuals and cultivated either by locals, indentured labour or slaves. To maximize productivity and quality, plantation owners invested in research on their particular crops. Of course, since these crops were grown in monocultures, experiments could focus on improving production, quality, processing and transportation of one crop without regard to local biodiversity, cropping systems or cultures. As had happened in Britain, many of these early activities were initiated by individual plantation managers looking to maximize commercial returns for shareholders rather than part of a formal research process.

When research did become more formalized, it followed characteristics and approaches that were familiar in British agriculture. Experimental stations devoted solely to improving specific crops were established across Britain's tropical territories. These stations had a strong commodity focus because many were owned, managed and staffed by commercial companies rather than government. Research institutions that began as private initiatives include the Tea Research Institute, Rubber Research Institute, and Coconut Research Institute, all in Sri Lanka, the Rubber Research Institute in Malaysia, the Sugar Experimental Station in Mauritius, the Coffee Research Station in Kenya, and Sisal Research Station in Tanzania. In each case, they were staffed mostly by British expatriates, with some locally trained staff and a large workforce of untrained labourers. By keeping things simple — one crop one system — research could focus on the management of large estates composed of uniform monocultures. Any additional crops grown on estates or by local communities were for their own subsistence and were not the subject of research interest or investment by plantation owners.

The examples given here relate specifically to Britain, but there are parallels with other European powers. The outcome of European colonialism was the *double whammy* that it represented for the agriculture of its colonies. Not only was there no research or development of local crops and cropping systems but an external system to produce cash crops for European markets was imposed on fragile tropical environments. Development of agricultural research in Europe had been built on local species and experience, but much of the colonial research was imposed using exotic species and foreign expertise or decided in faraway corporate boardrooms.

National Agricultural Research Systems

Soon after gaining independence (most in the 1960s), former colonies realized that the agricultural research and extension systems that they had inherited from the colonial powers were inadequate to meet national food production needs. In response, many established their own National Agricultural Research Systems (NARS) to focus agricultural research on emerging national priorities. Justification for the NARS was a need to make science and technology rapidly and directly available to transform local agriculture, which was not a primary objective of local universities. However, the leaders of newly independent nations had a dilemma: as well as reorganizing research systems to increase food production, they still needed to produce commodity cash crops for export so that they could earn desperately needed foreign exchange. Moreover, much of the agricultural expertise that remained in former colonies resided in their commodity research centres and with expatriate scientists who had chosen to remain after independence. Inevitably, funding, talent and effort continued to be channelled to cash crops, while local crops that had been grown for millennia were ignored or marginalized.

Unlike the commodity research institutions that each focused exclusively on improving a single plantation crop, no single model was adopted for NARS in developing countries.[18] Several NARS models exist, depending on the links between research and extension, level of independence from governments, and the connections between national agricultural research organizations and their former colonial powers. Some NARS have independent research institutes while others are associated with or integrated into universities. Not surprisingly, given their colonial and post-colonial experiences, the performance of NARS has been variable. Much of their research output has been disrupted by changes in government policies, political instability and the impacts of emerging regional and global challenges such as climate change.

However, the biggest constraint to the effectiveness of NARS has been the continuity, quantum and disbursement of funding from government and other sources for research, operations and infrastructure. In many former colonies, funding for national agricultural research has declined over decades which has created tensions between researchers and policy makers, challenges in retaining and rewarding high quality staff, and affected institutional and staff morale. Many local scientists working in national institutions have had to take on one or more additional jobs to support their family and use their own money to reach field sites. Perhaps predictably, talented scientists who find themselves working under such constraints have gone to work in the private sector or given up on research altogether.

The decline in financial support for national agricultural research has coincided with sustained reductions in global investment in agricultural research. Together, these cuts reflect a diminished status of agriculture in developing countries that has eroded human resources, research capacity and stability across the NARS. Of course, reductions in funding were not exclusive to agriculture. In the 1980s, governments in many developing countries were forced to limit public spending by the neoliberal policies of donors aligned the International Monetary Fund (IMF) and the World Bank's Structural Adjustment Programmes (SAPs). In return for loans, debtor countries were compelled to implement policies that encouraged privatization, liberalized trade and foreign investment, reduced public deficit, and balanced government budgets. By minimizing the ability of independent countries to manage their own economies, routes were created for multinational companies to enter states and extract their resources at minimal cost.

Official Development Assistance

For many former colonies, Official Development Assistance (ODA) has been a significant source of funding for agricultural research and development, often from the former colonial power. Rather than as direct assistance, this has often been provided in the form of funding for specific research activities. Initially, such funds went to dedicated programmes for particular commodity groups, e.g. cereal, root or tuber crops with allocations based not on formal analyses but by specialists from the donor country. More recently, structured approaches to research funding and management have been introduced through mechanisms such as 'Logical Frameworks' (LFs) in which universities and research institutes must bid to secure and manage research projects.[19] To meet the criteria specified in the LF, such research has to be 'demand-led' with activities that deliver specific outputs and defined outcomes. Whilst, in principle, demand-led research should ensure greater focus on impact and *value-for-money*, questions remain as to who decides what is 'demand-led,' how such research leads to longer-term solutions, and how it addresses complex issues such as climate change, loss of biodiversity and damage to ecosystems. Again, if ODA is directed to immediate challenges, from where will funding come for strategic, sometimes risky and curiosity-driven research that may provide the bases of longer-term solutions? Other criticisms of ODA are that explicitly or implicitly it is tied to human and capital resources provided by the donor, is vulnerable to short term political considerations and is aligned with the foreign policy interests of the donor country. A recent example is that, after over 40 years of independence, initially as the UK Overseas Development Administration and since 1997 the Department for International Development (DfID), the UK government has merged DFiD with the Foreign Office to create the Foreign, Commonwealth and Development Office. In its time, DFiD was seen as a global leader in international development and a flagship of the UK government's commitment to spend 0.7 per cent of Gross National Income (GNI) on ODA, which in 2019 amounted to over $26 billion dollars. It remains to be seen how its political independence will be protected under the new regime but it is clear that its funding has not, despite previous commitments.

Many studies have demonstrated that funding of agricultural research and development is not just morally justified, especially from former colonial powers, but also delivers high rates of return on investment. For example, a comprehensive analysis by the International Food Policy Research Institute (IFPRI) of over 300 publications showed a return of around 60 per cent per year for investment in research in developing countries.[20] Despite such evidence, national investment in agricultural research has continued to decline since the mid-1980s[21] In the face of shrinking public funding, greater private funding has been seen as the universal panacea for agricultural research in developed, emerging, and developing countries alike. However, despite the enthusiasm of policy makers and donors, private investment, has mainly flowed to those aspects of agricultural research that assure returns on investment, such as biotechnology, agrochemicals, veterinary products, seeds, and machinery. This leaves the public purse to cover novel or speculative research, operational and running costs of facilities, the building and retention of human capital, and research on local crops and cropping systems that are not part of commercial agriculture. The biggest consequence of these trends has been on salaries for permanent and contracted researchers. Where wages for national researchers fall disproportionately below those of their private and international counterparts, research staff have a financial incentive to seek alternative employment and potentially suitable staff decline appointments. This haemorrhage deters new talent from joining research institutions, discourages students from applying to universities, and diminishes the perception of agriculture as a career option. As a consequence, the age profile of agricultural researchers and farmers has become older and agriculture increasingly seen by government leaders as a sunset industry. In developing countries and emerging economies that see industrialization as the future, governments and policy makers often perceive agriculture to be a burden on public funds and not worthy of political support.

Global Agencies

During the Second World War, agriculture across Europe suffered with loss of crop and livestock production, labour shortages, lack of machinery, and shortages of seeds, fertilizer, and fodder, along with damaged infrastructure, supply chains, and trade. The agricultural system was broken. In many European countries, especially those devasted during the war, disruptions in trade and a soaring black market resulted in food shortages and hardships until the late 1940s. In response to these uncertainties, there was greater state intervention to ensure the production, processing, sale, storage, and consumption of food.

In contrast, without the effects of conflict, agriculture boomed in the United States. Significant increases in agricultural productivity were underpinned by high consumer demand and buoyant commodity prices. To help administer post-war food relief for Europe, the United States, Great Britain and Canada cooperated in a combined food board to distribute food and funding via the Marshall Plan. This ensured that recovery in Europe was rapid and effectively administered through the Organization for European Economic Cooperation (OEEC). Later, the OEEC expanded to include non-European member countries and became the Organization for Economic Cooperation and Development (OECD). The OECD pursued agricultural programmes that were tied to economic policies and standardization in return for development assistance. Agriculture became inextricably linked with politics, economic power, and the ideology and manoeuvrings of the Cold War. In this context, a single agency for global agriculture — the Food and Agriculture Organization (FAO) — was established under the auspices of the United Nations.

The United Nations Food and Agriculture Organization (FAO)

FAO was founded at a time when the Allied governments were reimagining the future of their nations while food shortages, rationing, and hunger were still commonplace. Many recognized that increases in agricultural production were critical to avoid political upheaval and possible societal collapse. These existential challenges to a world traumatized by conflict gave architects of the new world order an opportunity to set out a vision of hope based on international cooperation and effective governance, rather than a retreat to isolationism, narrow self-interest, jingoism, and xenophobia. As part of this international vision, FAO was set up as an overarching agency to increase food production and provide better nutrition to achieve social and economic development and higher standards of living.[22]

FAO was not the first attempt to establish a global agency for agriculture. In the late 19th century, the Polish-American agriculturalist David Lubin proposed that there should be an international organization for food and agriculture.[23] He also suggested an international agricultural congress where countries from across the world could share experience on agricultural production and economics. With his son, Simon, Lubin developed the concept of an international chamber of agriculture, and in 1896, he moved to Europe to implement his plan. In 1904, King Victor Emmanuel III of Italy gave his full support to Lubin's concept of an international agricultural congress and provided it with a building and an annual income of $60,000. In 1905, an international conference was held in Rome, which led to creation of the International Institute of Agriculture (IIA), also to be hosted in Rome. The aims of the institute were to help farmers share knowledge on improved production methods, establish a system of rural credit and have control over marketing their products. By 1934, 74 nations were represented at IIA gatherings and it became recognized as the first organization to deal with agricultural challenges on a global scale.[24] The IIA collected and disseminated data on agriculture, including production statistics and a catalogue of crop diseases, published the first agricultural census in 1930 and acted as a venue for discussion of new ideas about agriculture.

In 1943, President Franklin D. Roosevelt called a United Nations Conference on Food and Agriculture in Virginia, USA that included representatives from 44 countries. The Conference concluded with a commitment to establish a permanent organization for food and agriculture. In 1945, the Constitution of a Food and Agriculture Organization was approved in Quebec, Canada, with its initial headquarters to be in Washington DC, USA. The FAO soon became the recognized global authority for agriculture and its United Nations status made the IIA redundant. Therefore, also in 1945, the IIA ceased operations and was officially dissolved with its functions, facilities and mandate transferred to FAO. In 1951, FAO moved to its present headquarters in Rome.

In honour of David Lubin, the FAO named its library the David Lubin Memorial Library, which maintains his personal archives and the collection of IIA documents. Indeed, Lubin's ideas and legacy go beyond the IIA. In 1916, his proposal for a US Federal Farm Act introduced rural credit that contributed to the relief of American farmers during the Great Depression. In addition, his successful fight to lower freight charges led to development of the parcel post system. From his humble origins in Poland, emigration first to England and then America, and with no formal education, David Lubin's single-minded determination to deliver his concept of an international agency for agriculture left a lasting if largely ignored legacy. In his spare time, Lubin also wrote essays and treatises, and his novel, *Let There be Light*, proposed a universal world religion. He died at the age of 69 in the 1918 flu pandemic.

FAO is one of 17 specialized agencies that carry out various functions on behalf of the UN.[25] It is directed by the Conference of Member Nations through a council of 49-member states and a Director General who heads the agency with 197-member nations and associate members. From its headquarters in Rome, FAO maintains regional offices for Africa (in Accra, Ghana), Asia and the Pacific (in Bangkok, Thailand), Europe and Central Asia (in Budapest, Hungary), Latin America and the Caribbean (in Santiago, Chile), and the Near East, (in Cairo, Egypt). FAO's core Programme budget is supported through contributions of individual members set at the FAO Conference. This budget covers core technical work, cooperation and partnerships including the Technical Cooperation Programme, knowledge exchange, policy and advocacy, administration, governance, and security. FAO has departments for: Agriculture and Consumer Protection, Climate, Biodiversity, Land and Water, Economic and Social Development, Fisheries and Aquaculture, Forestry, Corporate Services and Technical Cooperation and Programme Management. Through these, FAO coordinates the activities of governments and technical agencies in programmes to develop agriculture, forestry, fisheries and land and water resources. It also carries out research, provides technical assistance on projects in individual countries, operates educational programmes through seminars and training centres, maintains information and support services, keeps statistics on world production, trade and consumption of agricultural commodities, and publishes a number of periodicals, yearbooks, and research bulletins.

Achievements of FAO

In almost three-quarters of a century since its establishment, FAO has evolved into the recognized global arena for negotiation, standardization, policy-making and research related to agriculture. Perhaps its greatest contribution has been to become the *go to* first point of reference for agricultural knowledge and expertise. FAO provides statistics on food production and consumption by each member nation and — with its sister UN agency, the World Health Organization (WHO) — the *Codex Alimentarius*, or *Food Code*, a collection of standards, guidelines and codes-of-practice established to protect consumer health and promote fair practice in food trade.[26] FAO is also responsible for updating and disseminating FAOSTAT, which since 1961 has provided free access to food and agriculture data on more than 245 countries and territories, along with AQUASTAT, which is its global information system on water resources and agricultural water management.[27]

Every year, FAO also publishes major reports on food, agriculture, forestry, fisheries and natural resources. These include: 'State of the World' reports for Agricultural Commodity Markets, Food and Agriculture, Food Security and Nutrition, Animal Genetic Resources for Food and Agriculture, Biodiversity for Food and Agriculture, Forest Genetic Resources, Forests, Land and Water Resources for Food and Agriculture, Plant Genetic Resources for Food and Agriculture, Fisheries and Aquaculture, and Soil Resources.

FAO has been at the forefront of many global initiatives and treaties. In 1974, it convened the first World Food Summit to address issues of hunger, malnutrition and food insecurity. The meeting issued a proclamation that;

every man, woman and child has the inalienable right to be free from hunger and malnutrition in order to develop their physical and mental faculties

This proclamation came with a global commitment to eradicate these issues within a decade. In 1996, FAO organized the World Food Summit, which concluded with the signing of the Rome Declaration. The Declaration established a goal of halving the number of people suffering from hunger by 2015. In 1951, FAO created the International Plant Protection Convention (IPPC) as an international treaty organization to prevent the international spread of insect pests and plant diseases in both cultivated and wild plants. In 1994, FAO also established an Emergency Prevention System for Transboundary Animal and Plant Pests and Diseases, focusing on the control of diseases such as rinderpest, foot-and-mouth disease and avian flu by helping governments coordinate their responses. Meanwhile, the FAO Locust Watch monitors the world-wide locust situation and keeps affected countries and donors informed of expected developments. FAO is host of the International Treaty on Plant Genetic Resources for Food and Agriculture (ITPGRFA), which entered into force in 2004.[28] FAO's Global Partnership Initiative for Plant Breeding Capacity Building (GIPB) is dedicated to increasing plant breeding capacity in developing countries. FAO's Globally Important Agricultural Heritage Systems (GIAHS) Partnership Initiative aims to identify, support, and safeguard traditional agriculturalists and their livelihoods, agricultural and associated biodiversity, landscapes, knowledge systems and cultures around the world. In addition, FAO is host to a number of other global partnerships and networks, including the Global Forum for Agricultural Research and Innovation (GFAR), a network of over 600 partner organizations working across the world in scientific research, education, rural extension, and advisory services; businesses and enterprise; international development agencies; and farmers' and civil society organizations.

Criticisms of FAO

There have been criticisms of FAO for at least 30 years. Many relate to its performance as the premier global agency for agriculture. In part, these criticisms resulted in the creation of two new organizations after the World Food Conference in 1974: the World Food Council (WFC) and the International Fund for Agricultural Development (IFAD).

The most telling concerns relate to whether FAO has maintained the lofty ideals and vision that created the organization in 1945. Across much of the globe, the Second World War left political turmoil, societal devastation, and human suffering that hampered reconstruction efforts and threatened further global conflict that, whilst largely avoided, culminated in the Cold War. At its inception, FAO had to face these challenges and provide leadership to a dislocated world and for a damaged planet. In response, it chose to set out a positive vision of a post-war world without famine that could be achieved through international cooperation and by putting science and technology at the disposal of forces for world peace. At the time, FAO envisaged a globally interconnected system of food governance supported by a holistic and long-term approach that would put farmers and social justice at the heart of development. As the global political climate strayed from idealism and hope to the cynical realities of Cold War politics, FAO's attempts at long-term structural transformation of agriculture stalled. Instead, it grappled with reconciling the conflicting goals of increasing agricultural productivity and markets, on the one hand, and the principle of food security as a shared global responsibility and human right on the other.

From the 1950s, FAO increasingly embraced an approach in which technical assistance, knowledge, and expertise were focused on increasing the yields and securing markets for specific crops as pathways to modernization and progress. Whilst in some parts of the world this technological approach spurred economic growth and industrial production, it left no room for alternative visions of agricultural development that embraced local knowledge and traditional systems as a basis for improvement. In practice, it provided few tangible benefits to farmers unless they abandoned their diverse agricultural systems in favour of top-down technologies on mainstream crops. Farmers increasingly came to be seen as beneficiaries of research rather than partners in it and recipients of innovation rather than the owners of it.

Perhaps this duality should have been obvious from the outset. An organization in which *food* precedes *agriculture* in its name is already making a statement about prioritizing products over systems.[29] The FAO logo makes the distinction even more stark. The text of the logo simply says *Fiat Panis* — let there be bread — with one food product — bread — in preference to a global system of agriculture. Indeed, not only is bread from wheat the only named food product the logo also includes an ear of wheat, a cereal associated with industrial and Western agriculture, as the embodiment of agriculture. Whilst not its intent, is this reduction to one product and one crop reflective of the purpose of FAO? This is after all an organization that covers diverse crops, animals, insects, forestry and fisheries and activities that span genetics to satellite mapping and simulation modelling, gender, natural resources, traditional knowledge and seed banks. These tensions became apparent when, in 2004, FAO produced a report called *Agricultural Biotechnology: meeting the needs of the poor?* The report highlighted the potential role of biotechnology to address global hunger without including the views of organizations representing the interest of farmers and NGOs. In response, more than 650 organizations from around the world signed an open letter stating that FAO had;

> *broken its commitment to civil society and peasants' organizations*[30]

The letter complained that FAO was siding with the biotechnology industry and that the report;

> *raises serious questions about the independence and intellectual integrity of an important United Nations agency.*

Despite attempts at reform, FAO is often seen as conservative, slow to adapt, mired in bureaucracy and red tape, and whose costs and procedures diminish its core competencies and weaken its ability to swiftly address urgent global challenges. Despite claims that FAO has decentralized — indeed, more than half its staff are now based outside Rome — there remain concerns that institutional culture and administrative and management systems remain unchanged. As the French theologian Alain de Lille, said in 1175 *"mille viae ducunt homines per saecula Romam"* — a thousand roads lead men forever to Rome.[31] Perhaps little has changed since.

International Agricultural Research Centres (IARCs)

Over recent decades, and across the developing and developed worlds, the number of national agricultural research institutes has been reduced and centres amalgamated. Cuts in public funding and changing national priorities are involved, but this consolidation is also associated with the emergence of international agricultural research centres (IARCs). Increasingly, IARCs have become recognized as the primary agents of innovation, new agricultural technologies, and plant breeding, with the NARS seen as vehicles for local adaptation and application. Established in 1971, the Consultative Group on International Agricultural Research (CGIAR), which is now the CGIAR System Organization, is the main globally recognized grouping of IARCs. As with the FAO, the genesis of the CGIAR is an outcome of the impacts of the Second World War.

Consultative Group on International Agricultural Research (CGIAR)

In 1943, the Rockefeller Foundation and the Mexican government established the Office of Special Studies, a partnership which laid the basis for what became known as the Green Revolution. In 1960, support from the Rockefeller and Ford Foundations led to the establishment of the International Rice Research Institute (IRRI) in the Philippines, and in 1963, the International Maize and Wheat Improvement Center (CIMMYT) in Mexico. The purpose of the two centres was to transfer Green Revolution technologies for widespread adoption of high yielding varieties of these three staple cereals: wheat, rice, and maize. Even with the remarkable successes of the Green Revolution, it was clear that these two institutions and their three mandate crops could not by themselves deliver the agricultural research and development impacts needed to feed a rapidly growing global population. In 1967, two new international centres were added — the International Center for Tropical Agriculture (CIAT) in Colombia and the International Institute of Tropical Agriculture (IITA) in Nigeria.

In 1969, the Pearson Commission on International Development advocated a broader international effort to support food production and tropical agriculture, and in 1970 the Rockefeller Foundation proposed a worldwide network of agricultural research centres under a permanent secretariat. This concept gained support from the World Bank, FAO, and United Nations Development Programme (UNDP). On May 19th, 1971, the CGIAR was established to coordinate international agricultural research aimed particularly at achieving food security in developing countries. The CGIAR had four strategic objectives: reducing rural poverty, improving food security, improving nutrition and health, and sustainably managing natural resources. As well as its strategic objectives, the CGIAR adopted six founding principles to guide research and resource allocation;[32]

1. Donor Sovereignty: Each donor is free to choose which centre/ research to support
2. Centre Autonomy: Centres have the autonomy to utilize their own resources

3. Consensus Decision Making: Decisions by the CGIAR will be done by consensus
4. Independent Technical Advice: An independent Technical Advisory Committee to set priorities
5. Informal Status of the System: The CGIAR does not have a formal legal status
6. Non-political, non-partisan, non-ideological: apolitical.

To deliver its strategic objectives, new international centres were added to the four already in existence. These would broaden the initial focus of CGIAR research on staple cereals to include cassava, chickpea, sorghum, potato, millet and other major food crops, livestock, fish, farming systems, conservation of genetic resources, water management, and policy research. The CGIAR centres also began to provide services and access to technologies for NARS in developing countries. By 1983, there were 13 CGIAR research centres around the world, and by 2001 the number had risen to 15;

1. AfricaRice (ADRAO/WARDA), Cotonou, Benin
2. Bioversity International, Rome, Italy
3. International Center for Tropical Agriculture (CIAT), Cali, Columbia
4. Center for International Forestry Research (CIFOR), Bogor, Indonesia
5. International Maize and Wheat Improvement Center (CIMMYT), Mexico City, Mexico
6. International Potato Center (CIP), Lima, Peru
7. International Center for Agricultural Research in the Dry Areas (ICARDA), Beirut, Lebanon
8. International Crops Research Institute for the Semi-Arid Tropics (ICRISAT), Patancheru, India
9. International Food Policy Research Institute (IFPRI), Washington D.C., USA
10. International Institute of Tropical Agriculture (IITA), Ibadan, Nigeria
11. International Livestock Research Institute (ILRI), Nairobi, Kenya and Addis Ababa, Ethiopia
12. International Rice Research Institute (IRRI), Los Baños, Philippines

13. International Water Management Institute (IWMI), Battaramulla, Sri Lanka
14. World Agroforestry Centre (ICRAF), Nairobi, Kenya
15. WorldFish Center (World Fish), Penang, Malaysia.

CGIAR Programmes

The CGIAR had grown from four to 15 research centres, linked by a set of founding principles that enshrined the sovereignty and autonomy of each centre. To encourage common actions, in 2001 the CGIAR adopted a series of Challenge Programmes to focus efforts across the centres along with external partners, each around a specific global challenge. By building critical mass with partners, it was envisaged that the impact and visibility of the CGIAR research agenda could be increased. The three Challenge Programmes hosted within a particular CGIAR centre are;

- The Challenge Program on Water and Food (CPWF)[33] to address the global challenge of water scarcity, poverty, and food security which brings together scientists, development specialists, and river basin communities in Africa, Asia, and Latin America. The CPWF's mission is: "To increase the productivity of water for food and livelihoods, in a manner that is environmentally sustainable and socially acceptable".

- The Biofortified Crops for Improved Human Nutrition Challenge Program (BCP)[34] (renamed HarvestPlus) to improve the health of poor people by breeding staple food crops that are rich in micronutrients, a process referred to as *biofortification*. HarvestPlus focuses on three micronutrients that are widely recognized by WHO as limiting in human diets — iron, zinc, and vitamin A (beta-carotene). Breeding programmes were initiated to increase the level of a specific micronutrient in six staple foods that are consumed by many of the world's poor in Africa, Asia, and Latin America — rice, wheat, maize, cassava, sweet potato, and common bean.

- The Generation Challenge Program (GCP)[35] aims to increase the use of crop genetic resources to create a new generation of crops that meet the needs of farmers and consumers. Its purpose is to apply molecular biological advances to crop improvement for developing countries to create an *integrated platform* of molecular biology and bioinformatics tools. This will be freely available to enable researchers, particularly in developing countries, to use elite genetic stocks and new marker technologies in their local breeding programmes.

Underpinning the three Challenge Programmes, most CGIAR research and development activities are centred on eight CGIAR Research Programmes (CRPs). These are structured in two interlinked clusters. The first cluster is the "innovation in agri-food systems (AFS)" which includes eight commodity-based CRPs;

- Grain Legumes and Dryland Cereals Agri-food Systems
- Fish Agri-food Systems
- Forest and Agroforestry Landscapes
- Livestock Agri-food Systems
- Maize Agri-food Systems
- Rice Agri-food Systems
- Roots, Tubers, and Bananas Agri-food Systems and
- Wheat Agri-food Systems.

The second cluster includes four cross-cutting "global integrating programs (GIPs)";

- Agriculture for Nutrition and Health (A4NH)
- Climate Change, Agriculture, and Food Security (CCAFS)
- Policies, Institutions, and Markets (PIMS), and
- Water, Land and Ecosystems (WLE).

The intent of the commodity-based CRPs is to operate within an agro-ecosystem context and link with multiple stakeholders along the research-development-end-user continuum, but by their very nature they remain CGIAR centric. In contrast, the cross-cutting GIPs have the flexibility to build partnerships with stakeholders across all of their activities that require collaboration on complex issues that go beyond the mandates of individual centres.

In 2016, the CGIAR Consortium was replaced by the CGIAR System Organization which, for the first time, gave a single legal status to the entire CGIAR system with a System Council to provide a governance function. As a single entity, the CGIAR is now free to enter into "treaties, agreements, arrangements and contracts, acquire and dispose of movable and immovable property and institute and respond to legal proceedings". From September 2020, leadership of the CGIAR System Organization was entrusted to an Executive Management Team (EMT) comprising three Managing Directors who are charged with its transition to *One CGIAR*. Amongst the roles of the EMT is to lead development of the One CGIAR science agenda until 2030 and prepare for consolidation of CGIAR Centers and Alliances into a few operational units to deliver common support services across multiple countries.

Successes of CGIAR

The CGIAR system now employs 8,000 scientists, researchers, technicians, and other staff, and hosts vast infrastructure and facilities in much of the world.[36] In 2021, the system will be 50 years old and its oldest centre (IRRI) was founded more than 60 years ago. It has contributed to a world that is now more food secure than when it was launched. Much of CGIAR's impact has been through crop genetic improvement and breeding programmes. Between 1965 and 1998, CGIAR plant breeders were involved in developing varieties that accounted for 65 per cent of the global area planted to 10 of its mandate crops — wheat, rice, maize, sorghum, millet, barley, lentil, bean, cassava, and potato. The CGIAR centres host 11 gene banks that conserve over 760,000 plant accessions. In 2006, the centres with gene banks signed superseding agreements that placed their collections under the auspices of FAO through the ITPGRFA, and adopted its standard contract for exchanging genetic materials. Through the CGIAR's Systemwide Genetic Resources Programme, the centres and their partners share knowledge about germplasm and its discovery and conservation, conduct joint research, establish common policies and practices, and contribute to international debate on how best to protect and equitably share genetic resources.

CGIAR scientists advise on how best to protect and equitably share genetic resources and the intellectual property derived from them. Seed contributions from CGIAR gene banks have also helped agriculture to recover after conflicts in countries such as Afghanistan, Angola, Cambodia, Mozambique and Somalia. For example, in Cambodia's *killing fields*, an estimated two million people died by execution, forced labour, and famine. As well as its fellow citizens, the Pol Pot regime also destroyed all of the nation's rice varieties, but after its overthrow, all of these varieties were replaced from IRRI's gene bank, leading to Cambodia again becoming a net rice exporter.

It is a sign of IRRI's impact that one centre based in the Philippines is responsible for the genetic resources of a crop that produces over 700 million tonnes of grain annually on over 158 million hectares of land in over 100 countries.[37] Nearly 640 million tonnes of rice are grown on over 200 million farms in Asia, representing 90 per cent of global production. Rice research is done throughout the world (even where it is not grown), but researchers look to IRRI for genetic material, research publications, and global statistics. This is not surprising because IRRI holds over 132,000 accessions in its gene bank, and its research has developed the high yielding varieties that form the basis of the world's paddy rice cultivation.

Another remarkable initiative in which the CGIAR is instrumental is the Global Crop Diversity Trust, a public-private partnership with an endowment of $260 million to conserve agricultural biodiversity.[38] In 2008, the Trust opened the Svalbard Global Seed Vault on a Norwegian island above the Arctic Circle, with a capacity to preserve seed material for thousands of years. Svalbard is the *Doomsday* repository for humanity's agricultural heritage. At its opening, 21 national and international institutes deposited nearly 300,000 duplicate seed samples. More than 200,000 came from CGIAR gene banks, which plan to deposit another 300,000 samples in the coming years. Among the first to systematically conserve biodiversity, CGIAR centres now maintain over 650,000 samples of crop, forage, and agroforestry genetic resources.

In addition to conservation and improvement of genetic resources, the CGIAR centres have also contributed to increasing the nutritional content of staple crops, breeding insect- and disease-resistant varieties, integrated pest management and improvements to biological control measures in livestock and fish production systems, better natural resource management, and improved policies in numerous areas, including forestry, fertilizer, marketing, and genetic resources conservation and use.

Criticisms of CGIAR

Vision: *A world with sustainable and resilient food, land, and water systems that deliver diverse, healthy, safe, sufficient, and affordable diets, and ensure improved livelihoods and greater social equality, within planetary and regional environmental boundaries.*[39]

Despite its undoubted and manifold achievements, the CGIAR has attracted criticism from a number of sources, many of which raise concerns about the role of the private sector, the influence of politics, and the technology-driven focus of its research on particular elements of its vision and purpose. Rather than at the executive level, accusations remain that at the operational level the CGIAR's partners are not the rural poor but rather the private sector that will only support research projects that are of commercial benefit. Critics question the increasing role of multinationals and the private sector, both in funding and benefitting from CGIAR activities. Despite its commitment *to produce public goods for the benefit of poor agricultural producers in developing countries*, CGIAR research is critiqued as promoting the dependence of farmers on costly intensive inputs such as pesticides, fertilizers, and other chemicals. Arguably, these are technologies that relate primarily to increasing the yields of calorie-rich staple crops grown by male farmers with access to cash or credit.

CGIAR's significant role in strengthening technical capacity in international agricultural research is undeniable. Since its inception, the consortium has produced improved varieties of major crops, improved farming methods and policy analysis, and created new knowledge. However, CGIAR's priorities remain firmly focused on scientific and technological solutions to societal challenges in which agriculture plays a part. This is not surprising because the foundation of the CGIAR was the Green Revolution. It has continued to promote agricultural intensification, biofortification, and technocratic solutions to environmental issues and increased food production. Its research continues to prioritize scientific solutions to increasing yields, as well as crop genetic improvement and breeding of staple crops. These have remained CGIAR's stock-in-trade. The accusation is that rather than trying to overcome the political and social factors that control access to food, CGIAR research has focused on increasing food production *per se*. The concern is that by declaring its preference for technocratic solutions, the CGIAR has increasingly aligned itself with agribusiness conglomerates and donors who promote scientific and technological research for yield increases of major crops. In fact, food policies do not just affect food production and consumption, but have enormous impacts on local ecologies, relative hunger and poverty, and access to nutritious, affordable, and culturally appropriate food for all.

Alignment with the private sector risks a top-down, one-size-fits-all commercial approach to agricultural research — one that ignores the knowledge and experience of farming communities, their crops and farming systems in favour of a single model of industrial agriculture. The criticism is that CGIAR research programmes are still designed and evaluated from a linear perspective based on a *commodity* vision of research. It is unclear how, by itself, a model designed around input technologies for major commodity crops can ensure the management of natural resources in the face of climate change and other challenges. To fulfil its vision, critics argue that the CGIAR must look beyond technological solutions to a new socioeconomic environment in which agricultural communities are partners, not clients, and are themselves sources of innovation. For this, it will have to become a facilitator of interactions between global research and local innovation networks, and a trusted source of information and expertise to help communities access new technologies and facilitate their own change process.

The history and culture of CGIAR institutions is primarily centred around commodity crops. In some cases (IRRI, CIP), the title of the institution includes the crop or crops for which it is responsible. In others, the title does not include the names of particular crops, but rather the mandate of the institution. For example, the International Center for Research in Dry Areas (ICARDA) is responsible for barley, lentil, and faba bean and the International Crops Research Institute for the Semi-Arid Tropics (ICRISAT) is responsible for six crops — pearl millet, finger millet, sorghum, pigeonpea, chickpea, and groundnut. One of the 15 CGIAR centres, Bioversity International, has a broad mandate for agrobiodiversity. Others, such as the International Water Management Institute (IWMI),

Center for International Forestry Research (CIFOR), World Agroforestry Centre (ICRAF), and International Food Policy Research Institute (IFPRI) are not restricted to specific mandate crops and their cross-cutting remit allows them to work closely with non-research development partners. Of course, there are many advantages to working on specific commodity crops — research, breeding, marketing, and socioeconomic and policy issues can be highly focused on specific targets and sponsors can decide which crop and where in the value chain to direct funding. Funding agencies know what they are paying for and researchers can focus to achieve their research objectives. The danger is that such immediately attractive options deter any serious considerations of more complex, long term and radical alternatives.

The most recent criticism of the CGIAR is not on attempts to integrate its research activities but in its decision to further consolidate operations and governance under a *One CGIAR* vision for international agricultural research. In July 2020, an open letter to the CGIAR, signed by the International Panel of Experts on Sustainable Food Systems (IPES-Food) included this statement;

"Reform of the CGIAR is long overdue. But on the eve of a major restructuring, we believe that the reforms on the table fall short of the fundamental change that is required, and risk exacerbating power imbalances in global agricultural development. We are concerned that the 'One CGIAR' reform process appears to have been driven forward in a coercive manner, with little buy-in from the supposed beneficiaries in the global South, with insufficient diversity among the inner circle of reformers, and without due consideration of the urgently-needed paradigm shift in food systems."[40]

In the open letter, its signatories contend that representatives of the global South were generally against the merger, while major funders and scientific institutions were supportive. It goes on to note that only seven of the 22-member group responsible for managing the transition process are from the global South and that the voices of farmers, civil society, and independent researchers are still largely absent. However, its most searing indictment is that the consolidation of the CGIAR system will not equip its members for the paradigm shift in food systems that is urgently needed.

For this, it will have to move from the limitations of the Green Revolution approach to one that builds build resilience through crop/species diversity and the mainstreaming of agroecology to deliver economic, environmental, and social co-benefits. It is hard to see how further consolidation of the CGIAR will lead to greater diversity or innovation in its actions.

Association of International Research and Development Centers for Agriculture (AIRCA)

AIRCA is an international coalition of seven research and development organizations established in 2012 to sustainably address the global challenges of food and nutritional security.[41] AIRCA's vision is: *Diverse food systems for people and planet.* Its mission is *to deliver innovations contributing to diverse, safe, nutritious, and climate resilient food systems for health, livelihoods, and environment.*

With population growth, urbanization and climate change, pressure on agriculture and trade-offs among different land uses will intensify. Identifying and optimizing these trade-offs means looking at the entire landscape and using a food systems perspective to develop options that pursue positive and lasting health, environmental, and socio-economic outcomes. AIRCA centres work in partnerships towards food systems that support livelihoods and health, in particular by;

- producing nutritious, safe, and affordable food that goes beyond calories
- devising inclusive employment and income opportunities, especially for women and youth
- designing approaches to optimize human, plant, animal, and environmental health

and that support the environment, in particular by;

- enhancing agricultural biodiversity with currently underutilized crops
- introducing landscape and agroecological approaches and
- optimizing input use efficiency, reducing waste and promoting circularity.

AIRCA members have skillsets to address complex challenges through complementary expertise, partnerships and global networks, and use an agile coordination structure for rapid response. AIRCA's efforts are complementary to those of the CGIAR, but its members focus more on marginal environments (e.g. mountains, saline environments) and diversification (vegetables, insects, underutilized crops). A strong focus is on environmental stewardship — protecting soil and plant health — and knowledge management. AIRCA members have offices in 33 countries, employ over 2,000 staff from 82 nationalities and have engagements in over 100 countries.

The seven AIRCA members and their respective missions are;

- **World Vegetable Center (WorldVeg).** Research and development to realize the potential of vegetables for healthier lives and more resilient livelihoods.
- **Centre for Agriculture and Bioscience International (CABI).** Improving people's lives worldwide by providing information and applying scientific expertise to solve problems in agriculture and the environment
- **Crops For the Future (CFF).** Generating, synthesizing, and promoting knowledge on underutilized crops for the benefit of the poor and agricultural sustainability
- **International Center for Biosaline Agriculture (ICBA).** Working in partnership to deliver agricultural and water scarcity solutions in marginal environments
- **International Centre for Integrated Mountain Development (ICIMOD).** Enabling sustainable and resilient mountain development for improved and equitable livelihoods through knowledge and regional cooperation
- **International Centre of Insect Physiology and Ecology (ICIPE).** Helping to alleviate poverty, ensure food security and improve the overall health status of peoples of the tropics, by developing and extending management tools and strategies for harmful and useful arthropods, while preserving the natural resource base through research and capacity building
- **International Fertilizer Development Center (IFDC).** Bringing together innovative research, market expertise, and strategic public

and private sector partners to identify and scale sustainable solutions for soil and plant nutrition that benefit farmers, entrepreneurs, and the environment.

AIRCA members have developed a number of initiatives. The most ambitious of these is the Global Action Plan for Agricultural Diversification (GAPAD), launched in December 2015 alongside the UN Climate Change meeting in Paris.[43]

Revolution Number 8 Revisited

After the end of the Second World War, enormous changes occurred across the globe. Some countries fought and won independence from their colonial powers, often at huge human cost (Algeria), sometimes only after their bloody partition (India), and occasionally without great conflict (Malaysia). Nowhere was the dizzying speed at which nations gained their freedom more remarkable than in Africa. Between 1951 (Libya) and 1993 (Eritrea), 56 African countries became independent or were born. No fewer than 35 African countries achieved independence during the 1960s; 7 gained self-rule from Belgium, Britain, and France in 1960 alone. African and other nations were taking control of their own destinies, and their leaders were agreeing to respect the borders that they had inherited.

These arbitrary creations of colonial powers often used rivers or other geographic features as borders, with little regard for ethnic groupings, cultural identities, and pre-colonial history. In many cases, including the Anglophone and Francophone countries of West Africa, colonial borders remain an obstacle to trade and cooperation on common interests such as agricultural development and biodiversity conservation. The political and social consequences of colonialism for individual nations and entire continents have been extensively reviewed and discussed in both academic and popular literature. What has not been discussed as widely are the impacts of colonialism on agricultural and natural ecosystems, sometimes over several centuries, and its legacy since the independence of former colonies over several decades.

Agricultural research has been conducted around the world for centuries, but has usually been for the benefit of particular groups in specific locations. Such research, often informal and anecdotal, was aimed at improving the performance of crops and animals on a single farm or even in a single field — often only for the farmer and his or her immediate community. Even formal research, for example that practiced on tropical plantations during the colonial period, was aimed at one community — a colonial power thousands of kilometres away. Research has been done in many places, but results were rarely shared between participants, thus general principles could not be elucidated or recommendations made about the most suitable agricultural practices across locations. To reduce this fragmentation in agricultural knowledge, diffusion of purpose and duplication of effort, the United Nations established FAO as the single agency with a global responsibility for food and agriculture.

Universities also began to compare themselves against global benchmarks and agricultural research outputs came to be evaluated both among institutions and with other disciplines. Whilst not streamlined to the extent that CGIAR centres had been for particular species, agricultural research in universities began to focus on core activities in which the university had particular expertise, facilities, laboratories, glasshouses, fields or farms through which it could secure additional funding. Academic research gravitated from exploring the whole agroecosystem of a field, farm or region involving many disciplines, to studying discrete bits of it in one specialism.

Winners

Revolution Number 8 has produced significant outputs that favour technological and business solutions to food production in some parts of the world and with some species. In this, it has been spectacularly successful. The commercial production of major staples grown as industrial monocultures now dominates much of the world's agriculture and has achieved productivity levels that have outpaced population growth. Winners of Revolution Number 8 are those who have directly benefitted from the new technologies. These include farmers, traders, commercial companies and researchers, societies for whom food security is now a cost-effective reality, and a global agrifood industry that has made almost limitless products from so few ingredients.

Losers

By focusing on a handful of species grown as monocultures, agricultural research has ignored or discarded other agricultural systems and species, traditional and indigenous knowledge, and alternative models of agriculture that go beyond mainstream production systems. The losers in Revolution Number 8 have been those 'forgotten' communities that have managed diverse and often complex agricultural systems for millennia, the forgotten species that they have cultivated, the biodiversity that has been destroyed collaterally in highly industrialized monocultures, the waterways and near-shore coastal areas affected by agricultural pollutants, the global climate change from greenhouse gases produced in highly fossil-fuel-dependent industrialized agricultural systems, and the loss of knowledge and cultures associated with crop species that are no longer cultivated and systems that have been replaced. But perhaps the biggest loser has been agricultural research itself, not in what it has found but in what has not been explored, valued or supported. This oversight is not simply a matter of lost opportunities for novel research but missed solutions for a world in crisis.

Is the Agricultural Research System Fit for the Future?

Over recent decades, there has been an unprecedented explosion in education, research, knowledge, and data about agriculture. Almost all of this effort has been associated with the winners of the agricultural research revolution. As the challenges facing society become ever more complex and entangled, educational and research systems become simpler and more streamlined. We now grow fewer crops and animals for more people than ever before in human history. Our monocultural model of agriculture and our modular model of education represent a duopoly of practise and method that rejects or at least disparages innovation beyond narrowly defined boundaries, avoids risks, deters new ideas, and diminishes diverse thought and action. Agricultural research in universities has never been as focused on familiar species, simplicity over complexity, technological solutions to societal and environmental challenges, and high-impact peer-reviewed publications. This narrow vision of scholarship leaves academics with few career options other than to undertake research on favoured and funded topics rather than on alternatives that challenge convention. It is not hard to see why.

On 28 February 1953, customers having their lunch in the Eagle pub in Cambridge were rudely interrupted by a man called Francis Crick who walked in and declared: "We have found the secret of life."[44] The *secret* for which Crick (a biologist) and his collaborator James Watson (a physicist) were later awarded the Nobel prize was the discovery of the double helix structure of DNA (deoxyribose nucleic acid). This is the fundamental molecule that carries genetic instructions for the development, functioning, growth, and reproduction of all known organisms. Its *publicly funded* discovery is a milestone in the history of science that gave rise to modern molecular biology and the development of techniques such as recombinant DNA research, genetic engineering, rapid gene sequencing and monoclonal antibodies, methods on which today's *private sector* multi-billion-dollar biotechnology industry is founded.

In 1953, Watson and Crick's discovery of DNA led to a one-page landmark paper in the journal *'Nature'* with the modest title "A Structure for Deoxyribose Nucleic Acid." In fact, Watson and Crick are credited for the discovery of DNA in the 1950s, but it was first identified in the late 1860s by Swiss chemist Friedrich Miescher. Following Miescher's discovery, other researchers revealed more details about the DNA molecule, including how its main chemical components link together. Without the scientific foundation provided by these pioneers, Watson and Crick may never have reached their ground-breaking conclusion that the DNA molecule exists in the form of a three-dimensional double helix. Science is less about sudden *discoveries* and more about evolving *ideas*. For these, we need an academic environment that allows for speculation and testing of hypotheses that go beyond conventional models, quick returns on investment and wealth creation.

In 2015, Ronald Vale, at the University of California at San Francisco, argued that Watson and Crick's seminal paper would have been rejected by *Nature* if they submitted their work today.[45] He provides a spoof letter from the editor to the prospective authors. "*A double helix? Bit speculative. I regret to say that we cannot offer publication at this time. While your model is very appealing, referee three finds that it is somewhat speculative and premature for publication.*" Vale goes on to say that in the past 30 years there has been an estimated four-fold increase in the amount of data required by major journals, largely because of the competition to publish in them. He argues that prestigious journals increasingly insist on authors having a "mature" story. Reviewers "fall in line" with such "escalating expectations" and often demand extra experiments, making it "harder to publish just a key initial finding or a bold hypothesis." Vale says this means that "crucial results are being sequestered from the scientific community." This both retards the rate at which new ideas can be "tested and advanced further" and delays early career researchers from gaining independence because they need high-profile publications for grants and tenure.

In 2020, James Lovelock, who developed the Gaia hypothesis, said;

> "*I have felt for some time that the universities are getting dangerously like the early church. They have dozens of different sects and they are quite proud if you belong to one of them: if you are a chemist you often don't know anything about biology and so on. This is why ordinary university science is not really helpful because the department looking at seaweed would not be the same as the one looking at methyl iodide. It is a division into bits. It's time universities were revolutionised and had much more common thinking.*"[46]

The power of modern data systems and communications technologies makes the production of rankings for almost anything ever easier. Once a ranking is produced, rarely is the method of its calculation challenged. In the case of universities, as well as league tables for their teaching and research, a quick internet search can display university rankings for *student satisfaction, graduate employability, best student cities, hottest universities* (presumably by temperature), *best student life, best party schools, best college dorms, most Astronaut alumni* and various *richest-graduates* lists. In a world of rankings, academia inevitably looks for research activities and publications that will improve the status of a university, department, or individual researcher and avoid the more complex challenges facing humanity and the planet.

The common and most sought-after metric for success is the *research impact factor* or citation of peer-reviewed publications. The underlying assumption for this metric is that peer-reviewed journals publish only trustworthy science and that trustworthy science is published only in peer-reviewed journals. The natural extension of this argument is that the higher the impact factor of a journal, the greater the quality and validity of the research papers within it. For agricultural sciences, this presents a dilemma because the impact factor of a journal rarely equates to the impact of the research itself. Agricultural sciences are caught between the challenge of producing work that will generate high *research* impact, which is the domain of a university, and doing research that leads to high *societal* impact, which is the domain of development agencies and NGOs. There is potential for breakthroughs at both ends of the spectrum (new technological innovations at one end and social transformation at the other), but there are few quick breakthroughs in understanding an entire agricultural system.

Of course, some will argue that not all research in universities and research institutes is on mainstream systems, which is true, and that alternatives have been investigated but each found to be unworkable, which is false. In the era of league tables and transparency, an appropriate question to ask is, how much money has been spent on research, advocacy, incentives and subsidies for mainstream agricultural species and systems and how much has been invested in their alternatives? The answer is not difficult to guess.

Final Thoughts

The challenges facing humanity are existential, at least for our species, and we will need bold and urgent responses if we are to survive, let alone flourish. The climate crisis is a present-day reality as are pandemics, along with the concomitant rupturing and polarization of political, social, and

economic systems and disruption of supply chains and trade. These dislo-
cations of the existing world order will only become more intense and
more frequent as the climate crisis unfolds, countries retreat into isolationism and societies become less equal. We are very short of the Paris target
to reduce CO_2 emissions enough to keep mean global temperature
increases to within 1.5°C. Current predictions are closer to 3.5°C of global
heating by the end of this century, which would effectively destroy agricultural systems in much of the world and destabilize all others. Not one
country is on target to meet all of the 17 SDGs of the 2030 Agenda for
Sustainable Development. Arguably, agriculture relates to all of them but
certainly SDG 1 (zero poverty) and SDG 2 (zero hunger), cannot be
achieved without transforming agriculture and the agricultural research
that underpins it.

Agriculture sits at the nexus of the climate crisis and sustainable development. Its success can be our main ally in meeting these challenges and
its decline a major cause of our failure to do so. Better business-as-usual
is not the solution; we need more radical options. Agriculture and the
research that supports it will have to be different — not simply better. An
example of better business-as-usual is HarvestPlus, one of the three flagship programmes of the CGIAR. The programme is ambitious and is
already having an impact on addressing micronutrient deficiencies in
many parts of the developing world. HarvestPlus is a great example of
putting CGIAR scientific and technological expertise toward breeding
varieties of different mainstream crops with higher levels of critical
micronutrients. The underlying principle of HarvestPlus is that, because
poor people depend on nutrient-poor staple crops, why not increase the
micronutrient content of these and other familiar and popular crops? In
this way, agricultural production, transport, trade, and consumption can
continue undisturbed with only technological interventions for biofortification of the mainstream crops that people already eat.

The logic of this argument is obvious, alluring, and elegant, but it is
flawed and an example of the linear thinking of incremental improvements through conventional research. There are many hundreds of currently underutilized crops that are more climate resilient than favoured
staples, richer in micronutrients and for which there is a wealth of indigenous knowledge on cultivation and consumption. With targeted research

efforts, many of these crops could produce nutritious and marketable products to complement expensive breeding efforts on major staples. They could also provide a livelihood and income for the very marginalized people on behalf of whom international research institutions were established. But for this we need to diversify not consolidate our research.

Over only a few decades, the unprecedented aggregation of agricultural research into global and national centres and universities and its disaggregation into discrete disciplines represent a revolution in agricultural research. Revolution Number 8 has set the background and framework for the future of agriculture — or has it? The Agricultural Research Revolution evolved over three quarters of a century and its outputs can be seen from outer space as monocultures of major crops and plantations, which now cover much of the Earth's land mass. These serried ranks of uniformity exemplify the reductionist and one-size-fits-all approach of modern agricultural science and the dangers of its present path. The next quarter of a century will determine the future of humanity and the 8.7 million other species with which we share the planet. We must now decide whether Revolution Number 8, and better versions of the same kind of agriculture, and the research that supports it, will be the solution for our common future, or whether we need Revolution Number 9.

6

UNDERUTILIZED CROPS

A Personal Journey

Context

This chapter is about how and why I came to work on underutilized crops and the approach that I and colleagues have taken to their study. It relies heavily on my work with one such crop, bambara groundnut. The final section compares bambara groundnut (*Vigna subterranea*) to groundnut or peanut (*Arachis hypogaea*), and asks why the former remains a subsistence crop grown mainly by women in Africa while the latter is now a major world commodity. It considers what we can learn from the global expansion of groundnut and the mistakes to avoid if we wish to make a real impact through wider adoption of underutilized crops such as bambara groundnut.

Since the start of agriculture (Revolution Number 1) there has been a continuing decline in agricultural biodiversity — not only the number of crop (or animal) species grown, but also the range of systems in which they are cultivated.[1] The Empires of Power (Revolution Numbers 2–5), the Green Revolution (Revolution Number 6), and the Agrifood and Research Revolutions (Revolution Numbers 7 and 8) that underpin the Green Revolution have accelerated this decline by focusing almost exclusively on a single model of agriculture based on the monoculture of a few species. Does it matter? Is this successful model that we now have fit for the future?

There are various ways to look at biological systems, including *empirical* approaches and *mechanistic* approaches. For agriculture, empirical approaches are guided by practical observations, sometimes gathered over many years, of what works best for a farmer at a specific location. Empiricism assumes that knowledge is based on experience. For example, the right amount of fertilizer for a particular crop is derived from the farmer's own knowledge of the soil, weather, and other characteristics of the field on which the crop is to be grown. Experiments at various fertilizer rates can confirm which fertilizer level produces the highest grain yield and can therefore be recommended for that location. The results can be verified by physical evidence of crop yield and its reliability tested through various statistical techniques. Any explanation of *why* a particular fertilizer rate is the best is less important than the physical evidence that confirms it.

A mechanistic approach to the same question involves an attempt to understand the *process* by which fertilizer is used by the crop. It requires an explanation of the various steps required for the uptake of fertilizer and how these may lead to the highest yield. In the absence of physical evidence, a mechanistic approach must build a conceptual model (a sort of *mind map*) of the system, and where necessary, fill knowledge gaps by assumptions or evidence from other crops or locations. An empirical approach will work best at any location based on experience alone, but is less valuable when conditions change beyond those that are familiar to the farmer, or if we want to predict what would happen at other locations, now and in the future. For example, local knowledge is of limited value for predicting the likely impacts of climate change on a crop or for the introduction of a new crop about which the farmer has little or no knowledge. In such circumstances, a mechanistic framework may be more useful because we neither have the time nor resources to test each crop at every potential location and under every climate scenario. In the particular cases of agricultural biodiversity (different crops) and climate change (different conditions), a mechanistic approach provides an initial framework on which to base our understanding, whereas empiricism is of limited value.

John Monteith — A Mechanistic Approach to Biological Systems

My exposure to the mechanistic approach to agricultural research began in 1979 when I embarked on my PhD at the University of Nottingham. I was a student in the Tropical Crop Microclimatology Unit of the Environmental Physics Section under the supervision of Professor John Monteith. In 1988, I was appointed as a new lecturer in the Department of Agriculture and Horticulture, also at the University of Nottingham. The experience of working with John Monteith set the foundation for the rest of my career.

Instead of empirical correlations of patterns of crop behaviour, Monteith built simple *mechanistic* models that related crop performance to independent environmental factors. Each could then be linked through physiological processes (physiology is the branch of biology that deals with the functions of living organisms). He then used field measurements of crops and their environments to elucidate general principles and physiological mechanisms to test and improve his model.

In this way, Monteith established *causal* relations rather than *correlations* between crops and their environments.[2] For example, field measurements linked with models of how light moves through crop canopies confirmed that the rate at which crops produced dry matter was proportional to the rate at which they intercepted sunlight (the more light that was captured, the more growth was produced). This relationship has formed the basis of our current understanding of crop performance when other factors, such as drought, are not limiting. The relationship also allowed practical applications such as the use of remote sensing of vegetation from satellites or aircraft to estimate crop performance based on interception of solar radiation over the growing season.

John Lennox Monteith was born in 1929 in Fairlie, a village on the Firth of Clyde in Ayrshire, Scotland.[3] He died 82 years later in Edinburgh. Extending his interests and impact well beyond his Scottish roots, Monteith is best remembered for the huge contribution that he made to environmental science and tropical agriculture, especially in Sub-Saharan Africa and semi-arid India. As a physicist trained in meteorology, Monteith transformed how biologists examine the relationships between living organisms and their physical environments. To do this, he pioneered a new discipline, Environmental Physics, and established novel techniques to measure and analyse how plants and animals interact with environmental factors such as heat, soil water, humidity, and carbon dioxide. Monteith's guiding principle was that physics provides a framework that allows scientists to cut through the clutter and complexity of biology — that to make progress we had to focus our understanding of the system on its bare functional elements. He was fond of the quote attributed to Lord Rutherford (who established the nuclear structure of the atom, discovered alpha and beta rays, and proposed the laws of radioactive decay): *"all science is physics or stamp collecting"*.

It is no exaggeration to say that Monteith established a new understanding of how the environment controls the growth and yield of crops. Traditionally, crop scientists and agronomists had sought to find correlations between crop growth at any location and the weather — a constant topic of discussion among farmers worldwide. The weather changes within and between seasons and across locations, so agronomists often establish *empirical* relations based on observed patterns of crop behaviour under different weather conditions. Empirical relations can be useful for any crop at one location, but they often fail when extrapolated across locations. To improve their usefulness, crop scientists use multilocational field experiments and statistical techniques such as regression analysis to correlate their observations at different sites. Monteith was dismissive of empiricism and scathing about statistics. In his annual presidential address to the Royal Meteorological Society in 1981,[4] he said;

"The statistical blunderbuss is a very clumsy weapon for attacking the problem of crop-weather relations; but it is also very uninstructive because it ignores the interaction of physical and physiological mechanisms. Field experiments to capture all the multi-year and multi-location variability are nearly impossible..."

The single achievement that Monteith is most closely associated with is the Penman-Monteith equation. Monteith was a student of Howard Penman, who in 1937 was appointed to the Department of Physics at the Rothamsted Experimental Station, in the UK. There he studied the transport of gases through porous materials and from that evaporation from bare soils. Penman's research into the evaporation process led to his seminal 1948 paper[5], *"Natural evaporation from open water, bare soil and grass"* where he presents an equation (later known as the *Penman Equation*) in which he used meteorological variables of air temperature and humidity, solar radiation, and wind run to estimate *potential evaporation* from an open water surface. He then demonstrated that the *potential transpiration* from a vegetated surface was very similar to potential evaporation as long as the vegetation was well supplied with water. In other words a crop behaved almost the same as if it was a body of water — say a lake. *Penman* estimates of potential evaporation based on meteorological data were very close to actual evaporation from wet surfaces, but they overestimated actual evaporation when the evaporating surface was not completely wet. This is the case for all vegetation, some of the time and some vegetation, almost all of the time, for example in dry environments. If the soil was drying or leaves were not saturated with water vapour (as is the case for the arid and semi-arid tropics) the Penman equation failed.

Monteith recognized that vegetated surfaces can restrict their rate of transpiration below that of a completely wet surface.[6] To do this, they close *stomata* on their leaf surfaces when access to water by their roots is insufficient to meet the demand for water imposed by the atmosphere above their leaves. To allow for the influence of the crop surface (i.e. its *canopy*) on transpiration, Monteith introduced the concept of a canopy *resistance*.[7] He treated the whole crop canopy as one *big leaf* and vegetation as part of an electrical circuit in which the crop canopy acted as a variable resistor — minimal resistance when the crop was well supplied with water and maximum resistance when the crop was stressed from lack of water and suffering from drought. By elegantly introducing the concept of a canopy or physiological *resistance* to theoretical or *potential* evaporation, Monteith modified the Penman equation based on physical principles, to include the influence of biological systems on evaporation. First published in 1965, the Penman-Monteith equation remains a global standard to estimate evaporation from vegetated surfaces and an elegant example of how physics can be applied to agricultural challenges.[8]

The UN Food and Agriculture Organization (FAO) recommends the Penman-Monteith equation to estimate crop water requirements for different vegetation types and to predict how changes in land use, for example deforestation, affect soil hydrology. The equation demonstrates how all vegetation is *coupled* to the atmosphere and has been incorporated into general circulation models to predict the effects of climate change on the hydrological cycle.[9] Its development and use illustrate how Monteith, a physicist, was able to modify physical principles into practical applications in biology that are now part of agricultural, ecological, hydrological, and meteorological sciences. Biologists have sought comfort in describing the complexity of the natural system, but Monteith believed that "*to make progress we must simplify*". Interestingly *physics* is derived from the word *physis*, which means `nature' in Greek. Monteith championed a new branch of science, *Environmental Physics*, that now forms the basis of our understanding of how biological organisms interact with their environment.[10]

By using a mechanistic rather than descriptive or empirical approach to problem solving, environmental physics can be applied across biological systems. Most of Monteith's research was on crops, but he also demonstrated how environmental physics could be applied to human health. In the 1970s it was extremely difficult to keep premature babies at a constant temperature in normal hospital incubators, with consequences on their likely survival. Monteith and his colleagues observed that evaporation through the skin of premature babies was much greater than that of full-term babies. Since evaporation causes cooling of a surface, the group recognized that this extra evaporation meant greater heat loss in premature babies. The group took the principles that Monteith had established for evaporation from crop surfaces and applied them to the surfaces of premature babies. Their analyses showed that commercial *evaporimeters* severely underestimated skin evaporation and therefore heat loss in premature babies.[11] Following from this study, there have been improvements to incubator design and much closer attention is now paid to the heat balance of babies in incubators and their internal environments.

In 1967 at age 38, Monteith became the first Professor of Environmental Physics at the University of Nottingham. Over the next 20 years his multidisciplinary teams studied the growth of crops and the heat balance of animals. His work made Nottingham, especially its Sutton Bonington campus, into a global powerhouse for teaching and research in Environmental Physics. Monteith used his intellect, wisdom, humility, and humour to encourage, inspire, and motivate those around him, many of whom went on to become research leaders in their own right.[12] Whenever the many eminent scientists and dignitaries came to visit him at Sutton Bonington, he would introduce even the most junior of us as his `colleagues', not his students. His comments were always constructive and invariably witty even when critical. I remember when reading the draft of my thesis his terse comment *"this chapter wasn't up to your usual high standards — maybe you should invest less time on the cricket pitch"* — his house overlooked the college cricket ground.

For me, two examples illustrate the brilliance of Monteith's inquisitive mind and ability for lateral thought. In his funeral oration, his son Graham described one of the many family holidays in the Scottish countryside in which Monteith, his wonderful and gracious wife Elsa, and five children crowded into the family estate car. On a particularly early start, the children were alarmed by a slamming of brakes and their father jumping from the driver's seat to take pictures of the frosty road surface. A scientific paper later appeared in the journal *Weather* entitled "Drying patterns on roads in the Scottish Highlands" by JL Monteith.[13] I also recall an occasion in 1984 when Monteith came to visit me while I was working as a crop physiologist in India. We had a habit of going for a walk at the end of each day around the campus lake. Over a series of evenings, Monteith stopped at the same place to photograph the sun setting over the peaks of a distant mountain range. Sure enough, a paper duly appeared entitled "Crepuscular rays formed by the Western Ghats", which included the statement "visiting Central India in the latter part of 1984, I was surprised to see that spectacular displays of crepuscular rays appeared at exactly the same time on many successive nights and that they were taken for granted by the local inhabitants."[14] The same could be said of how Monteith, a physicist and an outsider to the conventions of traditional agricultural sciences, could see a pathway that was missed by his contemporaries and counterparts. Sir Isaac Newton said;

> *"If I have seen further it is by standing on ye sholders of Giants . . . this is not at all because of the acuteness of our sight or the stature of our body, but because we are carried aloft and elevated by the magnitude of the giants".*[15]

Those who benefitted from their interactions with John Monteith, as a man and as a scientist, would say the same.

ODA Tropical Microclimatology Unit

In 1976, Monteith secured funding from the UK Overseas Development Administration (ODA) to establish a multidisciplinary team of scientists, technicians, and postgraduate students based at the University of Nottingham's Sutton Bonington campus. Their purpose was to study the relationships between tropical crops and their environments, particularly in hot, dry regions of the tropics such as Central India and Sub-Saharan Africa. The approach of Monteith's ODA Tropical Crop Microclimatology Unit remains a case study in novel, ground-breaking crop research and scientific collaboration. First, and most unusually, the team included biologists, physicists, environmental scientists, and engineers working at scales between the physical properties of soils to the turbulent characteristics of the atmosphere. Crops were treated as the conduit between the soil and the atmosphere in which the potential of their genetic characteristics was limited by the conditions of their environments. Second, this diverse multidisciplinary and multinational team had to work, not only across their own specialisms, but also in the variable conditions of tropical fields and in experimental glasshouses at Nottingham. Third, the impacts of their research had to extend to other locations not just under current but also future climates. Long before the terms were commonplace, the ODA Unit was unravelling the impacts of climate change on the growth and performance of tropical crops, developing simulation models for such crops under current climates, and predicting their performance in future climates.

The *jewel in the crown* of the ODA Unit's facilities at Sutton Bonington was a suite of five experimental glasshouses. In these, *stands* of tropical crops could be grown from sowing to harvest in naturally structured soil whilst air temperature, humidity, and soil moisture could be independently controlled throughout the lifetime of a crop. Over the ensuing decade, these glasshouses were used to investigate how crops respond to specific environmental factors in a controlled and closely monitored environment in which other factors were set to levels that represented their range in the tropics. In this way, the effects of say, drought, could be *uncoupled* from other environmental factors such as air temperature or humidity — something that couldn't be done in the field. This allowed the Nottingham group to establish an experimental understanding of how specific environmental factors control crop performance without the confounding effects of other factors influencing their measurements. Between 1976 and 1987, the Sutton Bonington glasshouses were used to grow crops in stands large enough to represent a complete crop *canopy*.

The progress of the crop in each glasshouse could be measured by sequentially sampling a small number of plants throughout its lifetime and relating growth and development to capture of environmental resources such as water, light, and heat over the same period. By having five identically fitted glasshouses, a *response plane*, say from wet to dry soil, could be tested whilst other environmental factors were kept constant. However realistic they may be, studies in experimental glasshouses 52° north of the equator cannot mimic the complexity of challenges faced by crops growing in tropical fields. To address this limitation, the Nottingham team conducted field experiments with staff at the International Crops Research Institute for the Semi-Arid Tropics (ICRISAT), in India, and at the World Meteorological Office (WMO) Agrhymet Centre in Niger. This allowed *real world* experiments at field scale with larger areas for crop sampling and environmental and physiological measurements. Experiments in fields exposed to the vagaries of seasonal weather and impacts of insect pests and diseases also provided a robust test of the relationships being developed at Nottingham between crops and environmental factors. Over 11 years, the Nottingham team published around 100 papers and helped train many postgraduate students (including me) and institute researchers at Nottingham and ICRISAT.

The mid-1970s were an exciting time in environmental physics — many of the now familiar concepts relating crop growth to microclimate were just being developed. By building on these and introducing new approaches and experimental techniques, Monteith and his colleagues provided a common framework in which agronomists, crop physiologists, soil scientists, microclimatologists, and crop modellers could describe crop responses to the environment.[16] Their analyses have enabled breeders to identify desirable crop traits, and crop managers to target agronomic practices based on an understanding of general principles of crop/climate relations rather than the formulaic application of inputs. To generate a sufficient body of evidence that could be applied across environments, the Nottingham group focused on how specific environmental factors, solar radiation, temperature, and atmospheric humidity affected the growth and yield of two tropical crops — pearl millet (a cereal) and groundnut (a legume that is also known as peanut). Both pearl millet and groundnut are mandate crops of ICRISAT, based near Hyderabad, India, with a Sahelian centre in Niger.

Pearl millet (*Pennisetum glaucum*) is a drought-tolerant cereal that has been grown in Africa and the Indian subcontinent for over 7,000 years.[17] It is the most widely grown type of millet (others include finger millet, proso millet, and foxtail millet). The likely centre of diversity of pearl millet is the Sahel zone of West Africa, where it is used principally as a food source amongst poorer households. In the US it is planted mainly as a forage crop for livestock. Groundnut (*Arachis hypogaea*), is a relatively drought-tolerant legume grown mainly for its edible seeds. Its centre of

diversity is Central and South America, but it is widely grown in the tropics and subtropics by both small-scale and large commercial producers.
Asia and Africa account for 95 per cent of the global groundnut area
where it is cultivated under rainfed conditions with low inputs by
resource-poor farmers.[18] Like other legumes, groundnut can *fix* atmospheric nitrogen through nodules on its roots. This ability reduces its
demand for fertilizers, and the residual nitrogen that it leaves in the soil is
made available to the roots of companion crops in mixed cultivation or
succeeding crops in rotational systems.

Keith Scott

In 1988, I was appointed as a new lecturer in the Agronomy section of the
Department of Agriculture and Horticulture at Sutton Bonington. I was
fortunate that soon after my appointment, the University of Nottingham
chose Professor Keith Scott to become the Chair in Agriculture. Keith
Scott, an agronomist, had previously worked with John Monteith and
applied many of his concepts to temperate crops, especially during his
time as Head of the Brooms Barn Experimental Station, which was
responsible for UK research on sugar beet. Keith was not just an ally and
friend, he was also a champion of the direction that we both wanted to
take experimental agronomy — away from empiricism and trials to
mechanisms underpinned by experimentation and robust field testing.

At the time, (and for many still) agronomy was viewed as a dull subject
of limited intellectual value. The stock-in-trade of agronomists was often
the response of a crop to varying levels of fertilizer, agrochemicals, pesticides, or growth regulators. By fitting a function through data points
across a range of application levels, the optimum for a particular crop,
variety, or location could be discerned with little understanding of the
underlying mechanism. Funding was readily available from commercial
companies for field trials, producing papers from them was straightfor-
ward, university departments and research institutes were happy to carry
out such research, and publications were plentiful if not illuminating.
A common description of this approach was *spray and pray* agronomy in
which researchers could easily demonstrate the impacts of agronomic
inputs with little understanding of underlying causes. Outsiders like
Monteith were seen as a bit of a nuisance with talk of multidisciplinary

approaches, crop models, physical processes, and mechanisms of resource capture without a practical background in agriculture. However, Keith realized that Monteith's approach was critical if agronomy was to emerge from its traditional silo to link with disciplines such as physiology, genetics, and environmental sciences and be able to meet emerging challenges facing agriculture and research.

Keith Scott was a force of nature, heavy of foot, with a Geordie accent and snowy white hair that had once been red. His attitude to life was to play hard, work hard, and challenge convention. He had strong likes and dislikes. He was not popular with some colleagues, but we became close friends and he used to stay with my wife, Sue, and me during the week, returning to the family home in Suffolk each weekend. To accommodate his itinerant lifestyle, he carried his belongings and breakfast supplies in a Volvo estate vehicle, and would frequently be seen delving into the back of his car for immediate requirements. To his friends Keith was *Ginger* and to him everyone was a *pal*. His humour was infectious, aided by an eye for detail, mimicry of accents, and exaggerated mannerisms. He had a repertoire of anecdotes that could keep his audience in hysterics with head in hands. But for all his quirks and clowning, Keith Scott was a serious and often obsessive academic. We spent many hours writing and rewriting documents until the exact phrase was finally wrought from the original text. Keith had grown up in South Shields near Newcastle with his cousins Ridley and Tony Scott who went on to become renowned film makers and producers of such blockbusters as *Blade Runner*, *Top Gun*, Alien, *Gladiator*, and *Thelma and Louise*. Keith was also an ambitious man. I sometimes felt that his many successes were driven by an expectation of self-achievement that anyone would have found onerous. Sadly, like his cousin Tony some years later, in 2000 Keith Scott's untimely death left a void amongst his family, many friends, and the academic world. I owe Keith a lot on a personal level, and his early belief in me was his greatest contribution to my academic career.

Tropical Crops Research Unit (TCRU)

In 1988, the University of Nottingham Department of Agriculture and Horticulture set up a Master of Science (MSc) in Agronomy, and because most of the students were from Africa or South Asia, I was asked to become Programme Director since I had worked in Sub-Saharan Africa and India. This was one of the most exciting times of my career — a mix of students from Africa, Asia, and Europe from different backgrounds and cultures working together. The European students gained from the wealth of experience brought to the programme by African and Asian colleagues, who were often coming back into academia from senior institutional positions in their own countries. Some of the European students graduated to work on development projects and careers in the tropics. Many of our best students came from Southern and East Africa, especially Zimbabwe, which was suddenly receiving scholarships for international studies as the country emerged from years of conflict and sanctions. Most were focused on how the course would be of practical use when they went home. I believe the experience that the students gained from working and socialising with each other was more valuable than the academic content of the course. Many have remained part of my network of friends and colleagues around the world, gone on to complete PhDs with me, and become academic partners in several projects.

The MSc in Agronomy was unusual in that it lasted for 18 rather than the normal 12 months of most UK-based Master's degrees. This extra six months allowed students to complete a substantial research project that often covered the full lifecycle of an annual crop — usually a European crop growing on the University's farm in the English East Midlands. The challenge for me was to find suitable projects for students from the tropics that would be relevant when they returned home and use the legacy of the Monteith era to build on their practical knowledge.

In 1988, funding by the UK government for the ODA Unit had come to an end and the suite of controlled-environment glasshouses lay empty. This coincided with my appointment in the Agronomy section. With Keith Scott's support, we secured funding from the university to refurbish and upgrade the facility, which we rebranded as the University of Nottingham Tropical Crops Research Unit (TCRU). In each glasshouse, air temperature and humidity control systems were upgraded and new data loggers were installed to control and monitor the internal environments. In the years since the establishment of the ODA Unit in the mid-1970s, the world had increasingly become aware of the term *global warming* and the wider implications of what has since become known as *climate change*.

There was increasing concern about how heat stress, and its close relative drought, would affect agriculture in already hot and dry regions of the tropics, but there was little evidence for their effects alone or in combination at elevated levels of atmospheric carbon dioxide (CO_2) under current or future conditions. To address this emerging challenge, we installed a CO_2 injection system into each of the five glasshouses. This allowed detailed and season-long studies of the effect of CO_2 elevation for tropical crops growing under various levels of soil moisture, air temperature, and humidity.

As with the ODA Unit, we assembled a multidisciplinary team of physicists, biologists, and engineers for the new TCRU. Again, we studied only two tropical crops — this time sorghum (*Sorghum bicolor*; another drought-tolerant tropical cereal) and groundnut. The experiments within the glasshouses looked at the effects of elevated CO_2 and drought on sorghum and groundnut at 375 ppm and 700 ppm of CO_2. The results from controlled environments in the UK again provided invaluable supporting evidence of the impacts of CO_2, drought, and heat stress individually and in combination on growth and yield of tropical crops. Research by the TCRU has been invaluable in building an understanding of how higher CO_2 levels with and without the effects of drought and heat stress impact two important tropical crops. The team also expanded certain computer modelling principles established earlier at Nottingham and elsewhere, and used their combined knowledge to develop and test new crop simulation models for the likely impacts of climate change on these two important tropical crops. For 12 years, Nottingham had worked on major tropical crops. It was time to be more ambitious.

Starting on Underutilized Crops

By establishing the TCRU, I hoped that students, especially those from Africa and Asia, would benefit from the principles, techniques, and methods established through research at Nottingham and with its partners in the tropics. At the same time, I wanted their research to contribute to a new body of novel research in scientific agronomy that would be relevant to their careers and countries in the future. The question was, on what topic? Rather than adding to the research on crops that we had worked on for so long, I felt that the TCRU facilities provided a perfect opportunity to test potentially useful but currently *underutilized crops* that had not benefitted from the intensive research that had already been dedicated to major crops through the network of international agricultural research centres established in the 1960s and 70s.

Initially, my suggestion to colleagues that I wanted the TCRU system to test hitherto underutilized crops was met with bemusement. Why use such a powerful system for crops that have already been shown to have such little potential? — *if they were any good we would have discovered them by now.* The advice, sometimes from exasperated but well-meaning colleagues, was that working on underutilized crops would be a waste of time and effort and a setback for my fledgling career — *underutilized crops are underutilized for a reason — they are of no value to the agrifood industry.*

I was told (correctly) that it would be difficult to publish papers on such crops because journal editors would struggle to find suitable referees with any knowledge of specific underutilized crops. This advice was usually well intentioned, but it was guided by the pragmatic realities of being a successful researcher — publish or perish, establish a subject specialism, plough a lonely furrow, focus on topics for which research funding is readily available. I am not quite sure why I rejected such sage advice from experienced senior academics. Perhaps it was my naïve enthusiasm combined with Keith Scott's belief in me that overruled the objections. I found it hard to accept that in Africa and India there weren't far more than just a handful of introduced staple crops such as maize that seemed worth investigating. Maybe it was also my belief that agronomic research was meant to be about new challenges rather than confirming what we already knew or what other researchers were already working on. Were underutilized crops really as *useless* as we had been told, or had we simply ignored them because they were mainly grown by local communities, often on a small scale by women and were therefore not attractive to funding bodies, fellow researchers, governments, and development agencies? The argument then, and still now, is that a dollar spent on researching an important and widely grown staple crop would have greater, faster, and more tangible impacts on more people than a dollar spent on an underutilized locally grown crop that had never been heard of by most people.

Of course, especially when funds are so scarce, limiting research to mainstream crops makes good sense. The problem is that without research and funding, *underutilized* crops remain *under-researched* crops. Without dedicated effort they would eventually disappear from agriculture through lack of interest — not from those who continued to grow them but from those who chose not to work on them. Wasn't the argument that *they are underutilized for a reason* simply a circular justification to do nothing about them? If we haven't evaluated underutilized crops, how do we know that they are of no value now or in the future? This argument carried the day, and so it was decided that the TCRU would become an internationally recognized centre for research on underutilized crops. To identify exactly which underutilized crops would be of most interest, I wrote to around 15 colleagues and contacts in Africa and Asia with the question, "if you had access to the revitalised TCRU facilities, which under-researched crop would you study?"

Bambara Groundnut

I first came across bambara groundnut in 1980 as a PhD student at the University of Nottingham. I had been sent with Dr Peter Gregory of the ODA unit to work on pearl millet at the WMO Agrhymet Centre in Niamey, Niger on the edge of the West African Sahel. Peter, tall, thin and bespectacled, had arrived in Niamey a few weeks earlier to set up the experiment. His previous experience in India and Morocco had given him the skills and fortitude needed to conduct field experiments on the edge of the Sahara Desert. He had already used his working knowledge of French to organize the local field staff, engage with the expatriate scientists, and secure the equipment that we would need to do the experiment. Ever the Englishman but also an internationalist, Peter had an easy way of engaging across cultures, and as a soil physicist, working across disciplines. Peter was to remain a close friend, ally, and champion throughout my scientific and management career; I owe him a great deal over a long time.

In Niger he taught me about the reality of working on field experiments in the arid tropics where the daily temperatures reach above 40°C and relative humidity can fall to 10 per cent. Peter quickly taught me the importance of strong boots (scorpions), a wide-brimmed hat (sunstroke), and plenty of water throughout the day (sudden dehydration). He also taught me the importance of meticulous preparation, daily record keeping in a field notebook, and the use of field instrumentation. Unlike those of us who concentrated on what we could see and measure above-ground, as a soil physicist Peter was interested in the hidden half of crops — their root systems. This required a new level of detailed observations of the length and extent of the root system and allied measurements of soil structure and water content. The experience of measuring crops from dawn to dusk without technical support has left indelible memories that I fear have made me rather less sympathetic to lab-based students than I should be. Constantly breathing the dust blown across from the Sahara Desert has left an indelible mark on my lungs as a souvenir from West Africa.

Even 20 years after gaining its independence from France, Niamey, the capital of Niger had an unmistakably colonial feel about it. Not only was there a virtual monopoly of French goods and services, there was an attachment to French culture. I was especially struck by the experience of travelling through fields of local crops such as millet, sorghum, and fonio to find that locals were queuing in the main supermarket to buy French baguettes that had just arrived on the flight from Paris. At weekends, I would travel on my moped through villages near Niamey for a better introduction to local culture. Here, I met communities that I had never heard of — Tuareg, Hausa, Zarma, each distinctly dressed in traditional clothing, whose houses, fences, and matting were often made from the leaves and straw of the millet that we were growing, and from whom I learned about local foods, crops, and livestock. One such crop was bambara groundnut, a small tough leathery-leaved legume that produced its seeds below ground (hence *groundnut*). A modest plant, but amongst its peculiarities was that the crop was grown almost exclusively by women ('*the groundnut of the women*'), usually on a small scale in their villages rather than in the fields where men grew staples such as millet, sorghum, or maize. The women grew bambara groundnut to feed their families when more favoured staple crops failed in dry seasons.

Early Faculty Work at Nottingham

To be honest, I never thought much again about bambara groundnut until I returned to Nottingham in 1988 as a new lecturer. There were various proposals for new research to be done at the TCRU, mostly on crops that were already the mandate of CGIAR centres, but interestingly, a contact at the University of Zimbabwe suggested bambara groundnut — the crop that I had encountered in Niger. He wrote, "locals claim that this crop can grow and yield in soils that are too dry for groundnut — why not find out if that's true?" A few weeks later a packet of bambara groundnut seeds arrived at Nottingham. I then visited the University of Zimbabwe and the Department of Research and Specialist Services in Harare, which was responsible for government agricultural research. I was taken to see field crops and visit a seed company that was distributing bambara groundnut seeds across Zimbabwe. I heard positive comments about *jugo beans,* the local name for the crop. However, I was told that there was scarcely any government-supported research on bambara groundnut because it wasn't a priority for the Ministry of Agriculture or international funding agencies, both of which favoured staple crops such as maize.

Back at Nottingham we found that there was hardly any published research on the crop and most papers described the botanical features of bambara groundnut, with little information about its potential. What we could find about the crop was that, like groundnut, it is a legume, and that also like groundnut, it produces its seeds below ground — hence the similar names. We also learned that although bambara groundnut has its centre of diversity in West Africa, it had been increasingly displaced by groundnut (also known as peanut in the US), the favoured cash crop of the colonial powers that had partitioned much of Africa. Despite this marginalization, bambara groundnut was still widely cultivated by local communities from Senegal to the Ethiopian lowlands, and across Sub-Saharan and Southern Africa and the island of Madagascar. However, unlike groundnut, bambara groundnut has not been adopted by any of the international research centres, although the International Institute for Tropical Agriculture (IITA) in Nigeria holds a collection of genetic material of the crop.

Not surprisingly, without national support, international investment, sustained research, advocacy, or champions, there had been no serious attempt to improve bambara groundnut. It continued to be maintained by communities as *landraces* that were grown, traded, and protected by local (mainly women) farmers in Africa and on a small scale elsewhere in the tropics. Despite the lack of formal breeding and improvement, there were local forms of the crop that were adapted to cooler, moist regions and also to areas that were too arid for other crops. In addition, the crop was reported to yield on poor sandy soils that were marginal for groundnut and other legumes. Of all the factors that we considered to justify research on bambara groundnut, the greatest was probably its morphological and botanical similarity to groundnut with which we had substantial experience. If it really was more drought resistant than groundnut, why not compare the two under experimental conditions to see whether bambara groundnut had any potential under the drought-prone conditions of the semi-arid tropics where groundnut often failed?

Initial Comparisons with Groundnut

In 1988, we began by growing bambara groundnut in the TCRU glasshouses under varying levels of soil moisture. Our initial results were encouraging and confirmed the observations of local farmers in Africa. The crop, or at least the landrace that we had received from Zimbabwe, was more drought tolerant than the variety of groundnut that we had used under similar drought conditions. Buoyed by this early success and fuelled by enthusiasm, especially that of some very dedicated African students in the MSc course who knew the crop, we grew bambara groundnut again the following year in the TCRU glasshouses. To our disappointment and intense frustration, the crop that had grown so successfully the previous year in the same experimental system produced virtually no pods at any level of soil moisture — not even under full irrigation.

Daylength Enigma

In our search for published literature on bambara groundnut, we had contacted the small community of researchers who had experience with the species. Most were in Africa, often in West Africa where the crop was popular as a snack food. We also came across Dr Anita Linnemann, based at Wageningen Agricultural University in the Netherlands, and Dr Frank Begemann in Germany, who were two of the few scientists in Europe with any experience of bambara groundnut. I met Anita Linnemann in Wageningen and was impressed that, in her house she had created a bambara groundnut library with carefully listed papers that at the time probably represented the global scientific literature on the crop. Frank Begemann had edited a series of 30 FAO monographs on underutilized crops of potential, of which bambara groundnut was one.[19] Looking back, Anita and Frank were early pioneers when underutilized crops were not a subject for serious investigation and for which neither has been properly recognized. Anita agreed to visit Sutton Bonington to see why a crop that was well supplied with water and optimal temperature had produced no pods in one year and plenty of pods the previous year under similar conditions. Her observations were a revelation for us. In her own controlled-environment system in the Netherlands, she had observed that pod development in bambara groundnut may be influenced by photoperiod, which is the daylength from sunrise to sunset at any location.[20] Trials on accessions — a group of related plant material from a single species collected at one time from a single location — from different African locations, exposed to several daylength treatments (12 hours or less and 14 hours or more) showed that longer photoperiods delayed or even prevented pod filling even where plants had already flowered.[21]

Further experiments showed that photoperiod had no effect on pollen fertility of bambara groundnut and fertilized embryos (pods) were initially found from both long and short photoperiods. The embryos developed at the same rate until 18 days after flowering. However, in long photoperiods, embryo growth stopped at this stage. The ovaries remained on the plant for another two to three weeks but eventually aborted over time. When plants exposed to long photoperiods were transferred to photoperiods of 12 hours or less, some of the ovaries went on to develop mature seeds under the short photoperiods. This confirmed what Anita Linnemann, working in controlled-environments in the Netherlands, had shown that photoregulation of pod filling is an important trait for bambara groundnut growing in Africa.[22] The photoregulation mechanism provides the plant with flexibility to adapt to fluctuations in the length of the growing period — often determined by the onset and end of the rains. Linnemann also observed that, at least for West African accessions, photosensitivity decreases with distance from the equator.[23]

Could Anita Linnemann's work showing photoregulation of pod filling in bambara groundnut in controlled-environments in the Netherlands explain the divergent performance of the same crop in controlled-environments in the UK? Armed with her evidence we went back to look at our own data from the Nottingham glasshouses. Yes, the conditions and control systems in both years were similar, but one factor was different — the sowing date. In the UK daylengths in the summer can be for as long as 16 hours. Since the crop had flowered in every year regardless of sowing date, we assumed that the crop was insensitive to daylength for pod filling. We had never encountered a crop that could flower and produce fertilized ovaries but then fail to expand into mature seeds without a photoperiod signal. Perhaps if we imposed a fixed daylength on crops in our glasshouses we could uncouple this photoperiod requirement and ensure that pods were filled to produce mature seeds? In 1990, we did the same experiment again across five irrigation levels (one in each glasshouse). However, this time we imposed a 12-hour daylength in each glasshouse by covering the crop at 8 pm each evening and uncovering it again at 8 am the following morning. What we found was remarkable. In each glasshouse, the same treatments that had failed to yield any pods the previous year produced numbers of pods. Even more exciting was that in the well-watered treatment, the yield of our landrace of bambara groundnut was as great as that of a known high-yielding variety of groundnut.

Out of Africa

So, having observed that daylength control of pod filling can be regulated under controlled environments in Europe, would the same mechanism apply in the field environments of Africa? For this, we contacted Dr David Harris who had previously been a colleague and friend in the Nottingham ODA Unit. Dave was now working as an agronomist in Botswana as part of a long-term project on sorghum and maize in arid environments. He agreed to our request to test the hypothesis that daylength is critical for pod filling in bambara groundnut in a field study at the Sebele Experimental Station near Gaborone, Botswana where his own research was based. The rainy season in south-eastern Botswana extends from October to May. There is no distinct start to the season and dry periods are common. Crops are generally planted after heavy rainfall before the soil becomes too dry for seeds to germinate. Periods of heavy rainfall are unpredictable and crops might be planted at any time between October to February, beyond which they are unlikely to mature because temperatures fall rapidly later in the season.

If the evidence from Nottingham and Wageningen was true, sowing date and not just seasonal rainfall should influence the yield of bambara groundnut in Botswana. To test this hypothesis, seeds of a bambara groundnut accession from Southern Africa were sown every week for 20 weeks from 16 October 1990 until 26 February 1991.[24] During this period, daylength went from 12.7 hours to 13.7 hours and back to 12.6 hours. The number of days to first podding varied between 52 and 98 days, with the fastest development at the shortest daylength, confirming the observations from controlled-environments in Europe. Plants sown early produced plenty of leaves, but they did not pod until the longer days of full summer; therefore, they produced the lowest pod yields. Conversely, the plants sown in mid-summer did not pod until early autumn when the days were becoming shorter and the pod yields were higher.[25]

The implications of photosensitivity on pod production for the practical management of bambara groundnut are very important and may go some way to explaining why it has not been more widely adopted. If we assume that pod filling takes around 40 days, then the crop duration for the same plant could vary from 85 days (planted in late December) to 135 days (planted in October). For a rainfed crop growing in the hot, arid climate of Botswana, the extra 50 days for crop maturity is time that it can ill afford and almost certainly means complete crop failure. It also explains the frustration of researchers who sometimes observed viable pod yields in dry years and complete crop failure when rainfall was plentiful. This anomaly can now be explained elegantly by looking at sowing date. All scientists hope for a *Eureka!* moment in their careers. If not exactly a Eureka moment, the results from Botswana, UK, and Netherlands at least represent an *Aha!* moment for our understanding of bambara groundnut. After these initial studies, we found similar effects of the influence of sowing date on the yield of bambara groundnut landraces in Tanzania.[26] These results further confirmed Anita Linnemann's original hypothesis that daylength controls pod filling in bambara groundnut.[27] They also provide a powerful argument for collaborative research that links controlled-environment and field studies based on a scientific and mechanistic approach rather than conventional agronomic trials alone.

Mapping Global Potential

The next question was, "if we now understand the critical role that day-length plays in the productivity of bambara groundnut, could we find or develop genetic material that was insensitive to the daylength requirement for pod filling?" In 2002, whilst we were developing genetic approaches, FAO approached us to map the agroecological potential of the crop using the new modelling and geospatial mapping tools that were becoming widely available in the late 1990s. This involved travelling to FAO head-quarters in Rome and, ever since, I have found FAO and the city that hosts it to be one of the most fascinating and frustrating places I have ever visited. FAO occupies a building in one of the most scenic areas of Rome overlooking the Baths of Caracalla and the Circus Maximus with the splendours of the Eternal City beyond. The challenge is not to be distracted by the views of Rome from the roof (which would be impossible), but to actually find the roof. The building that is now occupied by FAO was originally built in the 1930s under the Fascist government of Italy to be the seat of the Ministry of Italian Africa. After World War Two it was repurposed to become the headquarters of FAO as the new United Nations agricultural agency. Its purpose may have been changed, but its design remains the same. I have since visited FAO many times but can hardly recall an occasion when I have not had to ask how to find a particular office or at the end of a meeting how to reach the exit. I am told that this is a deliberate Mussolini feature; the building was planned to confuse its occupants and deter social gatherings to avoid the risk of plotting and intrigue. FAO is housed in three identical blocks within which labyrinthine corridors lead to small offices for staff working in various programmes in agriculture, forestry, fisheries, and land and water resources. FAO also supports research, technical assistance, education and training programmes, and collects data on agricultural output, production, and development.

Our project was commissioned by WAICENT, the World Agricultural Information Centre, and its head, a flamboyant Venezuelan with the largest handle-bar moustache I had ever seen. He introduced me to his team in a room full of people pouring over maps or entering data on computer screens. Even back in 2002, FAO was using its vast data resources and mapping tools alongside new technologies to develop digital maps for any purpose related to agriculture. However, their datasets did not include those underutilized crops that were not part of the annual statistics provided by Member States. Our task was to map potential areas of the world that could produce bambara groundnut as a case study that could then be applied to other underutilized crops. We applied this methodology both for Africa, where the crop is widely cultivated but where there is little experimental evidence, and for new regions where bambara groundnut is not cultivated but where environmental factors should be conducive to its growth. As well as producing a global map of where it could be grown, we developed a computer-based crop simulation model of bambara groundnut (BAMnut),[28] which when incorporated into a Geographical Information System (GIS) could predict, for the first time, the potential yields of bambara groundnut across the world. That the project achieved more than its objectives is due to my Iranian colleague, Mohammed Bannayan Aval. His intelligence, hard work, and unlimited enthusiasm to think big led us into activities that have since justified a generation of computer simulation models for underutilized crops.

A gridded resolution of 50 km × 50 km for global land areas (excluding Antarctica), between 1961 and 1990, was used to generate crop yield estimates using BAMnut. The methodology did not account for the specific effects of soil type, insect pests, or diseases on the potential productivity of bambara groundnut at any location. Nor did we estimate the influence of daylength on pod filling. Essentially, we asked, "where are the most suitable places for growth of bambara groundnut, what is its likely yield threshold, and over how much surface area are yields achievable?" Results showed that bambara groundnut could produce significant pod yields in much of the world beyond where it is now cultivated, with suitable areas in America, Australia, Europe, and Asia, as well as Africa.[29] In fact, locations within the Mediterranean region, where there is no history of bambara groundnut production, showed the highest potential yields. Interestingly, the global map indicated significant potential for bambara groundnut in Southeast Asia. Some years later, I found that the crop has actually been grown in this region for centuries.

With all the limitations mentioned above (no allowance for insect pests, disease, soils, and even daylength) the study identified potential areas for production based on the agroecological requirements of bambara groundnut. The study was important because it meant that before committing to expensive field trials and research, we at least now have a preliminary baseline of where to start testing any bambara groundnut crop. If we could identify the best (suitable temperature and rainfall) and worst (too hot/cold/dry) locations then we could focus on other limitations (daylength, responses to soil type, insect pests, diseases) that could prevent the crop from reaching its potential. This mechanistic approach based on simulation modelling and mapping (a desk study) can serve as a useful prelude to detailed field investigations on bambara groundnut to identify appropriate management practices and breeding targets. It also provides a basis to assess the potential of many other underutilized crops and for comparisons with those currently being grown at any location.

Since 2002, much has happened in crop modelling and available technologies to improve on the approach that was taken for one underutilized crop almost 20 years ago. However, that early study is important beyond for what it told us about bambara groundnut. It has even greater relevance to current issues of food security because the methodology can be rapidly extended to assess the potential productivity of many subsistence food crops that contribute to the diets of the world's poorest people living on the front line of changing climates.

Identifying the critical role of daylength in pod filling and mapping the global potential of traditional bambara groundnut landraces gave a new impetus for further research at Nottingham and with partners around the world. The next big question was, "what factors limit its wider adoption in areas where it is currently produced?".

Bambara Groundnut — The Groundnut of the Women

In 1999, through support from the UK Department for International Development (DFID), TCRU researchers completed a study of the factors that affect the marketing and use of bambara groundnut in Southern Africa, specifically in Zimbabwe and Swaziland.[30] There were individual, and no doubt unique, factors for each country and each crop, but the study helped us to understand the challenges facing the wider adoption of bambara groundnut in terms of awareness and perceptions, markets, products, gender factors, agronomy, and institutional support. The study, led by Karen Hampson — at the time an MSc student and now doing important work for Farm Radio International in Africa — was particular to bambara groundnut in these two countries but its conclusions have wider implications both for bambara groundnut grown elsewhere and for other underutilized crops around the world. The study recommended that farmers should be involved in agronomic trials and that training was needed for cooking, processing, product development, and recipes. The study also identified the need to establish market links and support mechanisms to promote final products for retailers, facilitate promotion and research, and establish links with regional export markets.

Despite the challenges for women growing bambara groundnut in Africa, our field studies identified that all street vendors, market traders, and farmers selling wholesale bambara groundnut in markets were women. Some had been particularly successful, for example in Botswana, I once visited a village where the only brick building was paid for by the proceeds of the sale of bambara groundnut by a mother who had also provided funds to cover the cost of five children at primary school. Indeed, for some women, bambara groundnut was their only livelihood because they were either single parents, or their husbands were absent or unsupportive of a *woman's crop*. In fact, as a household crop, women are responsible for passing down bambara groundnut seed through the generations, for saving seed yearly, and for putting some dried bambara groundnut away as food security. At present, many women farmers grow bambara groundnut because it is a good supplement to household food and provides some direct income. One implication of increasing the sale of bambara groundnut would be that as a cash crop, men may become more involved. Were this to happen, the control that women have at present over this crop and its profits would risk being diminished or lost.

EU Support for Research on Bambara Groundnut

Much of the early research at Nottingham on bambara groundnut was achieved through the dedication of PhD and Masters' students with some support from agencies such as the ODA and FAO. However, such research was difficult to sustain without external funding and collaborative effort.

Without the long-term commitment of the European Union (EU), our promising but inconclusive research on bambara groundnut would have eventually come to an end, as no doubt it had for many other underutilized crops at the same stage. This history of sporadic effort on underutilized crops explains why research is often anecdotal, repetitive, and not widely disseminated. Examples of such piecemeal effort on many underutilized crops abound and have earned a reputation for research on underutilized crops as *hobby science* or a cottage industry not worthy of serious academic enquiry. In fact, the initial enthusiasm of researchers who work on any crop usually dissipates when there is little prospect of funding, heads of department lose patience, and colleagues say "I told you so".

Support from the EU marked a turning point in research on bambara groundnut. Between 1992 and 2010, the EU funded three major projects exclusively on bambara groundnut, each co-ordinated at Nottingham — an astonishing commitment. These complex, multidisciplinary, and multinational research projects have revitalized interest in the crop. Bambara groundnut now has a critical mass of scientific research behind it and a growing body of literature. Underpinning these efforts has been the commitment and support of the EU for novel, sometimes risky, and unpredictable research on a crop with unknown potential and limited previous research. The three EU projects have linked a dozen partners in Europe, Africa, and Asia. Together, they have delivered coordinated research that spans ecophysiology, agronomy, crop genetics, breeding, entomology, pathology, marketing, processing and utilization, nutritional quality, socioeconomic characteristics, and crop simulation modelling of bambara groundnut.

In 1992, the EU funded the first international project on bambara groundnut, BAMGROW (1992–96).[31] To evaluate the agroecological potential of the crop, BAMGROW linked field experiments in Tanzania, Botswana, and Sierra Leone with experiments and analysis at Nottingham and Wageningen University in the Netherlands. In 2000, the EU-funded BAMFOOD project (2000–04)[32] linked molecular, physiological, and agronomic studies among partners in Africa (Botswana, Swaziland, and Namibia) and Europe (UK and Germany). In 2006, we secured a third EU-funded project, BAMLINK,[33] to evaluate the nutritional, ecophysiological, and molecular characteristics of bambara groundnut. BAMLINK was even more ambitious and linked Nottingham with African partners already familiar with bambara groundnut (Botswana, Ghana, and Tanzania), states in India that had never seen the crop (Gujarat, Rajasthan, and Karnataka), and fellow EU partners (Denmark and Germany). The long-term commitment of the EU to international collaboration allowed bambara groundnut to move '*out of Africa*' to regions of the world where it has the potential to become a crop for the future. Highlights of this sustained international effort on one underutilized crop are described in the following sections.

Operational Method of Hybridization

Despite its potential, no coordinated plant breeding programmes had previously been established for bambara groundnut before the EU-funded activities. A major challenge had been the failure to produce hybrids between two parents as a basis for breeding new varieties of the crop. Without established varieties, for centuries farmers have grown locally adapted *landraces* of bambara groundnut as a mixture of genotypes. This means low and unpredictable yields and variable responses to inputs such as fertilizers and irrigation. Compounded with daylength sensitivity for pod filling, the variability within landraces meant that early researchers were frustrated by the vagaries of the crop and deterred from continued research. However, we realized that if we could artificially cross (hybridize) promising parents, this would allow breeding genotypes with new combinations of desired traits through genetic recombination. It could also allow us to overcome challenges to wider adoption such as daylength control of pod filling and the *hard-to-cook* nature of its seeds that require long cooking times and abundant use of fuel.

The choice of parents for a crop breeding programme depends on breeding objectives, and each requires a reliable technique for hybridization. In 2001 an operational method of hybridization was developed at Nottingham with project partners in Germany and quickly tested and transferred to other partners in Africa and Europe. Around the same time, an expedition led by the French scientist Remy Pasquet collected seeds of the wild ancestor of bambara groundnut from the slopes of Mount

Cameroon. Five seeds of the wild accession were sent to Nottingham for investigation. Of these, only two germinated and when grown in pots displayed a very different spreading habit to the bunched habit of landraces that were cultivated by farmers. Following the new hybridization protocol, a *wide cross* was achieved between a male parent from the wild type from Cameroon and a female parent from a cultivated genotype from Botswana.[34]

First Genetic Linkage Map of Bambara Groundnut

The initial cross between the Cameroonian wild ancestor and cultivated landrace from Botswana provided a basis to generate a core collection of germplasm of bambara groundnut as a resource for plant breeding.[35] It also provided genetic resources for the first genetic linkage map of the crop, which was published in 2016.[36] Genetic linkage maps help identify the inheritance of genetic traits in a crop and targets for *marker assisted selection* (MAS) to breed for desirable attributes, such as seed colour and growth habit. Genetic linkage maps also provide an opportunity to identify genetic and physiological traits related to tolerance by different genotypes to abiotic stresses such as drought, heat, and cold. These efforts provided the basis for BAMBREED — the first international breeding programme on bambara groundnut.[37] Through the EU projects and links that they supported among the University of Nottingham and partners in Africa and India, we now understand much of the agronomy, physiology, and pathology of the crop and nutritional content of its seeds. Farmer knowledge and perceptions have been evaluated, as well as modelling and GIS mapping techniques established to identify ideotypes for breeding. We now have institutions in Africa, Asia, and Europe with a critical mass of research on bambara groundnut that is unrivalled on any other underutilized crop, and a methodology that can be applied to many other underutilized species.

Working in Teams

Despite being told that crop research is *a lonely furrow*, I have been fortunate to have had some outstanding colleagues on the bambara groundnut journey — one that has required tenacity and dedication to the ultimate goal as part of a team, rather single-minded focus on publications. These include three excellent postdoctoral fellows, Dr Sarah Collinson, Dr (now Professor) Festo Massawe, and Dr Shravani Basu who helped to coordinate the three complex EU projects. However, no academic colleague has been more committed to this long journey on underutilized crops than Dr Sean Mayes without whom the genetic linkage map of the crop and much of the EU project research would never have been completed. His leadership of the crop genetics research on bambara groundnut and its translation to other underutilized crops provides a role model both for the discipline and his many excellent postgraduate students who have gone on to become postdoctoral researchers and senior academics. Sean remains a close colleague and critical member of the research team described in Chapter 7.

Being part of a multidisciplinary team working on a longer-term challenge does not mean sacrificing individual or multi-authorship of peer-reviewed papers — sometimes it just takes a bit longer. The corpus of original research on bambara groundnut by colleagues from *four continents* is something that Nikolai Vavilov would have been pleased to see.

Underutilized Crops Revisited

Using the experience gained from bambara groundnut, here we ask, "what can we learn from the history of major crops that can be applied to underutilized crops?". We then reconsider an approach that we have taken on bambara groundnut and identify its strengths and limitations as a guide to future efforts on this and many other crops. Finally, we set the context for Chapter 7 — the need for a global effort on *crops for the future*.

Why have so many crops remained underutilized whilst others have become significant global commodities? The considerations are complex — properties, socio-political, cultural, historical, serendipitous, and no doubt many other factors that are particular or even unique to each crop. If we seek to diversify beyond the major staples that now dominate our agrifood system, we must look to explain the decline of agricultural biodiversity. We also need to learn lessons from the historical successes and failures of mainstream crops and apply these to the wider adoption of underutilized crops. Cynics will say that there are good reasons why underutilized crops have not become more significant and that we should not waste scarce research and development resources on them. In a changing world and facing new and unprecedented challenges, however, cynicism is not a good strategy. We may find that some crops that have been discarded by previous generations, ignored by scientists and denigrated by policy makers have the potential to be *crops for the future*. The question is, which crops? A useful example is to explain the histories and status of two legumes, groundnut and bambara groundnut, to understand why one is a major global commodity treasured by the agrifood industry and the other is not.

Groundnut — Case Study of a Major Crop

Groundnut (peanut, pig nut, goober, monkey nut) is native to Central and South America[38] where it has been cultivated for over 4,000 years.[39] In the 1500s, Spanish, Portuguese, Dutch, and German explorers took seeds of the crop to Spain and it was then taken to Asia and Africa. Today, groundnut is traded as a global commodity for its edible oil, food, and animal feed uses and is distributed across tropical, subtropical, and warm temperate zones.[40] China and India, far from its centre of diversity, account for the bulk of world groundnut production. The Portuguese first brought groundnut from Brazil to West Africa in the 16th century. There were already two African "groundnuts" — bambara groundnut (*Vigna subterrànea*) and Kersting's groundnut (*Macrotyloma geocarpum*) that were cultivated long before the arrival of groundnut. The familiarity of African farmers with their local types probably encouraged them to adopt the new groundnut. They did so rapidly, but rather than a single type, the Portuguese brought different groundnut accessions from diverse locations in South America to single locations in Africa. The crop is mostly self-pollinating and produces seeds that are genetically identical to the parent plant. However, if two plants are cross-pollinated, they produce seeds that are different from either parent. Some may exhibit new and desirable qualities and will be preferentially selected as the basis for new *varieties*.

Mixing of genetic material from across Central and South America created a secondary centre of groundnut diversity in West Africa with new ecotypes for different climates and uses.

An American Crop

The first journey of groundnut out of Africa was when enslaved people and their slave traders began to take the crop to the Caribbean and the southern states of the USA.[41] In the early 1800s, a more formal export trade for groundnut was established and it became a major source of income for the colonial economies of West Africa. The initial destination from West Africa was Great Britain, but soon America dominated the market. Slaves planted groundnut throughout the southern United States where it was used as a pig feed. As supplies from the American south began to reach the northeast of the country, affluent Americans shunned groundnut as *slave food* or *pig food* (hence *pig nut*), but poorer people took to it. Traders began to import significant quantities from West Africa, which was a source of cheaper supplies than from local growers. These imports ended in 1842 when an import duty was imposed to protect American growers. While Americans were keen on groundnut as a food, for most of Europe its main value was as a lubricant oil and an ingredient in the manufacturing of soaps and candles. In West Africa the French saw the economic potential of groundnut (and palm oil) as an edible oil, but French olive growers successfully lobbied the government to impose prohibitive duties on imported oils. By the 1830s, French merchants petitioned for a reduced tariff on groundnut which the government subsequently cut by 80 per cent. But the lower tariff only applied to whole groundnuts in their shells and not on oil pressed in Africa.[42] This protected oil processors in France and deprived African countries of important sources of income through value addition to their commodities. As groundnut imports increased, the French processors became richer and support for the tariff become even stronger.

In Africa, much of the groundnut crop was grown by migrants who travelled long distances from the African interior to groundnut growing areas on the west coast, such as Senegal and the Gambia. Not all came by choice. Caravans from the interior of Africa brought slaves, but rather than being sold into the trans-Atlantic slave trade, many were bought to work on groundnut plantations whose ultimate beneficiaries were the metropolitan French.

The adoption of groundnut as a major world crop is linked with its value to industrialized economies and their colonial rule, especially in Africa. In the early 19th century, a shortage of vegetable oil in Europe led to the development of the groundnut oil industry. The demand for feedstock led to the commercialization of groundnut plantations in Africa for export and the development of new value-added products. George Washington Carver, who was born a slave but became one of the greatest inventors and Black academicians in American history, developed over 300 hundred uses for peanuts including in chilli sauce, shampoo, shaving cream, and glue.[43] His innovations made peanuts a regular staple in the American diet but not for the first time in the Americas. In fact, the earliest reference to peanut butter can be traced back to the ancient Incas and Aztecs who ground roasted peanuts into a paste.

Modern peanut butter, its production process and the equipment used to make it are associated with at least three inventors. In 1884, Marcellus Gilmore Edson of Canada patented peanut paste, that was produced by milling roasted peanuts between two heated surfaces.[44] In 1895, Dr John Harvey Kellogg (the creator of Kellogg's cereals) patented a process for making peanut butter from raw peanuts. He marketed peanut butter as a nutritious protein substitute for people who had difficulty in chewing solid food. In 1903, Dr Ambrose Straub of St. Louis, Missouri, patented a machine to make peanut butter. Together, these innovators, their innovations, and the products that they developed ensured the future of groundnut as an American food crop that justified major investment, research, and mechanization. European markets relied on imports, but demand in the USA for peanut butter and other food products from groundnut was increasingly met by local production because the climate in much of the south was conducive to its cultivation. The result was the development of new cultivars, plant breeding programmes, agronomic packages, and improved agricultural and industrial machinery for cultivation, harvesting, and processing, as well as the manufacture of a range of value-added snack foods, condiments, and confectionery products.

Emergency Food Source

A more recent development is the use of groundnut as an emergency food for famine relief. Inspired by the widely consumed Nutella spread, in 1996 Plumpy'Nut was developed by two Frenchmen; André Briend, a child nutritionist, and Michel Lescanne, a food-processing engineer.[45] Nutella is made up of sugar, palm oil, hazelnuts, cocoa, skimmed milk powder, whey powder, lecithin and vanillin. Plumpy'Nut is a combination of peanut paste, vegetable oil, milk powder, sugar, vitamins, and dietary minerals. It is used as an emergency treatment for malnutrition since it supports a rapid weight gain in patients. The product is easy to eat and can be easily stored in packaging that has a two-year shelf-life since it needs no preparation or refrigeration it can be rapidly deployed for the mass treatment of vulnerable children at risk of famine. It also provides calories and essential nutrients that restore and maintain body weight and health. Traditionally, severe malnutrition has required hospitalization but Plumpy'Nut can be consumed at home and without medical support. The United Nations has recognized that Plumpy'Nut can be used to treat large numbers of children with severe acute malnutrition without being admitted to a therapeutic feeding centre. Plumpy'nut was used in 2007 by UNICEF and the European Commission's Humanitarian Aid Department in Niger in response to a malnutrition emergency and conforms to the UN definition of a Ready-to-Use Therapeutic Food (RUTF).

The East African Groundnut Scheme — A Case Study in Hubris

The future of groundnut is assured because its versatility, industrial uses, and research ensure that it meets many of the criteria associated with a major global commodity crop. However, it is also remembered for one of the greatest fiascos in agricultural history. Between 1922 and 1961, Great Britain administered the United Nations Trust territory of Tanganyika, formerly German East Africa and now part of Tanzania. The country covered an area of around 900,000 square kilometres, but due to tsetse fly infestation, only around 15 per cent was inhabited by a population of 6 million people. Much of the land was prone to drought, and other than rainfall, water supplies were limited. There was no history of mechanized agriculture.

In 1946, the post-war British government established a number of central bodies to contribute to colonial development. These included the Overseas Food Corporation (OFC) whose remit was to support the bulk buying of staple crops. One crop considered for investment was groundnut. The British government was under political pressure over post-war rationing, so groundnut as a source of edible oil was attractive. The government sent a mission to investigate the potential for groundnut production in Tanganyika, and it proposed a huge and ambitious project, which later became the East Africa Groundnut Scheme. The project would occupy 1.3 million hectares containing 107 mechanized agriculture units, each of 12,000 hectares — 80 in Tanganyika, 17 in Northern Rhodesia (now Zambia), and 10 in Kenya. As well as field production of groundnut, the scheme would involve massive infrastructure, including building a new deep-water port and railway in Tanganyika that was expected to create 32,000 jobs for African workers.

The project was originally suggested by the United Africa Company (UAC), a subsidiary of Unilever, but soon became a flagship of the British government.[46] It was estimated that investment of $33 million would cover six years of the scheme, which would decrease Britain's food bill by nearly $14 million per year over the same period by importing groundnut from Tanganyika to address a global shortage of edible oils at that time. The project envisaged that after vegetable oil had been extracted from the groundnut, the residue and groundnut shells could be converted into cattle feed and fibre for the manufacture of clothing. At face value, this was a brilliant concept. The humble groundnut would address the shortage of fats, clothing, and animal feed in post-war Britain and create livelihoods and economic development opportunities in Tanganyika, a particularly impoverished part of the British Empire.

To achieve these objectives, the project predicted that the scheme could produce 600,000 tonnes of groundnut a year within five years. This would require machinery for land clearing and adoption of large-scale mechanization. After their initial visit, the mission assumed that rainfall was adequate for groundnut production across the whole project area with potential to expand to new areas. They also recommended a two-year scientific research programme to include meteorology, soil science and mapping, crop disease surveys, and testing of crop varieties. Fully confident of the potential to grow groundnut on a vast scale, it was agreed that research would follow, not precede, the scheme. In October 1946, only a few weeks after receiving the mission's report, the British government agreed to proceed with the groundnut scheme. Additional funds were allocated for railway, port, and road construction to move a huge fleet of tractors to East Africa, provide workshops to maintain them, and create transport infrastructure for supply chains. Rather than the Colonial Office, the UK Ministry of Food managed the scheme, and in turn appointed UAC to manage the project. Since time was of the essence, it was agreed to start as soon as possible and dispense with the need for a pilot study, reconnaissance and survey, soil maps, rainfall data, or economic analysis.

Catalogue of Errors

Clearing vast areas of bush and preparing the ground for groundnut cultivation proved to be very difficult and there was a risk that drought could lead to major crop losses. However, the project team was confident that by applying good farming practices, the impacts of rainfall shortages could be avoided. In fact, widespread drought struck the region in 1948 and 1949. In 1948, the OFC took over the groundnut scheme, but by 1951 it was transferred back to the Colonial Office and reduced from the planned 12,000-hectare farms to units of between 600 and 2,400 hectares. The results from these much smaller projects were also very disappointing. In 1954, a new Tanganyika Agricultural Corporation (TAC) took over the scheme and the OFC itself was dissolved. The intention of the East African Groundnut Scheme was that 1.3 million hectares of bush would be cleared and cultivated to support the industrial production of groundnut in three countries (Tanganyika, Northern Rhodesia, and Kenya) which were each transitioning from British colonial rule to independent nationhood. By 1949, the area of the whole scheme was reduced to 145,000 hectares and two units in one country. By 1951, the grandiose project had been abandoned.

The road to hell is paved with good intentions (unknown). Started in 1947 by a well-intentioned British Labour government, the East African

Groundnut Scheme will be remembered as a fiasco. Whilst it is all too easy to see its demise simply as a failure of colonialism and the consequences of hubris, arrogance, and incompetence (which it was), there are several invaluable lessons for those seeking new options for the agrifood system.

Pride comes before a fall (Proverbs 16:18, King James Bible). The sheer ambition of the East African Groundnut Scheme was astonishing. Had it succeeded, it would have been a role model for post-war mechanized agriculture, contributed to the African and the British economies, and helped alleviate a world shortage of edible oil. Its proponents meant well, but it was a disaster in each of these objectives. In April 1947, it was anticipated that there would be 60,000 hectares (five *units*) under cultivation within the first year of operations in virgin terrain in a remote part of the British Empire. By the end of October only 4,800 hectares had been cleared, of which 3,000 hectares had been planted. That only 5 per cent of the initial area had been planted should have served as a warning to all concerned. Even on this restricted area, average yield was less than half the already paltry revised estimate of 0.34 tonnes per hectare. An initial allocation of 4,000 tonnes of groundnut seeds was ordered for the first season's planting, but because such little land had been cleared, only a small tonnage was sown. By mid-1949, the second year's harvest produced only 2,000 tonnes (against the 57,000 tonnes anticipated).

After two years, the yield was half of what had been provided for planting. By year two, the viability of the whole scheme should have been questioned, but its proponents remained upbeat, indeed bullish. As well as the daunting task of clearing and cultivating 1.3 million hectares of bush, infrastructure was to be provided with roads, a purpose-built port, and supply chains. Planners grossly underestimated the cost and complexity of building roads, railways, airstrips, and hospitals as well as providing water and sanitation. They had also badly misjudged the transport and supply problems, as well as the workshops needed for heavy tractors. Many of the stores and workshops for the tractors were built on an old lakebed, and in the rainy season some were washed away by flash floods. In addition to the projected availability of edible oil for the British market, other diverse benefits were envisaged from the production system. These included residue from oil extraction to produce cattle feed cake and conversion of the groundnut shells to make fibre. For the latter, Imperial Chemical Industries (ICI) constructed a factory in Ardeer, Scotland for the sole purpose of converting peanut protein into fibre. The initial output was projected at 9 thousand tonnes per year, but like the whole scheme itself, the fibre manufacturing process was a disaster. When garments made with groundnut fibre were washed or dry-cleaned, they snapped. Without testing a proof-of-concept prototype, the whole manufacturing operation turned out to be a waste of public money. Again, decisions were made from far away before evaluating the product, assembling the evidence, or testing the market.

Serious Flaws Ignored

It should already have been clear by the end of 1947 that there were fundamental problems with the groundnut scheme, but early setbacks and warnings did not persuade those in London of the folly of the entire project. Indeed, Sir Frank Lee, the permanent secretary at the Ministry of Food, wrote: *"Our standing as an imperial power in Africa is to a substantial extent tied up with the future of this scheme. To abandon it would be a humiliating blow to our prestige everywhere."*[47] As well as the eventual reputational damage that it caused to British prestige, there were huge costs before the government admitted defeat. This may have been a result of incompetence, hubris, or arrogance, but was probably the minister's attempt to save face. In total, the government had spent about $68 million on the groundnut scheme, of which only about $4 million might represent the value of capital assets handed over to the TAC. The net loss to the British taxpayer was about $50 million (around $600 million in current value).[48]

Measure twice, cut once (many people, but most memorably my father-in-law Jack Brough). From the outset, the scheme failed to consider the area's topography, soil, climate, and rainfall and imposed untested and ultimately unsuitable agricultural methods, including the wrong kind of machinery for the terrain. The soils contained much more clay than suitable for groundnut cultivation and many of its buried pods were left unharvested in the ground. After the rains, the hot weather baked the soil so that the digger-blades found it difficult to penetrate. The soil structure included quartz sand and compacted layers that were almost impenetrable and damaged ploughs and other implements. An earlier survey had shown that the soil was limited in phosphates and nitrogen. Undeterred by these warning signs, the project team continued to clear land and plant seeds. But problems multiplied. By the end of the summer of 1947, two-thirds of the imported tractors were unusable. Also, bulldozer blades were quickly ruined, and *shervicks* — part Sherman tank and part tractors — were wrecked. In the rush to start the programme there was no pilot study to test the production system, inadequate data on rainfall, soil type, slope, and the challenges they imposed on land clearance and cultivation. A scientific research programme was envisaged, but it was scheduled to follow, not precede, large-scale implementation of the scheme. There was no scientific basis to the project, no on-the-ground evidence to support it, and no experts to advise on it.

The greatest knowledge a man can possess is the address of the local library (Albert Einstein). Not only were management decisions made from London, there was no consultation with the local community on whose land the project was established. The original project team noted a lack of rainfall data, but they may have had a clue from local people who referred to Kongwa at the centre of the project as the *country of perpetual drought.* There was no attempt to gather local intelligence that could have guided the planners or warned of the pitfalls. The assumption that the land was untouched by agriculture ignored the scale of the indigenous flora and fauna and their effects on humans. Huge baobab trees were difficult to remove, not least because one was used as a local tribal jail, another was a site of ancestor worship, and many of the hollow trunks were occupied by bees' nests. Some of the workers needed hospitalization for numerous bee stings, others encountered angry elephants and rhinoceroses. Further problems were caused by the inaccessibility of the site and lack of water sources. Water had to be brought in and stored in a concrete-lined pool which locals insisted on using for swimming, despite protests by European workers. The scheme also had to train African workers from tribes with little previous experience with machinery and who found it difficult to use machines brought in to plant the seeds. Willing but poorly trained local drivers wrecked many of the tractors. The concept of mechanization (replacing human labour with expensive machinery) was inappropriate for East Africa — the notion of economies of scale under such hostile conditions was misplaced, and infrastructure needs were grossly underestimated. There was a high turnover of labour that cost time in training new staff. Disaffected workers went on strike for better pay and conditions, which caused local inflation that made provisions unaffordable for the local communities.

Success has many parents, failure is an orphan (Count Caleazzo Ciano). In 1946, a working party was appointed to review the East Africa Groundnut Scheme. Their report concluded that the cost of the scheme was six times greater than the value of the crops, and that the administration in Tanganyika should be "much smaller and more flexible". By 1949 it was clear that things were going from bad to worse. After the election in 1950, John Strachey was removed as food minister, and his successor, Maurice Webb, announced that the project would be abandoned. He accepted that the scheme had been implemented prematurely and before the methods used had been adequately tested. The accounts were in chaos, but oddly, he rejected the idea of abandoning the scheme or "retreating in any fundamental way". The best solution was to abandon "the purely food producing idea" and repurpose the scheme as "a broad project of colonial development with a wide and varied agricultural content". Debts amounting to $50.7 million were written off. No one was keen to take credit, blame, or even responsibility for the fiasco. The reputation of John Strachey never recovered but the damage had been done.

Bambara Groundnut — Towards a Research Framework for Underutilized Crops

Despite neglect by current administrations in Africa, hostility of colonial powers before them, and the ambivalence of research institutions, bambara groundnut is still around. In fact, it has been for a very long time. Ibn Battuta, the Moroccan scholar and explorer, observed the crop growing in Mali in 1352.[49] "They dig it from the ground, they fry it and eat it. It tastes something like fried peas". The food historian William Woys Weaver reported finding traces of bambara groundnut in the African American diet of the southern United States where it was called the goober pea.[50]

Tenacity

It is clear that bambara groundnut is a resilient crop; as are the people who cultivate it. Their tenacity has protected the crop and its associated knowledge and belief systems. The research so far on bambara groundnut owes much to the tenacity of the researchers who have made huge progress in our understanding of this much neglected crop. It is impressive that partners in Africa, Asia, Europe, and recently Australia, have contributed to world class multilocational and multidisciplinary research on an underutilized crop with little previous published literature. Over decades of research and personal commitment they have provided new insights into the ecophysiology, agronomy, crop genetics and breeding, entomology, pathology, marketing, processing and utilization nutritional quality, socioeconomic characteristics, and crop simulation modelling of bambara groundnut. Their approaches have been novel in that for the most part, their research has attempted to span a complete *research value chain* (RVC) from field to plate. This has required collaboration across disciplinary silos, information sharing, agreeing on priorities, and overcoming institutional barriers, disinterest, and occasional animosity. That they have achieved such a sustained effort for so long and produced such a sound body of published evidence is a credit to their determination and commitment to making progress on this crop.

The RVC approach adopted over the last two decades has taken bambara groundnut through a series of steps from field to plate, not always sequentially along the value chain nor continuously, but rather as funding and human resources have allowed. The approach has provided a framework for our understanding of one crop and its possible application to many others. This has been a daunting task, with an approach using conventional research methods organized into disciplines, each with their own sets of rules and practices. In that sense, research on bambara groundnut has been similar to that for any major crop — identify potential, overcome obstacles, find uses, and build acceptance. Research on this crop has been unusual in that it has linked disciplines and data along an RVC that mimics the journey of the crop from field to plate. For this, discrete subject areas such as crop genetics, plant and crop physiology, field agronomy, ingredient processing, product development, and market testing have each contributed to a single outcome — a greater understanding of what we know and what we don't know about one crop. Additional techniques to combine data on the environmental resources used by different ecotypes and simple simulation models have been developed of the crop that allow geospatial mapping of its potential across the globe. There have also been attempts (albeit as case studies) to identify socioeconomic and structural challenges to the wider adoption of bambara groundnut. Together, these activities have involved hundreds of researchers, students, and technical staff in more than 20 countries and over two decades.

Where Does the Framework Fail?

Although the framework has built an impressive body of quantitative research knowledge on bambara groundnut, is it robust enough to encompass the qualitative knowledge that has been curated by communities that have cultivated, consumed, and protected this crop for centuries? How can we understand the belief systems and motivations of the myriad of unconnected communities that have grown bambara groundnut across geographies for generations? That this knowledge is largely in people's heads and not in published literature, simply adds to the challenges facing researchers working on any underutilized crop. Here we consider a case study on how belief systems guide the actions of one community that has grown bambara groundnut for centuries. Without analyzing the belief systems that guide the decision making of communities involved with bambara groundnut, we cannot keep our promise to return the crop in a better state than we found it to those who have nurtured and protected it without the benefit of scientific study or investment.

Every time an African dies a library goes with her (African proverb). Bambara groundnut has a long history and a wide geography of cultivation. In the past few decades we have learned a lot more about its science, but what about local knowledge and belief systems that have been associated with this crop for at least 600 years? Where is this knowledge, how can we curate it, and is it relevant to the communities that still grow bambara groundnut and those who might? For rural communities, belief systems play an important role in the organization of society. Social taboos provide rules that guide community interactions with their environments and ascribe values to certain resources. These rules are highly structured around factors of social difference such as gender, age, and social status. Taboos operate as informal institutions, regulate social order and behaviour, and prevent disorder by penalizing transgressions. In rural Africa, breaking taboos is perceived to invite witchcraft or negative repercussions from the spirit world. This contributes to social pressure for individuals to conform to accepted norms. Mediation between the physical environment, society, and divinity falls within the jurisdiction of community leaders, particularly with spiritual authorities who are viewed as the link among the three. Taboos may provide a means to manage limited natural resources, but in the context of food systems, they can also regulate the distribution of food based on social difference. Taboos can even cause malnutrition if certain foods are prevented from being consumed.

Belief Systems in Rural Malawi — A Case Study

In the context of agriculture, the few published studies of local belief systems focus on major crops and link agricultural practices, belief systems, and social organization. Researchers working with farmer groups in rural Malawi found a number of beliefs and taboos associated with bambara groundnut that influence how and by whom the crop is produced, consumed, and traded.[51] The study also demonstrated the importance of food crops in the *worldview* of rural societies and their rules of engagement with the natural environment. The study concluded that belief systems and practices have important implications for development interventions and need to be considered at the design phase and not at the conclusion of a project. The study only described the relationship between one community in rural Malawi and one crop, but it raises issues that are relevant for the study of many other underutilized crops for which communities have been the custodians of knowledge and in which researchers are novices. Here are some of the findings from the study that illustrate why researchers working on underutilized crops will need to think beyond more familiar approaches and their own disciplines.

Custodians of knowledge: Rural communities in Malawi have an oral tradition through which the elderly transfer local knowledge, beliefs and taboos across generations. Elderly women said that the reasons behind the beliefs and taboos around bambara groundnut were seldom known but had been passed down without explanation or justification.

Cultivators: Community members reported that bambara groundnut requires little labour and is drought tolerant. Since few inputs are required, women are the main cultivators of the crop and sources of knowledge about it. Men are more likely to be in charge of crops requiring inputs as this involves greater control over expenditure and income.

Life/death symbolism: Bambara groundnut has a contradictory and symbolic significance of both life and death. It is used in traditional practices relating to medicinal healing as well as love and fertility rituals. The life/death symbolism is reflected in taboos around agricultural production, marketing and consumption, which are gender related. These taboos are perceived as a way to retain control access and benefits to specific groups, in this case, elderly women. The restriction is similar to other taboos that legitimize unequal access to food, such as preventing children and young women from eating eggs or drinking milk, which operate to limit food resources to men.

Traditional medicine: Bambara groundnut plays an important role in traditional medicines, preventing and treating illness, disease, spiritual attacks and other conditions. These medicines are often made by a traditional healer, who interacts with the natural environment and spiritual world, as poor health is often associated with punishments from spirits. The most common health and spiritual issues for which bambara groundnut is used is as a treatment by traditional healers are sexually transmitted diseases, infertility, epilepsy and spiritual attacks. Secret herbs, plants and sometimes animal parts are mixed together with the bambara groundnut for the patient to consume, or to place in cuts made in the skin by the healer. Other practices include placing a bambara groundnut on the head of a child when sleeping to prevent nightmares, treating abdominal pain and as a preventative medicine for measles. A necklace of three dried bambara groundnuts may be used to protect people from disease and black bambara groundnut is taken by those suffering from disease.

Ceremonies: Initiation ceremonies mark a rite of passage for individuals in their lives or as a means of entry into sects or societies. Bambara groundnut is used in ceremonies that include instruction in the behaviour and values of the community or sect and responsibilities in marriage and adult life. These occur at puberty for boys and girls or before a woman gives birth to her first child. Women participate in initiation ceremonies at the birth of their first child where new mothers receive instruction in childcare from elderly women who prepare a meal of bambara groundnut which is consumed by the new mothers, marking its association with fertility and love.

Protection of assets: Bambara groundnut is used to protect individual and household assets. It may be used as a talisman to protect the home and assets such as livestock against witchcraft that is aimed to harm household members or reduce their prosperity due to jealousy. Households use potions and ornaments containing bambara groundnut for protection from witchcraft. As part of this talisman, protected households cannot produce, cook or consume bambara groundnut while it is being used, as this may reverse the power of the talisman and invite harm.

Consumption: Bambara groundnut consumption is restricted when individuals or households may be vulnerable to death, such as during an illness or a threatening situation. People who work in dangerous positions, such as soldiers, police officers or hunters, should not consume bambara groundnut and its consumption is also restricted during travel since it is believed to cloud night vision. Women also are not allowed to eat bambara groundnut if their husband is travelling. Individuals taking medicine for an illness or spiritual condition are discouraged from consuming the crop as it is believed to dilute the treatment and bambara groundnut cannot be consumed by visitors to a home if they are not present when it is being prepared. This is due to the suspicion of tampering with the meal (witchcraft) if the visitor is not physically present.

Marketing and commercialization: Bambara groundnut is predominantly sold by women as part of societal norms that see them responsible for its production and preparation. Female producers have control over the income from the crop due to their role in cultivation and marketing and its perceived low value by men. However, gender roles can change over time, often linked with market potential. If profitability and status improve, there may also be changes in gender roles and possible household conflict.

Final Thoughts

Understanding the local context is critical when designing and implementing interventions that aim to improve agricultural value chains of underutilized crops. Belief systems impose additional factors in decision making that go beyond the production, yield, and income potential of a particular crop. Taboos that restrict people from growing a crop result in unrealized production potential and efforts to commercialize it may affect who produces it and who benefits from it. The past experience of women indicate that they are often excluded from benefits when crops become more profitable in the market — so why should they contribute to research efforts? It is also important to note that belief systems and taboos are changing, and young people increasingly question the traditional beliefs of their parents. Other taboos are weakening due to changes in religion, and exposure to other people, places, and information that contradicts long-held beliefs. The decline of taboos and traditional belief systems can also pose a threat to maintaining local knowledge of bambara groundnut and women's role in orally transmitting this knowledge. What is crucial is that we need to document traditional knowledge of bambara groundnut and similar crops and draw on women's knowledge and expertise in their cultivation and food preparation as a *basis* for research and not simply a corollary to scientific studies.

The Malawi study illustrates the wider challenges facing research on underutilized crops. We can integrate scientific data within a common framework for a specific crop, but how do we accommodate knowledge that is qualitative, vernacular, and often location-specific? At this stage, the main points discussed above provide guidelines about how we can approach these critical questions that are particular to underutilized crops. Chapter 7 describes an attempt to do.

Why Is Groundnut a Major Crop and Bambara Groundnut Is Not? A Researcher's Perspective

As with many other African crops, the *placement* of groundnut and the *displacement* of bambara groundnut are a consequence of centuries of colonial history. For good and bad, groundnut is now a major source of income and nourishment for millions of Africans, but bambara groundnut is not. In some locations such as Senegal, groundnut is a mainstay of the economy. It possesses genetic traits and biochemical properties that have made it an attractive option in many tropical and sub-tropical regions. Clearly the development of groundnut as an American crop and its rapid expansion in China, India, and Africa have consolidated its geographic potential and multiple uses. However, we should remember that the most important factor in its emergence as a world commodity crop is its oil content (between 44 and 56 per cent). This property attracted the interest, investment, and support of colonial powers eager to find a cheap source of edible oil for their own food systems that they could not easily produce locally (as with coffee, tea, cocoa, spices, and many other cash crops). Based on this initial demand for oil, everything else about groundnut fell into place for different markets — food, confectionery, animal feed, and industrial processes including paints, varnishes, lubricating oils, leather dressings, polish, insecticides, soaps and cosmetics, and even nitroglycerin. Research, advocacy and marketing followed. Its recent use as an emergency food provides a non-commercial market that will no doubt only increase as the population of malnourished and displaced people continues to expand.

Bambara groundnut is an indigenous grain legume grown and cooked mainly by subsistence women farmers in drier parts of sub-Saharan Africa. That one sentence is a powerful explanation of why bambara groundnut has remained underutilized, and unfortunately under-researched for so long. First, it is an indigenous grain legume, and as we have seen in Chapter 3, in the hierarchy of favoured crops legumes were not a priority of the Green Revolution nor have they ever been for the agrifood and research systems that have supported this revolution. In global terms, with the exception of soybean and groundnut, there are no legumes in the top 30 traded crops. Working on any legume (except soy and groundnut) assigns a researcher a lower place on the priority list for funding. The preoccupation of the global food system for high-yielding, calorie-rich, easily mechanizable, and readily processible crops has increased the discrepancy against favoured cereals that can achieve yields above 10 tonnes per hectare which is many times the typical yields of legumes. Based on yield and commodity price rather than, for example, nutritional content or contribution to soil health, makes a preference for staple cereals obvious for growers, processors, food companies, and consumers.

Second, it is *grown and cooked mainly by subsistence women farmers.* The reciprocal of this statement is that it is not grown, traded, or promoted as a cash crop by men. This has implications not only for justification to invest in it, but also for the status of men venturing into what is seen as *only a women's crop.* Along with other pejorative terms, it is clearly not an immediate attraction for men farmers, government policy makers, development agencies, or investors.

Third, *in drier part of Sub-Saharan Africa* (the most impoverished, marginal and forgotten areas of global agriculture) national governments have neither the money nor motivation to support research that is not seen as an immediate source of food, income, or foreign exchange.

For these reasons, if we were to select a crop for detailed investigation and significant research investment, bambara groundnut — a subsistence crop grown by marginalized women in Sub-Saharan Africa — comes close to the bottom of the list for donors, governments, and extension support or advocacy. For researchers, the easiest conclusion is the lazy refrain *if it was any good we would have discovered it by now*. However, is bambara groundnut really unworthy of even a fraction of the effort, money, patronage, and marketing expended over the past few hundred years on the world's major crops? Not if we have a different vision of agriculture. If we can emerge from the *yield-for-profit* paradigm that has driven the global agrifood system for decades to one that rewards *climate-resilient nutrition,* then bambara groundnut begins to look very different. By its nature, a crop that has survived the hostile climates of Sub-Saharan Africa for hundreds of years unaided by science and policy must be resilient. In the climate and health crises that humanity now faces, a good place for researchers to start looking for future options is to reconsider crops that have survived despite us not because of us. If we seek a crop that can grow in landscapes that are increasingly too hostile for major crops, provide proteinaceous nutrition rather than calorific sustenance, and improve the soils in which it is cultivated, then bambara groundnut looks a very different prospect for growers, consumers, and researchers. Of course, if edible oil is the main attraction, groundnut with its typical oil content of around 50 per cent is the obvious and valid choice (the oil content of bambara groundnut is around 6 per cent). However, if we wish to move from an oil rich diet, then bambara groundnut provides a healthy option to be processed into nutritious and tasty snacks, foods, and non-food products or by rediscovering the traditional foods that have been made from it for centuries.

We once saw groundnut as an underutilized crop that could become *a crop of the future* because of its rich source of edible oil. It did. If we now can see bambara groundnut also as a *crop of the future* because of its rich source of nutrients, all else will follow. For this, we need to look at the lessons learned from groundnut, now a popular mainstream crop, and apply them not just to bambara groundnut but more generally to underutilized crops. One lesson is that we cannot afford the enthusiastic pursuit of a new crop to cloud our judgement of its weaknesses and limitations. As said to me over 20 years ago *"it is underutilized for a reason"*. We must find the evidence for why as well as evidence to overcome the reason. We do not, however, have the time and resources for expensive failures. Groundnut was already a well-known and successful crop at the time of the East African Groundnut Scheme and could therefore survive this fiasco without permanent reputational damage. An objective, dispassionate evidence-base supported by high quality research on underutilized crops is the best way to avoid such damage.

CROPS FOR THE FUTURE

7

Work in Progress

Context

One of my favourite TV programmes of the 1970s was the American sitcom Mork & Mindy. It starred the brilliant Robin Williams as Mork, an extra-terrestrial being who comes to Earth from the planet Ork. Mork has been assigned to observe human behaviour by his leader Orson, who is looking to move to a new planet because Ork has just banned humour. Mork's spaceship happens to land near Boulder, Colorado, USA. The storyline for each episode involves Mork trying to understand human behaviour so that he can advise Orson on whether Earth would be a good place to live. At the end of each episode, Mork reports back to Orson, and there is a humorous exchange between them as they discuss social norms and human idiosyncrasies. Here's a fictitious dialogue on how Mork might have reported human attitudes on biodiversity to Orson.

Mork. *This is a really beautiful planet Orson, you'd love it. As well as earthlings, there are also half a million kinds of plants and millions of different animals and these funny little creatures called insects. The earthlings have already found that 30,000 of these plants and animals can be eaten and I've heard that some of the foods from them are really tasty but I can't seem to find any of them here in Colorado. In fact, these earthlings are really odd. Even though they have grown thousands of different plants, they only use three* special *plants for most of their food wherever they live on the planet and they just ignore all the rest. They don't seem to have much use for the insects either and are trying to kill them all.*

Orson. *Oh, I guess they only eat a few plants and animals so that they don't disturb all the others that they share the planet with. These earthlings sound like really kind creatures. I can't wait to meet them.*

Mork. *Well that's the funny part Orson. The earthlings remove all the other plants and animals from where they live so that they can have space to grow more of their own* special *plants. I asked them why they haven't even looked at the other plants but they keep telling me* "if they were any good we would have discovered them by now". *To be honest, the foods from their special plants can get pretty boring after a while — so all the others must be really awful.*

Orson. *Wow, you mean that they're destroying all the other plants and animals just so that they can have the whole planet for themselves and their* special *plants? Last time you told me there were signs of intelligent life on earth!*

This chapter is about the crops that are still around despite not being accorded *special* status by the global food industry, research system, and decision makers. In a world in crisis, we ignore them at our peril. Some could be *crops for the future.*

Plan A — The Case for a Special Plants Only Agrifood System

The case for Plan A virtually speaks for itself. The world's farmers have never produced so much food for so many people, and global food supply has outpaced population growth for many years. Of course, there are problems with its equitable distribution, waste, and accessibility for some, but food has never been cheaper or more plentiful on a global scale. This is largely because of the spectacular yield increases for major staple crops through the Green Revolution and associated technologies. The major crops are *special* for a reason — they are simply the best around and by further improving them we will have more than enough food to sustain humanity long into the future. Exciting new technologies (genetic and mechanical) will mean breakthroughs in the current yield ceilings of staple crops. Even without these, there is plenty of scope to increase crop yields through conventional breeding and improved management. In 2020, New Zealand farmer Eric Watson achieved the Guinness World Record for the highest wheat yield for the second year running.[1] Mr Watson produced 17.4 tonnes per hectare of wheat on his Ashburton farm. In 2020, the total amount of wheat that will be produced is predicted to be 758 million tonnes.[2] Given that the global average wheat yield is still only about 3.54 tonnes per hectare, the potential increase in wheat and other staple crops through crop breeding and better management is still very considerable even if every farmer cannot match the yields that Mr Watson can produce in New Zealand.

As well as greater yields, we have plenty of agronomic knowledge about cultivation of major staple crops and the supply chains that distribute their marketable products around the world. Limiting agriculture to just a few crops also allows large economies of scale and greater returns on investment in breeding, production, research, and marketing. Adding complexity makes logistics, quality control, feedstock, consumer education, and management all the more, well, complex. The argument that staple crops are primarily calorie-rich and nutrient-poor can be addressed by increasing the micronutrient content of specific crops by *biofortification* through conventional (non-GMO) approaches.[3] Genetic modification technologies have already led to significant yield increases and reduced use of pesticides.[4] The potential for such technologies is remarkable. And through monocultures, we now have the most efficient management system targeted to specific crops, and know exactly when to sow, apply inputs and control measures, and harvest their products. Introducing more species and cropping systems only complicates crop production and makes it difficult to mechanize and separate harvested products. There is simply no business case for diverse crops and multiple cropping systems.

In summary, modern agriculture has already achieved more than we could have asked from it since the end of the Second World War. If we continue to increase production of the main crops through better management and new technologies, we have the basis of a *second* Green Revolution — we must remain focused on the challenges ahead by improving our current agrifood system. At least in principle, agricultural diversification, greater biodiversity, agroecological and regenerative agriculture are all attractive, but they cannot achieve the doubling of food production that we will need in the next few decades. Wasting precious time and effort on agrobiodiversity is simply a distraction from the task at hand of feeding 10 billion people by 2050. We need to stay the course and produce more food for more people from the same amount of land.

Plan B — The Case for a More Biodiverse Agrifood System

Biodiversity includes all the variety and variability of life on Earth. Its variation is measured at the genetic, species, and ecosystem level.

Agricultural biodiversity (or agrobiodiversity) refers to that fraction of biodiversity that is used by humans for their material needs — food for themselves, feed and fodder for their livestock, and biomaterials for non-food uses such as fuel and construction. Agrobiodiversity includes both the genetic diversity of species that humans use, as well as the *agro-ecosystems* in which they are cultivated. Agriculture is a part of biodiversity both in its variety and its variation — the only difference is that we choose to manage it. Biodiversity and the planet will survive without human beings, but without biodiversity and its conservation the human species cannot survive. Agrobiodiversity not only supports food production but can help conserve global biodiversity, improve soil water management, sequester carbon in soils and biomass, and establish food production systems that are resilient to shocks.

Conserving biodiversity might seem desirable — even a moral obligation — but agriculturalists spend a lot of their time trying to reduce and even eliminate it. Insect pests attack agricultural crops and livestock, pathogens bring diseases to the species we grow, and weeds reduce those parts of the agroecosystem that we most value — such as seeds and veg-etative matter of crop plants. Agriculturalists then reduce the number and severity of insect pests, pathogens, and weeds through the use of agro-chemicals. The effects of agrochemicals on biodiversity and ecosystem services, such as pollinators, are familiar to both the agrifood industry and consumers who buy its products. Less well known however is the speed at which we have reduced agrobiodiversity in terms of the number and range of crop species grown and the agroecosystems in which we cultivate them. From the 30,000 edible species that have already been identified (there are countless more), humans have grown around 7,000 as crops but now farm only 30 *major* species for over 95 per cent of the food eaten by 7.8 billion people across the planet. The decline or demise of all the others means that only one in every 1,000 of the plants that we know to be edible is now considered as a useful food source, and only 0.0004 per cent of the crops that we cultivate now provide more than half of our food.[5] How has it come to this?

Those plant species that are not accorded *special* status have a number of other appellations — minor, orphan, neglected, underutilized, abandoned, lost, local, traditional, alternative, niche, women's, underdeveloped, or forgotten crops.[6] Their products are described pejoratively as *famine, poor people's*, or *subsistence* foods. The main argument against using any of these crops is that our modern, uniform food system is doing just fine, so why change a successful model? Implicit in the justifications for Plan A is that diversification offers no viable alternatives. The best we can do for these *forgotten* crops is to store their seeds in remote gene banks — just in case they might be useful in the future. But right now, there is no business case to use any of them as part of the global agrifood system — they are forgotten for a reason.

Perhaps the most frequent argument against the wider adoption of underutilized crops is the *if they were any good we would have discovered them by now* line of defence. Behind this stereotype are five *negative assumptions* used by the food industry, academia, and sponsors to justify why they reject underutilized crops and the wider diversification of food systems;

- No demand for these crops or markets for their products
- Even if a market exists, there's no supply chain to get products into supermarkets
- Even if a supply chain exists, quality seed and new varieties are needed to make desirable products
- Even with products and supply chains, there is no research expertise on these crops
- We can't decide where to grow them and how to use them without reliable evidence.

Answers to each of these statements must not only be acceptable but also *demonstrable* before any underutilized crop is deemed worthy of investment. The challenge is that without investment how can we demonstrate that any of them are useful? Even if all the statements above are valid (and there is some truth in each), the rational conclusion is that *Plan A* (staple crops grown as monocultures) will *by itself* be enough to nourish (not simply feed) 10 billion people on a hotter planet without destroying the ecosystems on which we all depend. It will not. On an ever-riskier and more hostile planet, mainstream crops will become increasingly vulnerable to climate extremes, their monocultures ever more demanding of higher inputs, and their carbon-heavy supply chains ever more susceptible to disturbance.[7] If breeding new, more climate-resilient varieties of staples doesn't outpace climate change, much of the world's current arable area will become unproductive for mainstream crops, and people who can no longer grow them will migrate to safer environments. By then, we will have removed all alternatives.

This exodus is not sometime in the distant future, but is happening now as we see in regions of Africa, the Middle East, and Latin America. But do we have a Plan B? Not really. Chapter 5 showed how the global research and education system has been finetuned to mainstream species, and even to particular crop varieties in specific locations, and for which the private sector sees short-term commercial benefits. Such a narrow vista means that alternatives to Plan A are unrealistic without disruptive innovations — not just in the number of species that we grow, but also in the systems in which we grow them — not only in our research, but also in the way in which we conduct it. Business- and-research-as-usual will not be enough.

Funding Crop Research

Universities, research institutes, international centres, and private companies have spent billions of US dollars to research the world's major crops. Such research has been of enormous global significance, but successes have largely been restricted to genetic and agronomic improvements of major staples grown as monocultures. The Green Revolution of the 1960s (two generations ago!) set the model for global agriculture in which high-yielding varieties of major species are supported by fertilizers, agrochemicals, irrigation, and mechanization. There are increasing challenges to the Green Revolution paradigm, but no alternative production model has been demonstrated on a large enough scale to be accepted by the agrifood sector. One reason why there is no alternative *production* model is that no alternative *research* model has been established to challenge the Green Revolution paradigm. Whether research funding is from public or private sources, it is limited and increasingly focuses on solutions that provide maximum impact and involve minimum risk. The outcome is that virtually all funding, expertise, advocacy, and subsidies are channeled to major crops grown as monocultures and their ubiquitous products. This means that resources are directed to better *business-as-usual* solutions rather than alternatives such as agricultural diversification on lands that are ill-suited to mainstream crops, products that are not part of the globalized agrifood system, or crops that are resilient to climate extremes but cannot compete with the calorific yields of major staples in mainstream systems.

The evolution of the international agricultural research system has reinforced the dominance of mainstream agriculture through its focus on species, (for example rice, wheat, maize, potato), groups of species (vegetables, marine creatures, livestock, forests, insects), landscapes (saline, mountain), systems (agroforestry), actions (policies), or resources (water, fertilizers). Only one centre, Bioversity International (which recently formed an alliance with the International Center for Tropical Agriculture (CIAT)), has a specific remit for global agricultural biodiversity.[8] Of course, individual centres can contribute to agrobiodiversity through research and breeding programmes that use the wild relatives and landraces of their mandate crops, germplasm collections, companion crops, policy interventions, or better use of natural resources. Although they can all make valuable contributions to greater agrobiodiversity, none has an explicit remit solely for the world's neglected and underutilized crops. If individual centres can be dedicated to a single crop, why can't we have just one dedicated solely to research on the other 7,000?

Conducting Crop Research

Major Crops

Crop research often involves a linear pipeline from basic to applied elements. In this approach, researchers usually test a hypothesis through experiments in laboratories, controlled-environments, or field plots where individual *treatments* can be applied within boundaries set by the experimenters. To be statistically valid, each treatment must be replicated so that the probability of a *treatment effect* is caused by more than just random variation. Experimental results are reported, and where treatment effects are statistically significant (usually a probability greater than

95 per cent), the published research may provide recommendations either for adoption by end-users, further research, or as evidence for policy makers. The research must be replicated (the same conditions imposed at different locations), which means that results must be quantitative and data provided in terms and units that are recognized for specific disciplines. Chapter 5 demonstrated how agricultural scientists face the challenge of spanning the divide between basic and applied research and at the same time demonstrating the practical applications of their experimental results. There is a plethora of research publications and reports for major crops, but it is often difficult to show how results obtained from research done under relatively well-controlled conditions in laboratories and field plots can be extended to the farm or region or used as evidence for policy makers. As well as being spatially replicated, the research must be replicated in time — over several seasons (to account for factors such as weather) or sites (to account for differences in environments) before general recommendations can be made.

Agricultural research itself has been a process of evolution in different places, at different times, and by different people and institutions, but there is no complete body of work even on major crops. Much research remains unpublished (because it is not statistically significant), incomplete (because it wasn't done over enough seasons), or is too speculative (where the conclusions are not robust enough for referees of peer-reviewed journals). Research that is unpublished is considered to be of no academic value and therefore not shared. Papers in expensive journals are often inaccessible to researchers in developing countries, and much agricultural research remains in the *grey literature* such as reports, working papers, government documents, white papers, and evaluations. With the proliferation of journals, data, and experiments, there is simply *too much stuff* to make sense of the entire agrifood system or provide detailed analyses, interpretation, or recommendations. To make sense of the welter of information and specialisms, agricultural science has retreated into silos of its constituent disciplines (for example, genetics, physiology, agronomy, food sciences, and marketing). Specialism is seen as the way forward for research breakthroughs and academic recognition through high-impact journals; there are few similar incentives to study the whole system.

Whether from public or private sources, agricultural research is increasingly focused on impacts that justify funding and the time involved. Funders are risk averse. There is also pressure on academics, journals, and sponsors to see results — a paper that shows a statistically significant effect of a particular *treatment* is easier to publish than one in which there was no effect. This sounds like common sense, but it skews research towards *confirmation bias* — the tendency to search for evidence that confirms or supports prior beliefs. In other words, we tend to do experiments that confirm our expectations rather than challenge them. This type of cognitive bias affects research activities not least because precious funds are spent on confirming the obvious rather than testing the alternatives. Evidence-based decision making requires evidence to do things differently — but we continue to provide evidence to do things the same way. The outcome of all of this restricted funding, short-term pressure, and confirmation bias is that money and research expertise is further channelled towards the usual suspects. The justification of sponsors is that a dollar spent on a major crop is likely to have more impact than an equivalent amount on an unresearched crop of limited significance. This may be true today, but will it remain the case when much of our agricultural areas are too hot, dry, volatile, or our soils too impoverished for mainstream agriculture? Nevertheless, a minister of agriculture is unlikely to be criticized for continuing to support attempts at national *food security* through the increased production of global commodity crops rather than championing their local alternatives.

Underutilized Crops

'Against this backdrop, imagine the challenge of working in agricultural diversification! Essentially, research on underutilized crops faces three obstacles beyond that on mainstream species;

- There is a paucity of peer-reviewed publications on any single underutilized crop. Often journal papers that do exist are thin on original data and rely on descriptive analyses of crop botany, distributions, and reported but unsubstantiated properties and end-uses. There is a danger that enthusiasts working on a particular underutilized crop will make claims about their favourite crop and fall into the *advocacy trap* of unsubstantiated assertions without independent and credible evidence

- Not only is there scant literature on any particular crop, it often duplicates that which is done elsewhere, is inconclusive because of insufficient time and funding, or doesn't meet the exacting standards of journals or editors who also cannot find suitably qualified referees for uncommon crops. Results may be contradictory since time and resources for more detailed studies are limited. Much original or preliminary research on underutilized crops remains consigned to the filing cabinets or shelves of researchers working in government institutes, often in printed documents that become unreadable with age. The outcome is that researchers often duplicate the work of others, the findings of different researchers and institutions working on the same crop remain unconnected, and money and time are wasted

- The biggest challenge is that the most precious research evidence on underutilized crops is often in the heads of those who have grown, protected, and used such crops for millennia — despite, not because of, agricultural science. Much of this evidence is vernacular (unwritten), often in languages that are not easily translatable for modern researchers and such knowledge is dying along with those individuals and communities who until now have curated it. Even when such evidence is written, it is considered to be inadmissible for publication because it is *qualitative* and based on the knowledge of individuals, passed down through generations rather than on *quantitative* and replicated measurements from a randomized block design experiment in a prepared field.

Given the above scenario and the growing gulf in funding, there are only two options for the future of underutilized crops. The first is to conclude that all of these crops are underutilized for a reason — they are of little value to the agrifood industry and for nourishing humanity. Trying to conserve these crops and the dying languages in which much of their knowledge is stored is futile — underutilized crops are of the past and should be left to those who still insist on growing them. The second option is to propose that they are underutilized because they have never been subjected to serious research or attempts to improve them beyond their current status. Instead of consigning them to history, with commitment from funders and the agrifood sector, as well as the knowledge and collaboration of those who continue to cultivate them and coordinated research, underutilized crops can be *crops for the future.*

Crops For the Future Research Centre (CFFRC) — An Experiment in Plan B

Between 2011 and 2020, Crops For the Future Research Centre (CFFRC) was based in Malaysia as the research arm of Crops For the Future (CFF), a global initiative for the world's neglected and underutilized crops.[9] Its guarantors were the Government of Malaysia, which agreed to provide funding for facilities and operational costs, and the University of Nottingham, which provided access to its campuses in Malaysia and the UK. CFFRC built on the legacy of the International Centre for Underutilized Crops (ICUC) that was hosted at the International Water Management Institute (IWMI) in Colombo, Sri Lanka and the Global Facilitation Unit (GFU) for Underutilized Crops that was hosted at what is now Bioversity International in Rome, Italy. By combining their resources, ICUC and GFU agreed that a single entity could best focus global efforts on the world's underutilized crops for food and non-food uses. The title *Crops For the Future* was coined by Dr Hannah Jaenicke, the Director of ICUC. In one elegant phrase, it replaced all the pejorative terms and negative associations with this vast array of crops.

As guarantors, the Government of Malaysia and the University of Nottingham in Malaysia provided a governance role for CFFRC, but it was free to build partnerships and secure support for global research on underutilized crops. At its launch, Professor MS Swaminathan, World Food Prize laureate and father of the Asian Green Revolution described Crops For the Future as *"the need of the hour"*. In 2011, I was appointed as CEO of CFFRC and remained in that post until May 2020 when the project came to an end; it was the most thrilling and challenging time of my career. As well as experimental research on underutilized crops, CFFRC was itself an experiment in *how to do research*. For this, it needed a plan on how to deliver its vision *"to be recognized as a world leader producing excellent, innovative research and knowledge dissemination on underutilized crops*" and its mission "*to develop solutions for diversifying future agriculture using underutilized crops*". For both vision and mission, CFFRC needed a research strategy.

Research Value Chain

One of the most frequent questions that I was asked, especially in the early days of CFFRC, was *what is the next crop of the future*? or *what is your list of crops*? We did not have a definitive list — *what would be the point of looking for crops of the future if we already knew which they were*? In any case, adding one or even 10 new crops to the list of major species was not our purpose. Instead, we set out to establish a framework that could be applied to any underutilized crop. By not being tied to a list of crops (*why is my favourite crop not on the list?*) we were free to focus on methods and approaches that could be applied across several crops at the same time. Rather than saying which crops we were working on, it was easier to say that we would not work on those that were mandate crops of the CGIAR and national or international agencies. Of course, this did not prevent us from making comparisons with these crops, and more importantly, learning from research and transferring technologies (especially in genetics, bioinformatics, and data systems) from major crops to accelerate progress on underutilized crops. Instead of a candidate list, we decided to use a small number of *exemplar* crops from which our expertise and learning experience could be transferred to others through a common approach.

From the outset, CFFRC took a transdisciplinary approach based on the concept of the 'Research Value Chain' (RVC).[10] Its principle is that research should transition along the typical agricultural supply chain from sowing and crop management to harvesting, transport, processing, marketing, and end-users. For this, academic disciplines must link along an RVC from genetic characterization and improvement of candidate species through their agronomy, ecophysiology, processing, and product development to the social, economic, and policy studies needed for their wider adoption. Although the RVC may seem to be simple common sense, it is not widely used.

With major crops, the sheer number of researchers and institutions involved deters integration across disciplines and institutional structures, and reward systems further discourage interdisciplinary engagement. Even if they are working on the same crop, it is difficult for researchers to engage with growers, processors, consumers, and end-users *as part of the research process*. Such *horizontal* integration may not be needed for major crops where much research already exists — researchers can simply go to the literature to support their own findings and those in areas of research beyond their own. In such cases, duplication of effort, in other words the number of similar findings, confirms the validity of the research and therefore justifies publication. However, for underutilized crops we have neither the luxury of duplication nor the comfort of specialization. To make progress, we need to escape from *disciplinary silos* and look for integration along the whole RVC — not simply as a series of multidisciplinary links but as a *transdisciplinary* approach to the grand challenge of agricultural diversification.

As well as needing evidence of connectivity along the entire value chain, sponsors will not wait for the typical sequence from basic to applied research for any underutilized crop. They may, however, support activities that perhaps demonstrate preliminary progress at each stage of the value chain — seeing a preliminary *working* version of the entire chain allows more time for detail and refinement later. Moreover, the purpose of the RVC is to avoid repeating any research that has already been done. As well as new research, we can also identify researchable gaps in the RVC that prevent particular underutilized crops from delivering social, economic, and environmental impacts and useful products. The RVC provides a mechanism to identify what we don't know about a particular crop as well as what we do know — the gaps in knowledge are usually the obstacles to progress for widespread cultivation of any species. Finally, research must address a problem to which underutilized crops can contribute to the solution. Academic research is often supply driven — an idea taken to application — but for underutilized crops it must be demand-led — *what does it offer that could not be achieved with major crops alone*?

CFFRC used the concept of the RVC to conduct research on underutilized crops simultaneously at each stage of the value chain from crop genetics, through development of product prototypes to markets and policy aspects. To deliver viable outputs from underutilized crops, research programmes were developed for different end uses such as nutritious foods for human consumption and feed for aquaculture systems. An additional programme built a global knowledge system for underutilized crops and a specific programme was designed to show how an RVC for one exemplar crop, bambara groundnut, could be used as a working model to improve many other underutilized crops. Each programme included discrete projects with a series of activities or tasks. A Programme Development and Monitoring Unit reviewed and evaluated the progress of each programme towards delivering *proof-of-concept* product prototypes that could be used as the basis for commercialization. The purpose of all of these monitoring and assurance mechanisms was to avoid activities that did not contribute to the "*problem statement*" and to demonstrate that funded activities, especially by research students, contributed to programme goals. As well as a mechanism to focus research on specific end goals (deliverables), the RVC provided an analytical tool that for specific underutilized crops could help identify;

- What we already know — to *coordinate* further research and avoid repetition
- How we can transfer relevant knowledge from major crops — to *accelerate* progress
- Which are the researchable gaps that we can fill — to *focus* only on necessary research; and
- Where we need to get to — to guide *strategic* thinking and set priorities.

Five Negative Assumptions About Underutilized Crops

Whatever approach, activities, resources, and commitment its staff dedicated to working on underutilized crops, the ultimate measure of the CFFRC experiment is how much progress it made in addressing the five *negative assumptions* used against Plan B. The activities below describe some of the progress made by CFFRC researchers in meeting these challenges. The results of their world-class activities are described elsewhere and their contributions are recognized here as well as in the Acknowledgements at the end of this book. First, I want to summarize the approach we took to each challenge.

No Demand for these Crops or Markets for their Products

Very soon after its establishment, CFFRC's researchers recognized that they had to develop products that were equivalent to or better than those already in use by the agrifood sector. It wasn't enough to show that these foods are healthier, use locally-sourced ingredients, have a smaller carbon footprint, or come from crops that grow on soils that are too hostile for mainstream crops. They must be *desirable* by themselves. As well as products from particular crops, the CFFRC team also had to show how their experience could be rapidly transferred to other crops to both avoid repeating efforts and to save time by learning from our mistakes.

FoodPLUS — Novel Foods and Functional Ingredients from Underutilized Crops

The first approach was for CFFRC's small food product development team to use ingredients from underutilized grains, vegetables, and fruits to develop a range of attractive and tasty food products such as energy bars, cookies, noodles, local snacks such as *murukku*, and dried ingredients for soups. These were not scaled up to commercial production, but the proof-of-concept prototypes for each product were tested on visitors and at events such as diplomatic functions in Kuala Lumpur. A more formal evaluation on variants of the products was provided through taste panels of individuals from different nationalities (mainly students) to provide qualitative assessments of taste, texture, flavour, and visual appearance. Laboratory measurements provided quantitative data on the nutrient and energy content of the foods and their shelf life. A second approach was to test the functional properties of ingredients from underutilized crops to see how they could supplement or replace those from major species. For example, how much could we replace the ingredients in, say, wheat noodles with ingredients from local underutilized crops without reducing the acceptability of flavour, taste, texture, or cooking properties? Various blends were tested to identify the maximum replacement capacity of ingredients from underutilized crops. The logic was that reducing the fraction of major crop ingredients in a food product with no loss in functionality would still stimulate a significant demand for underutilized crops without changing the eating habits and preferences of consumers. A great example of this replacement approach was with animal feeds, specifically aquaculture feed for farmed fish.

FishPLUS — Insect Meal Using Underutilized Crops as a Sustainable Source of Aquaculture Feed

Almost half the fish and other aquatic species consumed globally are farmed in land- or marine-based aquaculture systems.[11] Aquaculture is now the fastest growing sector of global agriculture and commercial feeds are the biggest economic cost of its operations. Not only are they expensive and imported, formulated feeds are unsustainable because they use fishmeal and fish oils extracted from wild-capture fish, (so called *trash fish* caught in fishing nets) and ingredients derived from major food crops, such as soybean and maize to provide vegetable protein and carbohydrates.[12] Bizarrely, unsustainably and immorally, to meet the increasing demands of aquaculture systems, we now feed wild fish to farmed fish and use other ingredients in aquafeed that could otherwise have been used in human foods. The vast overfishing of the oceans means that the cost of aquaculture feed has escalated dramatically in recent years, posing economic threats to the entire industry, not to mention the destruction of marine biodiversity. Because feed is their major cost, aquaculture farmers may take short cuts in using sub-standard feed and overstocking ponds with high densities of undernourished fish. Antibiotics are often used in an attempt to reduce death rates in these unhealthy environments. The result is that aquaculture is either an unsustainable and poorly regulated industry delivering unhealthy products, or formulated feeds have to be imported along global supply chains at great cost to the farmer and the environment. Just imagine, if rather than raiding the oceans for fishmeal and relying on imported staples, we could produce novel aquaculture feeds using locally-sourced non-food crops to replace imported staples and as substrates for protein using insects to replace imported fishmeal? Can we generate a sustainable, circular aquaculture economy that provides incomes for local farmers on marginal lands, healthier fish, and safer food for consumers?

The idea of using insects to replace fishmeal in livestock and aquaculture systems is not new.[13] In fact, various insect species have been found to be an effective partial replacement for fishmeal. The challenge is to find a cost-effective method of producing large enough quantities of insect meal to reduce our reliance on fishmeal. Various substrates such as agricultural and food wastes have been tried as a food source for insects, but each has had problems in scaling to commercial use either because of volume or quality control.[14] In 2018, CFFRC's FishPLUS team and its partners developed novel aquaculture feeds using insect meal from black soldier fly (BSF) larvae fed on local underutilized crops. Black soldier fly is a non-invasive insect species whose larvae provide a rich source of protein with a profile similar to fishmeal. Feeding trials at CFFRC on tilapia and prawns showed that BSF could replace up to 50 per cent of fishmeal in formulated aquaculture feed. The project also demonstrated additional economic and environmental benefits such as the use of *frass* from the excrement of insect larvae as an organic fertilizer.

A number of underutilized crops were tested as substrates to feed the BSF larvae. *Sesbania grandiflora* showed positive outcomes and nutritional analyses indicated that amino acid levels of the insect meal closely resembled those of fishmeal. *Sesbania* is a nitrogen-fixing legume that grows well on marginal soils and is not used as a human food crop. The use of underutilized crops as substrates for insect meal represents a potentially major breakthrough since it both reduces the demand for unsustainable fishmeal and provides economic uses for non-food crops. This disruptive technology can be scaled up for major aquaculture species and systems, and could be modified for poultry and livestock systems. Also, through this innovative approach, large areas of marginal lands could be used for small scale farmers to grow climate-resilient underutilized crops, both as substrates for insect meal and as primary sources of carbohydrates, lipids, and proteins to be incorporated into aquaculture feed. Instead of attempting to cultivate major crops in unsuitable environments, farmers could cultivate insects, or at least the substrate for them, as the first step in a value chain for high value, locally produced, nutritious, and healthy aquaculture products. The main reason given by the aquaculture industry to keep using fishmeal is that it is cheaper than other protein sources, including insect meal. This ignores the negative externalities of biodiversity, the environment, land use, and farmer livelihoods and the long-term consequences of not transitioning to a sustainable alternative. The costs of these externalities should not be borne by the aquaculture industry alone, but shared with those who enjoy its products. Like most other potential users of underutilized crops, aquaculture systems need to look beyond the bottom line of their operations and the usual options for their business models.

Forgotten Foods — The Future is History

The examples so far show how ingredients from underutilized crops can be used to make new versions of familiar foods or to develop feed formulations for healthier, more sustainable, and safer aquaculture products. In each case, consumers were not being asked to change their habits — they could choose snacks, fruit juices, and noodles with a novel twist or still eat familiar types of seafood. The fact that the foods or feeds included ingredients from unfamiliar crops might have been of interest but was probably not critical in their decision making. Our real challenge was to persuade consumers to explicitly *change* their habits and actively seek out unusual products from underutilized crops and other novel ingredients. Telling them that these were somehow *good for them or the planet* or tastier than familiar alternatives would not be enough — we needed a different approach. We decided that the future of food could lie in its history — our history, and indeed 'her story'. In February 2017, the director generals of members of the Association of International Research and Development Centers for Agriculture (AIRCA) held their annual meeting as guests of the FAO in Rome. Our schedule included a courtesy call on the then Director General of FAO, Dr José Graziano da Silva. At our meeting, each director was invited to describe the activities of their institution. When I mentioned CFFRC's work on underutilized crops, Dr Graziano noted that he had long wished to see a United Nations International Year of Forgotten Foods and that our forgotten crops and the interest of AIRCA members in agricultural diversification could build momentum for such an international year.

Back in Kuala Lumpur we agreed to launch a Forgotten Foods Network as part of a global effort to encourage wider interest.[15] To capture maximum impact, we decided to avoid a narrow definition of forgotten foods and instead include all those traditional foods that were increasingly being displaced by modern, uniform, and processed products. Forgotten foods can include ingredients from neglected crops, animals, vegetables, and even insects, as well as traditional varieties of the major staple crops. Many are derived from domesticated and wild plant and animal species that have been conserved and improved by farmers, or gathered wild for their food and medicinal properties. In the transition to a single western diet, forgotten foods have been neglected — their use has been reduced due to social perceptions, disinterest from research institutions, limited consumer awareness, and challenges in establishing markets for their commercialization.

As with their raw materials, there are always those who will claim that forgotten foods are forgotten for a reason — there is no interest in them — and that our modern diet of highly processed, takeaway and ready meals, rich in calories, sugar, and fat, and promoted by billions of dollars of marketing is the way forward because it is the preferred choice of global consumers. Perhaps. An alternative view is that before discarding our own heritage, we should rediscover the diverse food basket of our forebears — foods that until very recently were part of the diet of our communities but which have been neglected and forgotten by modern societies. From this treasure trove, we might identify those foods that could become part of a global effort to diversify human diets and nourish future generations.

In 2017, CFFRC's Forgotten Foods Network was launched as an online initiative to collect and share information on foods, recipes, and traditions that are part of our common heritage and which could again nourish us in the future. CFFRC's role was to provide an independent evidence base, without which we cannot identify and improve those forgotten foods and the many plants, animals, and insects from which they derive. For this, we need to link culinary history and recipes with scientific studies and new technologies on forgotten foods, measure their nutritional value, test their suitability for changing climates, and make novel products and cuisines that are nutritious and desirable to consumers and marketable by the food sector. The role of the Forgotten Foods Network was to become an on-line resource of foods, recipes, ingredients, images, stories, and scientific evidence on forgotten foods from ancient and modern crops, animal and marine species, and even insects. In November 2017, HRH The Prince of Wales launched the Forgotten Foods Network at CFFRC's headquarters in Kuala Lumpur, Malaysia.[16] As well as commending the value and importance of CFFRC's research, his visit, and a follow-up meeting that he hosted in London as patron of the global Crop Trust, strengthened the links between the Forgotten Foods Network and the Crop Trust's 'Food Forever' initiative. Food Forever is a wonderful initiative that is helping to secure support for the conservation of crop genetic resources by exploring and championing food diversity.

If a Market Exists, There is No Supply Chain to Get Products to Supermarkets

However desirable, nutritious, and marketable are their products, no matter how climate-resilient and environmentally-friendly are their ingredients or culturally rich and diverse their cuisines, there remains little interest from the food industry without a complete, continuous, and uniform supply chain of ingredients already in place from which foods from underutilized crops can be produced. This is quite a challenge given the history of marginalization and neglect by those now demanding an intact supply chain, but it is one that I am very familiar with! In 1989, I remember excitedly presenting a low fat, high protein, and tasty snack made from bambara groundnut to the executives of a UK food company (no longer in existence) to be told, *yes, this is very good and we could market it but we need at least 150 tonnes of flour for our factory operations.* Given the subsistence nature of the crop, this was unachievable. But the reality is that without at least a proof-of-concept supply chain from farm-to-plate, we cannot attract investors who are looking to their bottom line, which means a guaranteed supply of raw materials. The challenge with any underutilized crop is to integrate physically dispersed small-scale growers, transporters, processors, retailers, consumers, and markets at different stages of the agrifood supply chain to provide the working system that the industry demands. For researchers of underutilized crops, there is also the challenge to link distinct disciplines along the

RVC. Rather than develop and test a supply chain prototype elsewhere, we decided to develop one ourselves. We would link farm-to-plate case studies using CFFRC's facilities, and our staff would be the farmers, transporters, processors, product developers, and consumers of our own products. The project, called *CONNECT*, tested the acceptability of products made from two case study crops; bambara groundnut, a drought-tolerant legume, and *Moringa oleifera*, a fast-growing tree known as the 'miracle tree' because of its resilience to harsh environments. Moringa is also remarkable for the range of products that can be made from its seeds as a vegetable (its long 'drumstick-like' pods) and source of herbal medicine and from its leaves that are rich in protein, vitamins A and C, potassium and calcium.

Our interest was to use flour produced from the dried seeds of bambara groundnut and the dried leaves of moringa to produce high protein noodles. Our first objective was to grow bambara groundnut and moringa crops at our nearby field research centre. We transported harvested material to the CFFRC headquarters, made them into flour from which we developed food products, tested them on focus groups, carried out sensory evaluations, compositional and nutrient analysis, studied shelf life, and developed socioeconomic and policy recommendations. The second activity was to link data across the entire value chain from genetics through to consumers, and demonstrate a *block chain* of traceability using QR codes on all samples and integrate data into common databases. The third and most exciting challenge was to involve CFFRC staff at each stage from farm to market. In addition to our researchers and technicians, this meant involving our administration, finance, facilities, and human resources staff. In this way, we tested a prototype supply chain that linked planting materials, growing practices, harvesting, processing, and testing products and markets for bambara groundnut and moringa using our own staff at each stage of the supply chain. Through this experience, we identified seven *CONNECT* steps that could be used for the physical supply chain and parallel RVC for adoption of any underutilized crop in any location;

1. Crop selection and suitability; stakeholder analysis, crop suitability, potential yields, end-uses
2. Cultivation and harvesting; cropping cycle, field management, inputs, and harvesting

3. Post-harvest and processing; post-harvest management, supply chain, storage, and transport
4. Product prototypes; test product prototypes, recipes, and processing technologies
5. Nutritional/sensory analyses; composition, nutrient content, functionality, and acceptability
6. End-users and markets; end-user target groups, promotion, and routes to market
7. Impact studies; scaling up and economic impacts for consumers, partners, and society.

The *CONNECT* approach puts growers and sponsors in charge because they can compare diversification options against current practices for major crops and cropping systems. As well as the suitability of different crops, *CONNECT* provides a mechanism to develop and test novel crop products and identify their potential markets before going into large-scale production. *CONNECT* also provides an evidence base for environmental sustainability and economic viability of diversification options, livelihood opportunities for farmers, lifestyle options for consumers, and investment potential for sponsors and development agencies. Perhaps more than any of these tangible benefits, *CONNECT* taught CFFRC staff about ourselves and how research and support staff can become part of a common goal — all staff are consumers of food and can also be researchers without a formal qualification. Whilst the RVC approach had already removed the disciplinary silos between researchers — *CONNECT* reduced the institutional silos between researcher, technician, intern and administrator. There had been some concerns at the outset whether *CONNECT* was a sufficiently academic exercise to lead to research publications (it was) and that by linking researchers and administrators it would be unmanageable (it wasn't). However, its greatest contribution was in building a common identity and goal for all CFFRC staff.

If a Supply Chain Exists, Quality Seed and New Varieties Are Needed to Make Products

Much has already been said about bambara groundnut and how it became an exemplar species for CFFRC. As well as building on our earlier research and delivering a relatively complete body of knowledge on bambara groundnut, we wanted to see if our collective experience on this one crop could be developed into a framework that could be rapidly applied to other underutilized species. Is every crop so different that we must start from scratch, or can we share knowledge, technologies, and approaches among crops to accelerate progress on different crops at the same time and as a basis to compare with mainstream species? Could we save time, effort, and money by applying our learning experience across species? Through this iterative research and learning process, BamYield became both the first international programme for the genetic improvement and wider adoption of any underutilized crop and an exemplar for others. For this, we established five general principles;

- BamYield must span the entire RVC from genetics, physiology, and agronomy through to product development, supply chains, and policy. This meant that breeding new varieties had to include intended end-users from the outset, in fact, they had to guide selection of traits and characteristics for breeding objectives that met their requirements
- Alongside the RVC, BamYield must work with partners in the agri-cultural value chain, including small-scale farmers, processors, wholesalers, distributors, producers, and food importers and export-ers. Again, end users at each stage of the value chain were part of the research process and not passive recipients of its outputs
- To maximize its geographical impact, BamYield must integrate activi-ties and data across locations. As part of this international approach, BamYield established collaborations and coordinated experiments with partners in Malaysia, Thailand, Indonesia, South Africa, Botswana, Nigeria, and the UK
- BamYield could not be limited to researchers and practitioners, but must complement work being done in national organizations such as ministries of agriculture, health, rural development, NGOs, and those involved in sustainable agricultural production and fair and equitable trade
- As well as high-quality research and publications, the learning experi-ence gained through BamYield must be conveyed beyond a narrow

academic community. This book and one specifically on bambara groundnut that is now in preparation are part of this intention.

What are the *specific* activities that we learned from BamYield that could be applied to the genetic improvement and breeding of other crops? First, there was a clear objective — the production of higher-yielding and climate-resilient bambara groundnut varieties that could contribute to increased food security, better nutrition and improved incomes for subsistence and small-scale farmers. To achieve this, BamYield linked researchers, practitioners, and multiple agencies through a series of seven iterative *clusters* of activities to;

1. Identify uptake-limiting factors and initiate research on limiting genetic, agronomic, harvesting, processing, product, market, and socioeconomic factors along the RVC
2. Apply and test initial results from breeding, agronomy, and product testing with stakeholders
3. Identify the most promising genotypes from stakeholder engagement and initiate production, nutritional assessment, and market evaluation of promising genotypes
4. Develop a core collection of genotypes for multilocational testing and evaluate that core collection against desirable traits identified by farmers and other stakeholders
5. Establish farmer workshops and field demonstrations, host regional knowledge exchange workshops, test, and register varieties
6. Establish an online knowledge exchange system for partner interactions, disseminate best practices for management, harvesting, storage, and product development
7. Link farmers with each other, markets, and processors.

This approach meant that not only did we have to work across the entire value chain and involve non-formal research in the process, but we could not start at one end of the RVC and work our way along to the other — we had to take an entire food system approach that broke out of disciplinary silos and also from the typical sequence from basic to applied research.

Finally, BamYield established *transferable* tools and approaches that could be rapidly applied to the genetic improvement and breeding of any underutilized crop. These include;

1. A protocol for multi-regional trials and breeding systems
2. New processing methods and products to increase market demand and provide income
3. Knowledge systems to link researchers, processors, and end-users
4. Development of computer *'apps'* to put communities in charge of the innovation process
5. Ongoing exchange of knowledge between communities and researchers to sustain progress.

Even With Products and Supply Chains, There is No Research Expertise on These Crops

There are thousands of years of indigenous knowledge on thousands of underutilized crops, but most of this information is either inaccessible to researchers, inadmissible for publications, or buried in the grey literature of government reports and policy papers.[17] Academic research that has been published is scattered across institutions and geographies, and often relates to the particularities of a single crop in a single location for which raw data are usually not supplied. CFFRC provided an opportunity to bring together diffuse and uncoordinated research and expertise on underutilized crops. We set out to achieve three goals — to build a community of researchers based at CFFRC; to establish a global network of research and development partners; and to create a generation of researchers with competencies to lead future research and development of underutilized crops.

The first challenge was to build a community of scholars dedicated solely to research on underutilized crops and which spanned all disciplines of the RVC. The word *solely* here is important. Individuals and research institutions around the world have and continue to work on various aspects of underutilized crops, but they are rarely able to focus exclusively on such crops, either because of funding constraints, institutional priorities, or other responsibilities. Such crops may feature on national and international priority lists, but they rarely appear high enough to secure continued funding and advocacy. Unless underutilized crops are their *day job,* it is hard for scientists to dedicate more than sporadic efforts on them between other activities. The outcome is often wasted effort and reluctance to persevere.

From its inception in 2012, CFFRC recruited a team of researchers and support staff to be based at its headquarters and at a nearby field centre in Malaysia. The total number including field staff, administrators, and facilities management never exceeded around 50, of which fewer than 20 were researchers. As a small group of mainly young staff working on unfamiliar crops in a new organization, the CFFRC team faced challenges at the academic, operational, and at least initially, reputational level. How to establish projects and programmes, what equipment and facilities are needed, how should we collaborate with partners, when can we publish results, what monitoring and financial systems do we need, how can we communicate with the public and media? All daunting challenges for which help and guidance from wise heads was needed and provided.

As CEO, as well as appointing each member of this wonderful and dedicated young team, I was incredibly fortunate to secure the help of a number of world class colleagues to act as mentors, advisers, and leaders of the research that was being established at CFFRC. The late Professor Chin Ong brought with him vast research experience in tropical crops and agroforestry systems at the International Crops Research Institute for the Semi-Arid Tropics (ICRISAT) in India and the World Agroforestry Centre (ICRAF) in Kenya. Professor Aik Chin Soh, a fellow Malaysian, provided expertise and links to the Malaysian and Indonesian oil palm sectors. As well as academic links to the University of Nottingham, Dr Sean Mayes provided cutting edge research expertise in crop genetics and plant breeding of underutilized crops, leadership of the work on bambara groundnut, and joint directorship of our postgraduate programme. Professor Sue Walker led our agrometeorology and crop modelling activities and brought extensive agricultural experience and academic links from her native South Africa. Dr George Hall contributed his technical expertise in chemical engineering and vast experience within the food and aquaculture sectors to guide research on food products and aquaculture as well as mentoring in research proposal writing and publications.

As well as these leading academics, colleagues joined CFFRC from the private sector. Most notably, Mr Azizi Meor Ngah brought his vast experience as former CEO of the Malaysian Agrifood Corporation to the challenges of agribusiness, supply chains, and logistics. Mr Max Herriman used his expertise in marine and aquaculture systems to lead our FishPLUS programme as well as guide programme and project development, monitoring, and evaluation systems, and help me with strategic planning. Professor Peter Gregory agreed to advise and coordinate CFFRC's research strategy and mentor young staff in research activities, reporting, and scientific publications. Peter and I had first worked together in Niger, West Africa as part of the ODA Microclimatology Unit at the University of Nottingham. Since then he had gone on to lead research groups in Australia, direct research centres in England and Scotland, and become the Pro Vice Chancellor for Research at the University of Reading in the UK. Now he willingly provided guidance on research activities, proposal preparation, and scientific publications. To add to this guidance, we had the help of a number of academic colleagues from the University of Nottingham, most notably Dr Susan Azam-Ali who provided support in food and nutritional sciences and Professor Festo Massawe in crop genetics.

I was fortunate that Mr Bruce Fraser from New Zealand agreed to help advise and guide CFFRC financial management systems using his experience of the CGIAR system, especially in the CGIAR secretariat. To work with me on strategy and guarantor liaison, I had the commitment, loyalty and continuous support of two outstanding Malaysian colleagues, Ms Rossuraya Abdullah and Ms Zunita Zubir. Their fellow Malaysians, Mr Felix Miller, with a background in the oil palm and horticulture sectors, provided leadership of CFFRC Operations and Mr Gin Teng Ooi, with extensive experience in commercial field trials, was responsible for our Field Research Centre. Ms Gomathy Sethuraman took responsibility for our technical resources and Mr Steven Tan for our facilities.

Finally, at the board level we were honoured to have the constant support, guidance, and wisdom of Professor George Rothschild who, as its founding father, had been responsible for establishing CFF and guiding its transition to becoming a truly global initiative for the world's neglected and underutilized crops. As well as his knowledge, George brought with him his experience of leading international research institutions such as the International Rice Research Institute (IRRI) in the Philippines, directing Australian research agencies, advising countless organizations involved in agricultural research and, most importantly, sheer positivity and encouragement for me and younger colleagues.

It is impossible to describe the sheer thrill of establishing such a special organization with a unique mandate in a subject of global significance. Most who worked for and with CFFRC described the family atmosphere of the organization. Starting together helped to create that affinity, but I am convinced that a belief in the aims and common purpose of 'crops for the future' sustained that commitment through many challenging times. If I could choose one learning experience from a decade of the CFFRC experiment it would be that when we gave leadership responsibilities to young staff they invariably exceeded our expectations. We almost always underestimate the capacity of young people to take responsibility, learn from their experiences and provide imaginative solutions. For example, the flair and technical competence that Ms Tan Xin Lin showed in fashioning outstanding products from unfamiliar ingredients, Ms Advina Julkifle demonstrated in the evaluation of ingredients, Ms Hilda Hussin, under the direction of Dr Patrick O'Reilly, in developing socioeconomic and marketing studies, and the ingenuity that Giva Kuppusamy and his team employed in developing novel aquaculture feeds would bring credit to any institution. That they took responsibilities in a new field of studies is even more remarkable and that they were willing to cross their disciplinary boundaries to work together is an example for all researchers. In the same vein, young postdoctoral researchers such as Dr Wai Kuan Ho, Dr Maysoun Mustafa and Dr Ebrahim Jahanshiri took leadership roles for entire themes on underutilized crops and produced cutting edge research that is comparable with that on major crops. This provides a testament both to their individual abilities and a vindication of the approach that CFFRC took to staff recruitment and development.

The multinational and multidisciplinary team that was assembled at CFFRC bridged the traditional divide between academia and business and the unique combination of established research leaders and young talent created an exciting new dynamic for a young research organization. The willing and constant guidance and mentoring that eminent colleagues gave to our staff as well as the friendship, loyalty and support that they gave to me is a testament both to their own qualities and their commitment to the wider objectives of CFFRC. There are many other champions, colleagues and collaborators who contributed to the establishment and performance of CFFRC that are not mentioned here. I remain forever in their debt.

Having established our core team at CFFRC, our second goal was to build a global network of partners. This meant making alliances with institutions around the world that could support efforts in agricultural diversification. Our main links were with fellow members of the AIRCA consortium which included institutions that were all bigger and older than CFFRC. This allowed us to adopt common mechanisms, for example in monitoring and evaluation, a communication strategy, and a repository of publications. With AIRCA partners, we also developed joint initiatives such as the Global Action Plan for Agricultural Diversification (GAPAD) that we introduced in Paris alongside the UN Climate Change meeting. The purpose of GAPAD was to galvanize support for agricultural diversification to address specific Sustainable Development Goals (SDG's) of the United Nations 2030 Agenda for Sustainable Development. As part of the GAPAD process, CFFRC and AIRCA partners coordinated workshops in Nairobi, Kenya on SDG2 (Zero Hunger and Improved Nutrition), in Kuala Lumpur, Malaysia on SDG7 (Renewable and Clean Energy), and in Rome, Italy on SDG13 (Climate Action). As well as AIRCA, CFFRC established links with other international partners, Malaysian institutions, private companies, and its guarantors, and by 2019 had a network of over 50 partners on five continents represented in 300 global locations. None of these institutions worked mainly on underutilized crops, but each had disciplinary expertise, technologies, or facilities that could be applied to their research and development. They also provided the new CFFRC team with immediate access to world-class scientists working in contrasting agroecological environments and with complementary skills and expertise to build collaborations that extending the breadth and depth of CFFRC's activities and outreach.

Our third and perhaps most important responsibility was to create a new generation of researchers equipped with the skill sets and commitment to advance research on underutilized crops. For this, we worked with the University of Nottingham to establish the CFFPLUS Doctoral Training Partnership (DTP). In recent years, Nottingham and other universities have developed DTPs as a model for postgraduate training around PhD projects, each linked to an overarching theme or a *grand challenge*. In many cases, DTPs involve a number of research institutions and associate partners, including NGOs and private sector organizations. By building this research ecosystem, the DTP model allows complex issues to be addressed by cohorts of research students working together with academic supervisors and external partners, all working towards a common grand challenge. As part of the DTP model, students also receive training in professional skills and take part in collaborative activities that build future networks both among themselves and with prospective employers and sponsors. The historical lack of investment in underutilized crops meant that there had never been an incentive for PhD students to do their research on such species nor a mechanism to encourage cohorts of research students to work together on the grand challenge of agricultural diversification.

To address this limitation, in 2013 the CFFPLUS (*Providing Links to Underutilized Species*) DTP was established with the University of Nottingham. CFFPLUS allowed students to do their research as part of CFFRC's programmes and gain a prestigious PhD from the University of Nottingham. They also worked as part of a cohort of students, each at different stages of the RVC and with partner institutions and industry. By 2017, CFFPLUS had become the biggest research training programme ever on underutilized crops, with 50 individuals from many nationalities awarded PhDs on topics related to CFFRC research. As well as their own projects, CFFPLUS students were involved in workshops and training that linked them with other disciplines and programmes. The CFFPLUS DTP achieved more than its very substantial contribution to building human resources for research on underutilized crops — it serves as a model for expansion to other universities and related DTPs. To date, CFFPLUS students have authored dozens of peer-reviewed publications in international journals and helped to build international partnerships with supervisors from national and international institutions. As well as cutting edge research on underutilized crops, CFFPLUS students have developed transferable skills in quantitative analysis, experimental design, and transdisciplinary approaches to research problems, and are part of a generation of future research leaders that can address the grand challenge of sustainable diversification of global agriculture. The expertise and networks developed by the CFFPLUS students are already a making a major contribution to global expertise on agricultural biodiversity. Of equal importance are the extraordinary skills that the CFFPLUS DTP Manager, Dr Maysoun Mustafa and her colleagues at CFFRC under the guidance of Dr Sean Mayes and Professor Festo Massawe demonstrated in linking CFFRC research objectives on underutilized crops with the academic goals of students and their supervisors.

We Can't Decide Their Uses and Where to Grow Them Without Reliable Evidence

Through its staff, students, and partnerships, CFFRC has built a critical mass of expertise and talent for the wider adoption of underutilized crops. Its research outputs, proof-of-concept prototypes and publications have also contributed to a growing body of evidence for agricultural diversification. However, by themselves, these resources cannot easily be used to link research, make investment or policy decisions, or estimate benefits of agricultural diversification. For this, we need the best available evidence to guide decisions on the cultivation and uses of different underutilized crops in different locations and for different purposes. Until now there has never been such a global repository of knowledge. This has hampered both the wider adoption of underutilized crops and comparisons among them and their mainstream counterparts. Without independent and best available evidence, we cannot expect farmers, investors, and policy makers to take the risks of diversifying beyond their current options and preferences — why should they? In 2014, CFFRC's CropBASE team led by Dr Ebrahim Jahanshiri working with young staff such as Nur Marahani Binti Mohd Nizar and Tengku Adhwa Shaherah and outstanding programmers such as Ayman Salama, began to develop the first global knowledge base for underutilized crops.[18] By 2020, CropBASE contained over 9 million data points on over 2300 underutilized crops, their composition and nutritional properties, potential uses, and economic returns at any location now and in the future. For the first time, CropBASE combines community (qualitative) knowledge and scientific (quantitative) data on the world's underutilized crops in decision support systems and digital applications that can be used by individuals, institutions, policy makers, and investors. By integrating data from different sources, CropBASE provides an online gateway across the agricultural value chain — from genetics to consumption — for currently underutilized crops. To make data usable, CropBASE is built around four digital applications that link to a *decision support system* for underutilized crops;

- SELECTCROP — a geospatial tool for crop shortlisting and suitability mapping at any location
- ASSESSCROP — an evidence-base for nutrient and compositional analysis of products
- GROWCROP — an agronomy application to aid day-to-day decisions on different crops and;
- USECROP — a tool for market intelligence, accountancy, and cost/ benefits of crop options.

Both GROWCROP and USECROP are still in early development, but SELECTCROP and ASSESSCROP are already being used to identify where to grow these crops and for what purposes.

What Crop Can I Grow at My Location under Current and Future Climates?

SELECTCROP allows farmers, researchers, and the agrifood industry to make informed decisions and comparisons about cultivation and uses for different underutilized crops for different purposes. These are probably the first decisions that farmers will need to make before deciding whether to invest in agricultural diversification. By entering their location, farmers can use SELECTCROP to provide a listing of what crops can be grown at their site based on climate and soil characteristics. Global Information Systems (GIS) and mapping technologies aligned with SELECTCROP give predictions of crop suitability based on whether the grower's preference is for the highest yielding, most nutritious, most climate-resilient, or most profitable crop. By generating geospatial maps, SELECTCROP can be used to estimate the suitability of different crops on a local, national, regional, or global level under current and future scenarios.

What Is the Composition and Nutritional Value of Ingredients and Products?

As well as their yields, the agrifood sector is increasingly interested in the nutritional value of ingredients and their suitability for product development, shelf life, and supply chain logistics.[19] ASSESSCROP provides an online global database for the composition, nutrient content, and functionality of products from underutilized crops. It also provides protocols to measure;

- proximate composition (carbohydrate, protein, and lipid)
- energy (caloric content)
- macro and micronutrient content (vitamins, minerals, phytochemicals)
- functionality (quality of protein, starch, and lipid fractions), and
- anti-nutritional factors ((including protease inhibitors, lectins, total free phenolics, tannins).

These data provide the evidence-base for incorporation of ingredients from underutilized crops into food products, animal feed, and biomaterials for renewable energy. By using laboratory and industry standards, results from underutilized crops can be compared with those from other crops under laboratory conditions and from the published literature. They also provide a basis to develop cooking techniques, recipes, and products derived from underutilized crops that can optimize nutritional content, and verify microbiological safety, and quality.

A Food System Approach

As a small, new organization working in largely uncharted academic and organizational territory, there are several lessons from CFFRC's experience.

Perhaps the most obvious is that we took on more than we should and had to drop promising lines of research simply because we didn't have enough capacity. One example is our programme on renewable energy that investigated the potential of biomass from underutilized crops as feedstock for renewable energy — certainly an important topic but not one in which we had enough expertise along the RVC. Our programmes provided organizational structure, monitoring and evaluation systems through the RVC approach, but engagement across parallel programmes was limited and seen by colleagues as leaving little flexibility to deviate from planned activities. There was also concern that while the research programmes spanned the RVC of disciplines, colleagues found it difficult to identify the sequence of priorities for specific activities along the chain. Another factor was that although colleagues were keen to work with those from other disciplines, they also needed the critical mass of expertise within their own specialisms to be able to work across the *vertical* boundaries that emerged between horizontal programmes.

In 2018, CFFRC developed a revised research strategy that would allow it to move beyond specific time-bound and linear research programmes. The strategy combined subject matter expertise and research activities within four interacting *themes*;

- Genetic Resources and Field Studies; Crop Improvement and Production; Product Development and Quality; and Knowledge Systems, and three connecting *threads*;
- Bioeconomy; Nutrition, and Societal Development.

Together these reflect a *food system* approach that can rapidly build multidisciplinary teams to tackle researchable problems and provide the flexibility to approach emerging and unforeseen challenges in which agricultural diversification can be part of the solution.

Crops for the Future — Work in Progress

In less than a decade, CFFRC made significant progress in responding to each of the five negative assumptions used against the wider adoption of underutilized crops. The most encouraging development was how

CFFRC's activities coincided with a growing global interest in agrobiodiversity and the search for a new vision of agriculture that celebrates rather than diminishes diversity in all its forms. There are some good examples of how CFFRC research contributed to this broader movement.[20] In answer to a lack of products, the CFFRC team developed a range of foods using ingredients from underutilized crops. None of these are revolutionary (their recipes are freely available), but they demonstrate almost limitless possibilities for using currently overlooked local plants to make attractive, tasty, and nourishing foods. Here in Malaysia, we drive past local ingredients growing in plain sight on our way to the supermarket to buy apples imported from New Zealand, oranges from Egypt, and peaches from the USA. You want blue desserts? Why not use the flowers of butterfly pea? Green shortbread? Incorporate moringa leaves. Perhaps a naturally fizzy, high vitamin C and healthy fruit juice? Try the astringent fruit kedongdong. Can we make a local but authentic biscotti? Incorporate bambara groundnut flour. As well as developing these new versions of familiar foods, CFFRC tested and integrated information about their nutrient and micronutrient content, food safety, shelf life, and how production technologies could be transferred to make products from other crops. Each proof-of-concept product prototype is available for those food manufacturers and industry leaders who *get it* and are willing to grasp the opportunities of producing a multitude of nutritious, tasty, locally sourced, and marketable foods from diverse ingredients.

The best example of such visionary leadership came through our link with Chris Langwallner and his WhatIF team in Singapore.[21] In 2015, Chris and his colleagues, who had been working for large agrifood companies, decided that instead of using their expertise to formulate calorie-rich but nutrient-poor food products using artificial additives, flavour enhancers, sugars, and refined oils, they would set up their own company to make *healthy comfort foods* from what they called *future fit* crops — nutritionally dense, climate resilient, and resource efficient. Their WhatIF brand has developed familiar foods such as instant noodles using ingredients from bambara groundnut and moringa to enhance their nutritional profile and add natural flavours.

By linking these ingredients with proprietary technologies, WhatIF now supplies air-fried noodles with less than 50 per cent of the fat of conventional brands from crops that, as well as producing nutritious ingredients, can revitalise marginal lands and support livelihoods for small-scale growers in Asia and Africa. CFFRC's research on functional ingredients allows commercial partners to make novel foods that drive demand for underutilized crops by replacing conventional and imported ingredients. Of course, the same principles of blending functional ingredients can be extended from using insect meal to replace fishmeal for aquaculture feeds to comparable feed formulations for poultry and other livestock. Using insect meal and ingredients from underutilized crops, fish and livestock systems can also generate livelihoods for farmers tending marginal landscapes who can grow climate-resilient crops as sources of feed — the fish and fowl won't know the difference but the planet will.

Changing the composition of human foods and animal feeds whilst not asking consumers to change their behaviour is an achievable and desirable objective using novel ingredients, modern technologies, and imaginative promotion — the destination is in sight. However, persuading people to eat *unfamiliar* or *forgotten* foods requires a change in behaviour and in how people think about food — a journey without a clear route map. According to the famous Chinese proverb, *a journey of a thousand miles begins with a single step,* the Forgotten Foods Network was a modest first step to change attitudes to food. Its principle was that if we sought the knowledge of earlier generations, we could find forgotten foods that were once part of our own cultural identities but which we have somehow lost on our way to modernity. In our quest to rediscover forgotten foods, we soon found many allies who were already delivering this vision. Some of them were simply inspirational.

The Lexicon of Sustainability was founded by social entrepreneurs Douglas Gayeton and Laura Howard-Gayeton in 2009.[22] They are storytellers who use their expertise in visualization through film, imagery, and words to create narratives around how behaviour change can help build more resilient food systems and meet global challenges. Based in California, the Lexicon is an international collective that links public and private partners, entrepreneurs, food producers, researchers, activists, and experts from around the world. It uses storytelling to mobilize actions to tackle the grand challenges of the global food system. By showing the story behind powerful ideas, Lexicon encourages people to think about what they eat, what they buy, and how they can contribute to a healthier, safer food system. Storytelling is as old as humanity and a skill that can be learned. The Lexicon Lab provides master classes in storytelling as part of Slow Food's master's programme in food communications based in Pollenzo, Italy.

Since 2017, CFFRC has engaged with colleagues in the Global Forum on Agricultural Research and Innovation (GFAR), on a Collective Action on Forgotten Foods. GFAR is the largest global multi-stakeholder network of over 600 organizations involved in research and innovation in agriculture — an ideal platform for collective actions that are managed and delivered by its members. GFAR acts as a catalyst for partnerships, mentoring, learning, and knowledge sharing. Forgotten foods have immense cultural and consumer significance, many dietary and nutritional benefits, and can increase system resilience to climate change by maintaining agroecosystem and landscape diversity. For any truly collective action, however, we need to reframe agrifood research and innovation around the very communities that have conserved forgotten foods so that they become the agents of innovation. This is not easy. Large-scale collective efforts on forgotten foods are complicated by the nature of the many crops, species, and contexts involved, their individual scaling and processing issues, and the marginalization of many of the communities who produce and consume them. We need common actions that;

- raise awareness of the economic and nutritional value of forgotten foods
- make them accessible in attractive readily-transportable forms
- empower farmers and consumers to realize their benefits
- increase system diversity in farmers' fields
- encourage scientists and the media to demonstrate their benefits
- have adequate policy and legal frameworks to ensure their conservation and use, and
- promote their adoption as viable foods for both rural and urban consumers.

To achieve these ambitious aims we need an Action Plan for Forgotten Foods and a narrative of how they can contribute to better, more diverse diets, and global sustainable development goals.

Of course, it's not just a good narrative that will change behaviour — dietary choice is driven by a multitude of factors that include price, availability, convenience, taste, and familiarity. But our food choices can also be influenced by associations with cultural identity where these differ from a homogenized Western diet based on ready meals and processed foods. Modern educational systems no longer teach students how to cook, where to source ingredients, or how to trace the geographical and cultural origins of their food. Nor are these skills passed to younger generations who remain unaware of their own traditional food cultures. The result is that food is seen as a joyless quest for cheap calories consumed alone in front of a computer screen and not as part of a rich cultural heritage or a desirable way of life. An exception is the Mediterranean diet which, in 2013, the United Nations Educational, Scientific and Cultural Organization (UNESCO) included in its intangible cultural heritage of humanity list.[23] As well as wonderful foods, the Mediterranean diet emphasizes values of hospitality, neighbourliness, intercultural dialogue, creativity and a way of life guided by respect for diversity. It plays a vital role in cultural spaces, festivals, and celebrations, bringing together people of all ages, conditions, and social classes. Such expressions of social values beyond the food itself encourage the transfer of dietary behaviours and practices across generations and an understanding of culinary culture.

While the Mediterranean diet is now seen as the global benchmark for a sustainable and healthy diet, it should not be seen as a solution for all societies. Foods and ingredients that are eaten in the Mediterranean diet might not be available, accessible, or culturally appropriate elsewhere. A healthy, diverse diet is not a one-size-fits-all solution — by advocating a universal Mediterranean, or indeed any global reference diet, we risk replacing one form of cultural hegemony (the Western diet) with another. Sustainable and nutritious diets need not be the same, but can be built around common principles that are adapted to local agroecological conditions, ingredients, and cultures. ASSESSCROP provides an evidence-base that allows individuals and communities to choose local ingredients that achieve the same objectives of health and nutrition.

In response to increasing consumer interest, food companies, processors, manufacturers, retailers, and food service providers have begun to incorporate more biodiversity into their supply chains.[24] Consumers increasingly recognize that the products on their supermarket shelves are based on the same limited ingredients that are transformed by food technologists into countless permutations of shape, size, colour, and texture. They are marvels of product development and chemical engineering but they are not biodiverse — their diversity only starts in a laboratory or factory operation. The challenge for food companies is to demonstrate biodiversity from fields where crops are grown, supply lines, ingredients, and the products that they sell. Without biodiverse supply chains we can't have biodiverse products. Without transparency along the whole value chain, food companies, producers, and consumers remain unsure of what is in their products and, without tools, they cannot demonstrate the biodiversity of their products.

Launched in 2013, the African Orphan Crops Consortium (AOCC) and its partner initiative, the African Plant Breeding Academy (AfPBA), provide the most comprehensive crop improvement effort across Africa directed at national crop breeding programmes. They are supported by a global consortium of partners that includes 29 government organizations, scientific and agricultural bodies, universities, private and public sector companies, regional organizations, and NGOs. A sign of the growing interest in underutilized crops is that this consortium includes Mars Incorporated, University of California Davis, World Wildlife Fund, Beijing Genomics Institute, and the Center for International Forestry Research and World Agroforestry (CIFOR-ICRAF), which hosts AfPBA and AOCC laboratories in Nairobi.[25] The AOCC's goal is to sequence, assemble, and annotate the genomes of 101 traditional African food crops to allow scientists to improve their nutritional quality, productivity, climate resilience, and disease and insect pest resistance. The 101 crops, selected by surveying African women's groups, agricultural scientists, sociologists, anthropologists, nutritionists, policy makers, farmers, government representatives, universities and other stakeholders, are familiar in households across the continent. Three criteria were used to select the crops; they must be rich in vitamins, minerals, micro- and macronutrients; relevant to pan-African agriculture; and limited by a lack of genetic and breeding research. AOCC will publish all sequence information so that it can be freely used by any scientist or organization. As part of this effort, CFFRC worked with AOCC to sequence the bambara groundnut genome as an exemplar for the genetic characterization of other underutilized crops.

The CFFPLUS DTP built a critical mass of talent and future research leadership, however, the immediate challenge for CFFRC was to demonstrate that its own research performance on underutilized crops was comparable with that on mainstream crops and other international centres. We did not try to compete with the sheer number of publications produced by long-established institutions, but instead focused on the quality of our research. By 2017, CFFRC's 'Field Weighted Citation Impact' was 1.77 (the normalized average for any subject is 1.00), and by 2019, it had achieved research ratings that were comparable with leading international research centres. Despite the challenges of research and submission, CFFRC showed that publications on underutilized crops can be comparable in quality with the best in mainstream science.

Much of the engine for CFFRC's publications used CropBASE which includes genetic, agronomic, and nutritional data from external databases as well as CFFRC's own research. CropBASE has also been used to provide information on underutilized crops for decision-support tools for major crops and land-use systems. Its SELECTCROP tool is now part of the EU LANDSUPPORT programme in which CFFRC joined a consortium led by Professor Fabio Terribile at Federico II University, Naples, Italy. The consortium includes 17 EU partners in an ambitious effort to provide decision-support tools to evaluate trade-offs between agricultural, environmental, and policy challenges of different land uses.[26] Through SELECTCROP, CFFRC is building maps of the potential cultivation of underutilized crops across European locations. In return, through LANDSUPPORT CFFRC gained access to high-level EU land-use decision support platforms and databases to inform land use and crop diversification options that include underutilized crops.

The link between SELECTCROP with LANDSUPPORT shows how information on underutilized crops can be incorporated into land-use policies and decision-support systems for mainstream agrifood systems. What it doesn't reveal is how weak the available information on agrifood systems actually is — both in the quality of data and links between agricultural productivity, nutrition, and factors such as climate change. Lawrence Haddad, who in 2018 received the World Food Prize with Dr David Nabarro for their work on hidden hunger, leads the Global Alliance on Improved Nutrition (GAIN) and has recognized the dearth of trustworthy data on food and nutrition. Haddad and his co-author Jessica Fanzo of Johns Hopkins University have now launched the Food Systems Dashboard that links available data on factors such as food waste and greenhouse gas emissions to food security and agricultural productivity.[27] The dashboard includes 170 indicators that cover nearly every country in the world. The need and potential significance of the dashboard has never been greater, and so too is the opportunity to include indices such as the Agrobiodiversity Index that was launched by Bioversity International to measure biodiversity across nutrition, agriculture, and genetic resources and the EAT–Lancet report that seeks to provide the best available evidence to determine a universal reference diet that is healthy for both humans and the planet. Again, CFFRC's CropBASE is a powerful resource to link underutilized crops with these important global initiatives.

Perhaps the most exciting aspect of the last decade has been how CFFRC's work and knowledge systems on agricultural diversification have coincided with a growing enthusiasm for a different kind of agriculture. This new vision extends beyond agriculture in farms and fields to include urban and vertical spaces, digital technologies, Internet of Things (IOT) applications, novel products and disruptive business models. Its advocates and practitioners extend beyond traditional agriculturalists to include the tech-savvy, well-educated, globally-connected, and socially-conscious 'Millennials' and 'GenZs' who now make up the largest global demographic of humanity. Rather than formal qualifications and conventional approaches these generations seek real time guidance, evidence and technical support that empower them to innovate in the agrifood space and help them to make connections with investors and partners to scale their impact and build sustainable agribusinesses. Such support needs a different kind of research and innovation platform that goes beyond academia and commercial advice. Perhaps its best example is the Thought For Food (TFF) platform that was launched through the single minded vision, dynamism and inspiration of its remarkable founder Christine Gould. Whilst there are a growing number of initiatives to support the proliferation of agritech startups none has had the global impact of TFF. In less than a decade TFF has built a community of over 25,000 Millennial and GenZ innovators from over 170 countries, helped launch over 60 startups that have collectively raised $200M, and accelerated over 5,000 new agritech ventures.[28] In addition to the support and mentoring that it has given to fledgling agritech enterprises, TFF has transformed how information is provided, shared and updated for next generation agriculturalists. The TFF Digital Labs provide a mobile-optimized digital platform that shares knowledge across the TFF community freely, to anyone, from anywhere at any time. It is exactly the sort of partnership that can help take the research and development that CFFRC and its partners initiated to enable a new generation of agriculturalists to deliver commercial and societal impacts of agricultural diversification.

Final Thoughts

Without the financial commitment of the Government of Malaysia and academic support of colleagues at the University of Nottingham, the CFFRC experiment could not have been possible in Malaysia. Just imagine if every country followed Malaysia's example and contributed to a global effort to transform agriculture with forgotten crops, forgotten foods and forgotten and new knowledge. The funding provided by the Government of Malaysia was for seven years, but CFFRC continued for nearly 10 years. In that time, it can point to a number of achievements, but the most significant is that it contributed to a growing realization that our food systems have to change faster and more radically than we might imagine and many would like. The inspirational words of Professor MS Swaminathan that crops for the future was '*the need of the hour*' are never more relevant today. Indeed, it is the '*need of the moment*'.

This is not to say that the Plan B that agricultural diversification represents must immediately or indeed completely replace Plan A. Those who dismiss the wider adoption of underutilized crops often imply that its proponents are advocating the wholesále replacement of Green Revolution technologies and staple crops grown as monocultures. They are not. There are circumstances where mainstream agriculture practiced with strict adherence to environmental and ethical practices will continue to be the best option for much of the world's agricultural lands. However, attempting to impose top down uniformity on agroecologies and cultures to which they are increasingly unsuited will fail in the turbulent conditions that we and food systems face. What we need is objective evidence to equip policy makers and agriculturalists on what options are best suited to particular circumstances. These options will need to include a wider range of crops, cropping systems and business models. For such evidence we need to recalibrate our research priorities, for it to be applied we need a new generation of practitioners, for it to be facilitated we need political and institutional leadership and for it to have impact we need society.

The current pandemic is a final warning that we can't carry on consuming, transporting, and cultivating crops and farming exclusively in the manner of the last few decades. In a few weeks and across the planet, a virus 10,000 times smaller than a grain of sand has left the mirage of national and human exceptionalism in tatters. For some, the food system held up pretty well, but not for the poor, the vulnerable, and the sick — which is most of us. As we scramble for a technological fix to a societal problem, it is clear that we cannot vaccinate ourselves forever. Our best option is to increase our immunity to the pandemic by being healthy and living within planetary boundaries. The biggest comorbidity indicators for Covid-19 are all related to non-communicable diseases linked with poor diets — obesity, hypertension, and diabetes. We already knew that our food systems and diets were killing us, but we are now beginning to realize that they can also save us from the collapse of the human body and the societies that they inhabit. Amongst other behavioural changes, this will require a change in our attitudes to food, a planet with humour and diversity that can welcome Orson, and nothing less than the transformation food systems for good.

8

REVOLUTION NUMBER 9

Transforming Food Systems for Good

Context

Ode to Joy (All men shall be brothers)

First played in 1824, Ludwig van Beethoven's *"Ode to Joy"* is the final movement of his Symphony No. 9, considered by many to be one of the greatest pieces of music ever written. In Symphony No. 9 (also known as The Choral Symphony), Beethoven became the first major composer to include the human voice within a symphony. The piece employed an orchestra that was bigger than any other at the time and lasted for longer than any previous symphonic work. After Symphony No. 9, other composers started to use new forms of composition, experimented with large ensembles, introduced extreme emotion, and explored unusual forms of orchestration. The Ode to Joy text that Beethoven used was written by the German poet Johann Christoph Friedrich von Schiller in 1785. It was a poem that celebrated the unity of all mankind. Its first stanza translates from the German as;

O friends, no more of these sounds!
Let us sing more cheerful songs,
More songs full of joy!
Joy!
Joy!
Joy, bright spark of divinity,
Daughter of Elysium,
Fire-inspired we tread
Within thy sanctuary.
Thy magic power re-unites
All that custom has divided,
All men become brothers,
Under the sway of thy gentle wings.

This last chapter proposes a set of actions that illustrate the transformational role that the food system (and we) can play for the future of humanity and the planet. As with Beethoven's Ninth, some of these actions may seem unconventional at first, but they propose an approach to the transformation of food systems whose time has come. These actions are neither exclusive nor sequential, but I believe they are urgent if the agri-food system is to become a solution to the climate crisis rather than a significant cause of it. In so doing, it can become a force for good that goes beyond profit and a source of joy that goes beyond sustenance.

This chapter first examines how the global energy and transport sectors — that have until now been dominated by fossil fuels and the internal combustion engine — are being transformed, not because of, but despite their industries. It then considers options for the future of the global agrifood system, why *business-as-usual* is no longer viable in the face of the climate crisis and mass extinction, and why its transformation will most likely come from outside the modern farming and agrifood sector. It concludes with a call to action that proposes nine collective actions to transform our food systems for good — Revolution Number 9.

Transforming Energy, for Good

Energy Sector

Negative Externalities

The exploration, exploitation, production, and transportation of *fossil fuels* (gas, oil, and coal) are all considered to be extractive industries. Their purpose is to convert stored natural resources into energy through industrial processes that make them useful for human beings. Fossil fuels are derived through the decomposition of buried dead organisms that contain carbon formed many millions of years ago beneath the surface of the planet. Since these resources must be raised, often from deep beneath the Earth's surface, extractive industries disturb local ecosystems. Their products also disturb the planetary ecosystems far from where they are extracted or even used.

For example, a car built in Baden-Württemberg, Germany, driven in Beijing, China using fuel extracted from oil beneath the Gawar Field in Saudi Arabia adds carbon dioxide to the atmosphere of the entire planet — through atmospheric circulation — and at each stage of the process — through extraction, production, transport, and emissions. For as long as humans have engaged in extractive industries, they have considered each resource to be so plentiful, cheap, and accessible that alternatives were unnecessary. They have also considered the effects of exploiting fossil fuels of insufficient consequence to outweigh their benefits. After all, fossil fuels enable the manufacture and transportation of products to large numbers of people far from the site of their extraction or the factories where products are made. Fossil fuels are described as non-renewable resources because existing reserves are being extracted much faster than they are being formed. Extractive industries are therefore unsustainable beyond the exploitation of the resources that are already there. The era of fossil fuels that began with the Industrial Revolution is coming to an end — sooner rather than later, willingly or under consumer and investor pressure.

Extractive industries have underpinned economic growth, but often at a cost to both the health of human beings and that of the planet. These costs are sometimes called *negative externalities* because they are suffered by parties other than the producer or consumer.[1] For example, they might include the territorial and human rights of individuals affected by extraction, the ancestral rights of communities on whose land extraction occurs, or the consequences of environmental damage and loss of biodiversity to the planet.

Historically, extractive industries have been protected from the consequences of their actions either through military or colonial might, the economic power of their industry, or the political allegiances of those in charge of extraction locations. However, under pressure from civil society, non-government organizations, and more recently consumers and the wider public, extractive industries have increasingly attempted to justify their actions and reduce their harm. Much of the justification was initially championed through marketing campaigns and public relations exercises, but now these industries are increasingly required to meet more stringent operating standards by regulatory authorities. For this, they must demonstrate compliance with certain standards to satisfy investors and shareholders. There is growing private and public pressure to improve the environmental footprint of each extractive industry if they are to secure funding and subsidies and avoid negative publicity. Many private companies are now adopting less harmful methods, industry associations are disseminating best practices and external agencies are providing guidance on better environmental management. Meanwhile, governments and international agencies are working with environmental experts to refine policy and regulatory frameworks that govern the operations of the energy sector.

Environmental, Social, and Governance (ESG) criteria are increasingly being used to screen potential investments in extractive and other industries. Environmental criteria include how a company uses energy, manages hazardous waste and pollution, conserves natural resources and protects biodiversity. They may also be used to evaluate environmental risks and mitigation strategies to reduce carbon emissions and comply with environmental regulations. Social criteria include how a company looks after the health and welfare of its employees and deals with suppliers, customers, and the communities where it operates. These criteria may also include whether a company donates some of its profits to benefit local or disadvantaged communities, supports Corporate Social Responsibility (CSR) programmes, or encourages employees to perform voluntary work. Governance criteria examine a company's leadership, executive pay, audits, internal controls, and shareholder rights, whether a company follows transparent accounting methods, and the rights of stockholders to vote on important issues. Investors may also seek assurances that companies have no conflicts of interest in the appointment of board members, refuse to make political contributions for special treatment, and don't engage in illegal practices such as bribery, modern slavery and child labour.

Many mutual funds, brokerage firms, and financial advisors increasingly offer ESG-compliant products. The driving force for adopting ESG practices is often from younger investors who wish to see that their investments are based on shared *values* as well as profit. The speed of industry compliance with ESG criteria has been remarkable.[2] The term ESG was first coined in 2005 at the *Who Cares Wins* conference in Zurich that brought together institutional investors, asset managers, research analysts, consultants, governments, and regulators to examine the role of ESG values in asset management and financial research.[3] By 2018, ESG investing was estimated at over $20 trillion or around a quarter of all professionally managed assets around the world.[4] The rapid growth of ESG investing builds on the Socially Responsible Investment (SRI) movement and is sometimes known as *sustainable investing, responsible investing, impact investing*, or *socially responsible investing*.

For years, socially responsible investments were seen as less profitable because they restricted the types of companies that were eligible for investment and often performed worse in terms of stock price than companies that had little regard for ethical practices. More recently, however, investors have come to realize that ESG criteria have useful purposes beyond ethical concerns. For example, they allow investors to screen and avoid *risky* companies whose practices may lead to environmental impacts that damage reputation and stock value, and might potentially lose billions of dollars. Even beyond ethical considerations and risk reduction, as ESG-compliant business practices have become more widespread, investors are increasingly finding that they may be more profitable than conventional *yield-for-profit* companies. Green practices are no longer seen as just a form of *green washing*. They actually improve a company's bottom line as well as being a force for good.

While ESG-guided capital markets have had an important role in improving best practices, encouraging sustainability, and even generating greater profits, the reality is that in the future ESG financing will determine if and how much investment funding will be made available to extractive industries. Risk factors such as climate change are no longer viewed as financially immaterial, so the risk-reward ESG *tap* has become clear. When raising capital, higher ESG risk directly increases the cost of capital for firms pursuing unsustainable practices in the form of higher interest rates on loans and lower investment potential through raising equity. As large institutional investors are increasingly mandated to incorporate ESG factors into their fiduciary duties, there has never been greater shareholder pressure on extractive firms to add *sustainable value* to better align their actions with strict shareholder objectives.

Positive Externalities

So far, extractive industries, such as oil, gas, and coal, have reduced negative externalities through a focus on *doing less damage to people and planet* with improved practices, adherence to environmental, social, and governance criteria and values-based investments. By doing less harm, companies are already seeing that they can attract investors and increase rather than reduce their profitability. However, reducing negative externalities doesn't require them to change their core business — an oil and gas company can still extract oil and gas, but must do so more responsibly. Business practices may be improved, streamlined, or even replaced, but the core purpose, competencies, and markets of the company remain unchanged. Such companies are still extracting, selling, and using non-renewable natural resources for the benefit of their shareholders.

A *positive externality* exists if the production and consumption of a good or service benefits a third party not directly involved in the market transaction. For example, governments may choose to invest in education because it directly benefits the individual, has tangible benefits for their citizens, and potential benefits for society in general.[5] At least in principle, as societies become more educated they are less likely to damage themselves, others, and the planet. In this sense, education is a *global public good* with benefits and/or costs that potentially extend to all countries, people, and generations. For extractive industries such as fossil fuels and their consumers, their greatest positive externality would be to stop extracting and using fossil fuels and find viable and less polluting alternatives. By its very 'nature', the extraction of fossil fuels damages nature. By their very use, consumers of fossil fuels contribute to the harming of nature, themselves, and future generations.

Since 1970, global carbon emissions from fossil fuels have increased by about 90 per cent. Emissions from fossil fuel combustion and industrial processes contribute about 78 per cent of the increase in total greenhouse gas emissions between 1970 and 2011.[6] By economic sector, carbon emissions were from: electricity and heat production (25 per cent); agriculture, forestry and other land use (24 per cent); industry (21 per cent); transport (14 per cent); residential and commercial buildings (6 per cent) and 10 per cent from other energy uses that include fuel extraction, refining, processing, and transportation.[7] From these figures, we can see that replacing dirty and non-renewable fossil fuels with clean and renewable alternatives would have the single biggest beneficial impact for protection of the planet and its inhabitants. So, why don't we do this?

Renewable energy comes from natural sources or processes that are being continuously replenished rather than via the extraction of finite stored reserves. The principal sources of renewable energy include solar, wind, hydro, tidal, geothermal, and biomass energy that are used to generate electricity or directly drive machinery. Not only are they replenished, many of these sources are *free* — the sun shines, the wind blows, tides move, and water falls. So, if renewable energy sources are generally free, clean, and sustainable, why wouldn't we transfer all the investments, subsidies, and promotional efforts directed at traditional extractive industries into replacing non-renewable, increasingly expensive, dirty, and unsustainable energy sources with their renewable alternatives?

In response to calls for a transition to diverse forms of renewable energy, the traditional answer from the fossil fuel sector has essentially been, *"if they were any good we would be using them by now."* When shown that there are methods of generating electricity other than mining and drilling for fossil fuels in specific locations with declining reserves, the predictable response of the energy sector is that, *"they are not economically viable compared with fossil fuels"*. When shown that scaling up production dramatically reduces the cost of renewable energy, the industry response is almost always that alternative sources *"are unreliable"* since the sun does not shine and the wind does not blow all day and every day. Leading fossil fuel companies claim that they have pumped billions into *clean energy* projects, however, in practice their combined budget on green energy schemes has been a trivial fraction of their combined earnings. Even after public, investor, and regulatory pressure, the oil and gas sector has not fundamentally challenged its core purpose of extracting diminishing reserves of oil and gas from beneath the Earth's surface. Meanwhile, the world has moved on. By 2019, renewable energy investment hit \$288.9 billion, far exceeding fossil fuel investment.[8] Appropriately, the fossil fuel sector is now seen as a *sunset* industry.

As with the extractive oil and gas sector, the automotive industry has remained wedded to the use of fossil fuels until recently. Again, arguments of cost, scalability, and technology were used to justify why carbon-emitting combustion engines were preferable to cleaner alternatives, which were again said to be, *"not economically viable compared with fossil fuel using vehicles"*. In fact, electric vehicles have been around since the mid-19th century. However, the lower cost, faster speeds and greater range offered by 20th century internal combustion engine vehicles led to a worldwide decline in the use of electric motor vehicles.

Nevertheless, in 2004 Tesla Motors began development of the Tesla Roadster after Elon Musk joined the company and invested $6.3 million in Tesla stock. Subsequently, the company received a $465 million low-interest loan from the Obama Administration.[9] By March 2020, the Tesla Model 3 had become the world's best-selling electric car with more than 500,000 sold, and Tesla became the first manufacturer to produce one million electric cars. Senior leaders at several large automakers, including Nissan and General Motors, have acknowledged that the Roadster demonstrated that there is significant consumer demand for more efficient vehicles which they had noted but not themselves tested. In an August 2009 edition of *The New Yorker*,[10] the General Motors vice-chairman Bob Lutz was quoted as saying;

"All the geniuses here at General Motors kept saying lithium-ion technology is 10 years away, and Toyota agreed with us — and boom, along comes Tesla. So I said, 'How come some tiny little California startup, run by guys who know nothing about the car business, can do this, and we can't?' That was the crowbar that helped break up the log jam."

In July 2020, Tesla's stock valuation was more than $208 billion and in August 2020 it peaked at over $400 billion, which made it the most valuable car company on the planet — a pretty large crowbar and a pretty good investment.[11]

The above examples for energy and transportation illustrate how transformational changes, sometimes called *disruptive innovations*, occur from outside the industry. The fossil fuel sector didn't decide to transform itself into a renewable energy industry, nor did car makers choose to switch from combustion engines to electric motors. In each case, these industries had the skills, technologies, and money to do so, but external forces — often consumers and end-users — were the agents of transformation. In the same way, the hotel industry didn't invent Airbnb nor did the taxi industry invent Uber. Once offered, consumers decided that clean energy, electric vehicles, flexible transport, and accessible accommodation were preferable choices for reasons that include ethical concerns as well as economic and convenience factors. Industry initially dismissed, resisted, and then followed change. Public-relations-driven efforts to reduce negative externalities were no longer good enough. The transformations that we have seen in the energy and transport sectors are unlikely to be reversed — renewable energy and electric vehicles are increasingly seen as 'the new normal'. What about the global agrifood system?

Transforming Food Systems, for Good

Global Agrifood System

Negative Externalities

Agriculture is perhaps the best example of *renewable* energy in that it converts *free* natural resources, principally soil nutrients, rainfall, carbon dioxide, and sunlight into primary products (plants) and secondary products (animals) that are useful or desirable for humans. Agriculture is also a global public good in that its products benefit individuals, countries, and humanity in general. However, the global agrifood system is increasingly seen as an extractive industry designed solely for the purpose of generating profits for its shareholders. Why is that? It is extractive in that it uses fossil fuel at each stage of the value chain: production (fuels in the manufacture and use of machines, equipment, agrochemicals and fertilizers); transport (carbon-heavy supply chains); manufacturing (processing of ingredients into products) and consumption (preparation, cooking and heating or cooling of prepared products).

Agriculture is also extractive in that it *mines* the surface soil layers of the planet for mineral nutrients. These can be replaced by fertilizer applications (that use fossil fuel in their manufacture), but each operation that disturbs the soil during the crop cycle damages the biology of the local ecosystem. It is an industry in that it converts natural resources into foods, feeds, and fuels in specific locations and on a scale that allows them to be transported long distances on carbon-heavy supply chains for use elsewhere. The global agrifood system is increasingly seen as a factory operation both in the manufacturing of food (factory foods) and in the manufacturing of their raw materials (factory farms) on an industrial scale.

Industrial agriculture includes intensive crop management to maximize the yield of desirable parts of plants (usually those containing calories).

and the intensive rearing of animals to maximize the yield of desirable products (usually meat, milk, eggs, or skin). The terms used in agriculture also increasingly resemble those used in factories and industrial manufacturing. For example, *battery* cages are used for various methods of animal production, including: egg-laying hens; mink, rabbit, chinchilla and fox in fur farming, and the Asian palm civet for *kopi luwak* production of coffee. The term *battery* relates to the arrangement of rows and columns of identical cages connected with common divider walls, as in the cells of a battery. Similarly, animals such as cattle, pigs, and chickens are often raised in feedlots whose purpose is to increase animal body weight as quickly as possible. To enhance their production efficiency, animals are kept in confined spaces so that they will maximize weight gain, and in doing so deliver economies of scale, minimize costs, and maximize profits. Feedlot agriculture keeps livestock at high densities and uses modern machinery and intensive breeding programmes to optimize the configuration and properties of the livestock for large-scale production.

Like fossil fuels, industrialized agrifood systems drive economic growth at a cost to our health and that of the planet. Their negative externalities can similarly infringe the territorial and human rights of individuals affected by farming and the ancestral rights of communities on whose land farming now occurs, along with detrimental consequences for the environment and biodiversity. As with fossil fuels, there is also increasing private and public pressure to reduce the environmental footprint of agriculture, for example, by reducing the use of agrochemicals, fertilizers, and groundwater pollution. Many agribusinesses are adopting less harmful methods, and external agencies are providing more robust policy and regulatory frameworks to govern the agrifood sector. Moreover, modern farming methods are increasingly seen as unethical and even immoral, especially in relation to the welfare of farmed livestock and farm workers. In response to consumer pressure there has been a move towards less intensive livestock systems, although these changes and their interpretation have varied across the world. Take for example battery cages for laying hens. In 2014, around 95 per cent of eggs in the USA came from battery caged hens,[12] whereas by 2012 they had been banned in the European Union. This led to a reduction in the number of eggs from battery cages in Europe and created the false impression that all laying hens in the EU would either be free-range or housed in barns. In fact, illegal conventional battery cages have been replaced by slightly bigger cages fitted with *enrichments* such as perches.[13] The hens in these bigger cages are now called *ex-cage colony hens*. Animal welfare organizations such as Compassion in World Farming claim that such cages provide no worthwhile welfare benefits over conventional battery cages. Within the EU, standards also vary. For example, in 2007 Germany prohibited conventional battery cages. This was five years before the requirements of the EU directive, and the use of enriched cages from 2012. Its consumers have demanded better welfare standards and the industry and government have had to oblige.

Agrifood actors justify intensive livestock production by arguing that they have no desire to be cruel but that they have no choice. They note that the public has come to expect cheap food and is unwilling to pay a premium for practices that safeguard the environment, improve livestock living conditions, or enhance the employment and working conditions of farm workers. Supermarkets insist that cage-free hen eggs will cost more and that consumers will simply switch to cheaper eggs from competitors. Similarly, countries that enforce higher welfare standards for poultry will be swamped with cheap imports of poultry products from those with lower safety or ethical standards. Of course, similar arguments apply to products such as organic fruit and vegetables, fairtrade tea and coffee, and wild-capture seafood. We can continue to argue the benefits of cheap food against environmental and ethical costs, but until now, the global agrifood system has chosen to reduce its negative externalities (that is, to do slightly less harm to people and the planet) than to fundamentally transform itself.

Business as Usual

There are a few options that the global agrifood sector has identified to face the future while not disrupting its current model. The first is to continue with *business as usual* in which market forces determine any changes to current production systems and consumption patterns. The argument here is that *we have always found a solution to societal challenges before and with human ingenuity and new technologies, something will turn up — it always does.* Advocates say that intensive agriculture and industrial food systems have served us well until now and that rumours of their imminent demise are premature or even politically motivated. Food has never been cheaper, more accessible, and more plentiful, and as demonstrated during the Covid-19 pandemic, the global food system has remained intact, even if the long-term consequences of the pandemic are unclear. Nevertheless, the argument is that any changes in the food system should essentially be market-driven. The system will self-regulate in the face of new challenges, *don't panic, you can't buck the* super *market.*

A second option is to improve the current agrifood system through new technologies, incentives for better practices, and penalties for bad behaviour. *Better business as usual* includes: greater efficiencies in production and consumption through mechanization and reduced labour; use of digital technologies; traceability metrics; 'block chain' decentralized ledgers that record the provenance of each asset along the supply chain and better education for growers, processors and consumers to help reduce waste and eat healthily.

Perhaps the most significant gains will be by reducing supply-chain losses and food waste, which according to the FAO are estimated at 1.6 billion tonnes, with a global cost of $750 billion annually.[14] There are also exciting new approaches to develop *circular economies* in which raw materials, components, and products lose as little of their value as possible and renewable sources of energy are used. The entire system is considered as a series of closed loops that minimize resource leakage at each stage. We can nudge change by thinking about the entire agrifood system to see where we can reduce waste.

Option three is to maximize agricultural productivity per unit land area through a combination of higher-yielding crops and animals, and the use of greater inputs to meet higher potentials.

This *supersized business-as-usual* model argues that in much of the world, yield gaps are primarily caused by a lack of inputs and poor management practices that create yields far below current potential.[15] At the same time, novel genetic and digital technologies offer the prospect of new opportunities to break the current yield ceiling of conventional crop breeding, as well as secure new varieties that are more resistant to insect pests and diseases, and may even be able to withstand the effects of climate change. Proponents of intensive industrial agriculture argue that by increasing production on the same area of land and maximizing productivity in benign climates with fertile soils, we can feed the increasing global population without encroaching on the remaining areas of forests and natural ecosystems. Further intensification of the agrifood system will meet our future needs.

A fourth option is *novel business-as-usual*, which uses new technologies to develop novel foods that are more nutritious or biofortified with additional micronutrients, less harmful to the environment yet still desirable to consumers. An example is replacing animal protein with proteins from plants, algae, and insects to develop meat-free processed products that may taste as good or even better than animal-based products, as well as help reduce carbon emissions from livestock and release large areas from extensive grazing. These alternatives could also address the ethical issues of animal welfare and environmental pollution caused by intensive livestock production. However, there is the risk that under the guise of alternative proteins we are simply replacing extensive ruminant grazing by clearing new land for intensive cultivation of protein-rich crops such as pea and soybean grown as high-input industrial monocultures. Nevertheless, the potential to link ingredients from plants, insects, algae, fungi, and microbes with state-of-the-art technologies, such as fermentation, is enormous. The future lies in delivering new food technologies to recalibrate the food system and reduce the impacts of agriculture on the environment rather than replacing one industrial globalized food system with another.

The options described above are not exhaustive, mutually exclusive, or without merit. Indeed, many will say that they don't represent business-as-usual at all because they include innovative elements. Each option can lead to healthier outcomes for consumers, new business opportunities for food producers and processors, and environmental benefits. However, none of these options challenge the very essence of the current global agrifood system; they are not *transformational* in that they fundamentally seek technological solutions to societal challenges while maintaining the *status quo* to the maximum extent possible.

Climate Changes Everything

Climate Change, Food, and Population

The climate crisis, biodiversity destruction, global health crises, and further pandemics mean that all of our current assumptions about business-as-usual are either inadequate, obsolete, delusional, or fundamentally flawed. We will have to change our approach and strategies if our way of life, and indeed many of us, are to survive. The future of food is much more complex than we have assumed it to be. The changes needed are so significant that they will require nothing less than a revolution in the agri-food system. We need look no further than the existential threats posed by climate change to see why.

The beginning of agriculture was probably caused by a major change in climate that occurred after the last ice age (c. 11,000 BCE).[16] Its end — at least in its present form — will almost certainly be caused by the major climate changes that are happening now. The first occurred without human intervention; the second is occurring because of it. In the few thousand years between these two planetary events, human beings have changed the world more than at any other time in our previous millions of years on the planet. Before the age of agriculture, our ancestors survived through a combination of hunting animals and gathering plants to provide sustenance, shelter, and clothing. Perhaps without the external disruption caused by sudden changes in climate all those years ago, humans would have continued to live a peripatetic lifestyle in natural ecosystems with little cause to settle in one place and establish sedentary societies around agroecosystems — *settling down* appears to have unsettled us!

The biggest human impact on the surface of the Earth has been the spread of our agricultural footprint, an impression that goes far beyond those who farm and consume. For example, modern agriculture affects almost all other species. Wendell Berry, author of *The Unsettling of America; Culture and Agriculture* described eating as *an agricultural act*.[17] In other words, if we eat we are involved in agriculture, and that means all of us. For decades Berry warned of the dangers of rampant consumerism, industrial agriculture, and the dissolution of rural communities. His warnings were prescient even before the term climate change was coined; they are all the more important now for a world in crisis. Berry proposed a *50-Year Farm Bill,* setting out a long-term vision that would exactly reverse the present ratio of 80:20 annual cropping to perennial agriculture. By reversing this ratio, most agriculture would be based around perennial and regenerative systems in which soil cultivation would be limited to the least vulnerable land, soil erosion and chemical pollution would be radically reduced, and soil health gradually restored. The goal is to enhance the resilience of the agricultural system to climate shocks and changing weather patterns whilst retaining productive potential of land.

Perennial agriculture can build on the principles of regenerative agriculture as espoused prominently by the Rodale Institute, Pennsylvania, USA, among others. The underlying principles of regenerative agriculture are also demonstrated in a role model developed by the Land Institute near Kansas, USA, which is investigating how perennial cereals, legumes, and oilseeds can provide staple foods without disturbing soil ecosystems or increasing carbon emissions through tillage. As well as food production from perennial species, such *no-till* systems can increase carbon sequestration by adding to and preserving soil organic matter, maintaining soil health by retaining its microbial content, and enhancing soil nitrogen without mineral fertilizers through the use of leguminous *cover crops.*

Most countries, scientists, and members of the general public increasingly accept that climate change is a reality. Even in the USA, whose then leader withdrew from the Paris Agreement, most of the general population accepts that the world is getting hotter and agrees that human activities are the primary cause of climate change. Over 70 per cent of Americans think global warming is happening, an increase from 60 per cent since March 2015, and 62 per cent understand that it is mostly caused by human activities.[18]   However, only 46 per cent of the US population believe that climate change will affect them personally — a remarkable level of insouciance amongst the majority of Americans. It is likely that even fewer recognize how their own food choices affect not just themselves but the whole planet.

When linked with the use of fossil fuels across the whole value chain, the global agrifood system is now the largest contributor to carbon dioxide emissions. Rather than its major cause, agriculture could be our greatest ally against climate change. This could be through reduced carbon emissions, carbon sequestration, and mitigation of its worst impacts — all of which require a repurposing of our food system. For this, we will need radical solutions, not because we may *want* them but because we *must* develop them. These solutions need to include new options that make the global agrifood system more resilient and less destructive while still meeting the needs of a growing global population. We can no longer assume that the impacts of climate change are *far away, in the future, and affect other people*. They are *here, now, and existential* — at least for humanity. The extent and impact of extreme weather patterns and humanitarian crises are ever more frequent as more volatile climates push food insecure regions beyond their limits of endurance. We cannot simply reverse the impacts of climate change on agriculture; instead, we will have to adapt to uncertainties that none of the business-as-usual options can predict with any confidence but for which we need to be prepared.

Under current projections, the global population is estimated to reach 10 billion by 2050, levelling off at between 11–12 billion by 2100.[19] Within the next three decades, this means that we will need to increase food production by at least 70 per cent if we are to feed over 2 billion more people.[20] This represents a 56 per cent gap between crop calories produced in 2010 and those needed in 2050 if we pursue *business as usual*. In turn, this means a requirement of 593 million hectares more land (an area almost twice the size of India) between the global agricultural land area in 2010 and that expected by 2050. And lastly, we will have to close a 10 gigatonne greenhouse-gas-mitigation gap between expected agricultural emissions in 2050 and the target level needed to hold global warming below 2°C.[21] The Intergovernmental Panel on Climate Change (IPCC), which includes more than 1,300 scientists from around the world, forecasts a temperature rise of over 3°C by 2100.[22] Unless we take drastic actions — especially those involving food systems — we will not even come close to meeting commitments under the Paris Agreement of limiting atmospheric emissions of carbon dioxide.

Global heating and population growth will not be evenly spread across the world. Modelling suggests that they will occur concurrently in those areas that are already marginal for agriculture.[23] This means that we will increasingly need to grow more food, for more people, on a hotter planet. Indeed, much of the planet will be too hot, dry, or impoverished for major crops and livestock. However, these disadvantaged places are precisely where most people now live and where most children will be born.

How Did We Get Here?

Over the last 10,000 years or so, human beings have sought to establish complete dominance over the natural world. After the last ice age, much of the Earth experienced periods of drought and seasons of hot and cold weather rather than a stable and equable climate throughout the year. This meant that human beings had to cultivate crops whose stored products allowed them to survive when foods couldn't be gathered from the natural environment. Slowly, most human societies adopted a sedentary habit in which only some people were required to farm. To meet their needs, humans have increasingly replaced land occupied by *natural* ecosystems, within which humans moved in search of food, to *managed* agroecosystems in which they have corralled their plants and animals into designated structures and spaces. These plants and animals provide ingredients that can be harvested, slaughtered, processed, stored, transported, sold, cooked, consumed, or discarded.

Land for farming was increasingly demarcated from that for housing, transport, and other pursuits and the rest of the natural world was left largely undisturbed until such time that its resources were sought by human beings. After generations of encroachment onto natural ecosystems, agriculture now occupies almost half of the land surface of the planet that was once nature. A significant but difficult to quantify percentage of land has also been lost to nature in any meaningful sense, as well as to agriculture, because it was degraded through unsustainable agricultural practices. Such land lies idle as new deserts, salt flats, is disease infested, or barren wasteland.

Fragmentation of land did not occur only between farms and nature, but rather within farmed land itself. Diverse agroecosystems that resembled the complexity and functions of natural ecosystems gradually gave way to those designated exclusively for arable agriculture, livestock, and fish. From the almost limitless cornucopia provided by polycultures of plants (agroforestry, intercropping, relay cropping, compound cropping, and alley cropping, to name a few) and integrated agrisilvopastoral combinations of crops, trees, and animals, we now have one dominant system — monoculture. The outcome is that modern agriculture boasts the fewest species and least complex agroecosystems in our history. Typically, in modern agriculture, one crop variety or animal breed is grown without any neighbouring species of plants, animals, or insects, whether friend or foe. In a quest for efficiency, business, academia, and societies have followed a similar reductionist pattern of simplification by creating market forces, disciplinary silos, and political demarcations in which adjacent markets, subjects, and territories are seen as competitors if not enemies. These practices mimic the models and processes of industrialization — which, for example, almost never paid the price to society of atmospheric carbon emissions but considered them to be a necessary consequence of *modernity*.

The result is that we now rely on a handful of plant and animal species for most of the world's food. To achieve global *food security*, humans have selected and improved *the best* crops and prioritized their cultivation. This was done to optimize returns on investment of time and effort and to consolidate the area allocated for their uniform production through mechanized monocultures. In so doing, we have reduced the role of human beings, and some might say humanity, in the farming process in much the same way as the Industrial Revolution did to artisans through mass manufacturing. In industrialized economies, large capital-intensive farms of several thousand hectares are managed by around 1–4 per cent of the population.[24] In non-industrialized economies, small-scale farmers make up around 40–60 per cent of the population and they typically farm areas smaller than 2 hectares.[25] However, at least 70 per cent of the food humans consume is grown by the world's 1.5 billion small-scale farmers, and in many developing countries they produce up to 90 per cent of staple foods. While small-scale, often women, farmers are the backbone of global agriculture, the vast majority of funds and efforts to support research, development, innovation, marketing, investment, loans, and subsidies are focused on industrial scale agriculture in industrialized economies. There is also focus on the transfer of such systems, along with the aspirations that accompany them, to non-industrialized countries.

Complex Systems and Complicated Shadows

"It can scarcely be denied that the supreme goal of all theory is to make the irreducible basic elements as simple and as few as possible without having to surrender the adequate representation of a single datum of experience."

Albert Einstein

The above statement by Einstein is often paraphrased as *"Everything should be made as simple as possible, but no simpler."* Complexity has its limits not in its expansion but in its reduction, as it is with the industrialized agrifood system. How few crops and animals do we actually need? What is the minimum number of multinational food companies that can maximize economies of scale? How much can we expand the size of farms to minimize their number? How few farmers are required if we maximize mechanization and robotics? In this book I have argued that we have passed that critical minimal stage of simplification. Specifically, we now have *too few* crop and animal species, *too few* farmers, *too few* agricultural systems, *too few* supply chains and *too few* decision-makers. Beyond a certain level of simplification, something vital is lost to a system. In this instance, what we have lost includes trust, culture, taste, identity, knowledge, talent, nature, resilience, sustainability, and even joy.

The agrifood system has never been less complex, but for many consumers and the supply chains that serve them, its processes have never been more complicated. Imagine the complications for a supermarket that has to put thousands of products on its shelves *just-in-time* for consumers who, say, want a strawberry in December in the Northern Hemisphere. Not only does the supermarket have to source suppliers, ingredients, and availability, it must ensure that its supply chains work perfectly so that the strawberry arrives in peak condition and on time at the point of sale. Too early or too late risks wastage and lost profits as well as an unhappy customer who can choose to shop elsewhere. These *complicated shadows* mask the increasing simplicity and homogenization of the agrifood system and the research and education systems that underpin it.

To minimize risks, the agrifood sector uses sophisticated mechanisms to predict supply, logistics, shelf life, consumer preference, and a myriad of related factors. The system has increasingly replaced human beings, not just as a reprieve from the drudgery of stacking shelves and moving containers, but also in decision-making at each stage of the supply chain. Machines, sensors, artificial intelligence, and opaque traceability mechanisms all use algorithms to replace the judgment of human beings. In computer science, an algorithm is a *finite sequence of well-defined, computer-implementable instructions, typically to solve a class of problems or to perform a computation*. Algorithms work better than human judgment in complicated systems where there are *many moving parts*. For example, in a supermarket supply chain external forces such as transport systems and changes in supply and demand can be predicted through a set of interlinked processes, each of which can cause change but is predictable. Algorithms are less effective when the whole system changes in a complex and unpredictable manner. A competitor with a new business model — an Uber or an Airbnb — is a complex problem for which an algorithm is not ideal to determine the best response. Climate change is the most complex problem that humanity has yet confronted. No algorithms can tell us how to manage it; we will have to do that by ourselves. And climate change demands a different set of skills that equip us to adapt to *uncertainty* rather than those that we currently promote to manage *certainty*.

The Climate Crisis Requires Complex Solutions

We live in a world divided by wealth and poverty, over and undernutrition, peace and violence, disease and health, abundant and degraded natural resources, democracy and autocracy. Food production is central to each of these divisions and as a result it affects wealth, peace, nutrition, health, natural resources, and the stability of societies. People who cannot feed themselves are vulnerable to hunger, starvation, illegal migration, human trafficking, international drug smuggling, and terrorism. The failure of the food system in any country leads to increasing disparities and declining livelihoods, gender inequality, lack of education, poor health, decreasing labour productivity, weaker infrastructure, and possibly, insurrection. Food prices played a major role in the *Arab Spring* that started in 2010 but were part of a wider set of grievances leading to anti-government protests and uprisings against endemic corruption, low standards of living, and increasing wealth disparities.

In a fascinating study of the causes of food crises and political instability in North Africa and the Middle East, Marco Lagi and his colleagues plot the prevalence of food riots against rises in the FAO Food Index and show that spikes in food prices in 2008 coincided with over 60 food riots in 30 countries.[26] After a temporary drop, another price hike in 2010 and 2011 coincided with more food riots, widespread protests and government changes that led to the Arab Spring. The authors conclude that;

"in food-importing countries with widespread poverty, political organizations may be perceived to have a critical role in food security. Failure to provide security undermines the very reason for existence of the political system."

The problem is not that all of the above factors are *complicated* — which they are — but that they are *complex*. If they weren't, governments could employ clever data scientists to develop algorithms to predict exactly where and when, for example, food riots will occur. Such an algorithm might predict that a riot will occur at some time but not where or when since human behaviour is unpredictable and the causes of a riot are complex. In volatile times for the planet and humanity, algorithms are of little use unless communities and political leaders have the multiple skill sets to navigate unchartered territory and meet complex challenges.

Why the Yield-for-Profit Paradigm Won't Work in the Future

Like other extractive industries, our current food system is based on a *yield-for-profit* equation. In other words, the exchange of *calories for cash*. In oil, gas, and coal, the calories are in the form of fossil fuels, which through various industrial processes are converted by machines into *work* that is useful to human beings. In the case of the food system, the calories are in the form of foods that, through various metabolic processes, are converted by human beings into *work* that makes them useful. In each case, energy is lost along the chain from extraction to outputs. The *yield* is a measure of the conversion efficiency from extraction of calories to useful work; in other words, *how many calories are produced at the end of the value chain for the calories used throughout it?* In the case of agriculture 'crop yield' is often measured simply as the number of calories produced by a crop per unit of land area. In exchange for crop yield, each industry receives cash based on the value that markets place on the yield. After deductions for the cost of operations, this leaves *profits* (or losses), which may not be related to the number of calories produced against those used along the whole value chain. In intensive agricultural systems, the real energy costs are not only those directly involved in converting *fuel into food* but also include the indirect costs of energy such as that used to manufacture farm equipment, in delivery systems, or to produce fertilizers and other inputs such as pesticides and herbicides. Since the 1970s, Professor David Pimental at Cornell University has published ground-breaking research on the energy costs of industrial agriculture. In such systems, Pimentel calculated that the energy costs of producing food crops and biofuels were often greater than the amount of energy derived from them. In other words, the conversion ratio of extracted calories to useful work was negative.[27]

There is a substantive transformation of the energy sector happening now because renewable energy can provide the *yield* required by a growing population at a value that makes it affordable for customers and profitable for businesses. The speed of this transformation is astonishing. RethinkX, an independent think tank that analyses and forecasts the speed, scale, and social implications of technology-driven disruption, predicts that the world could achieve all of its energy requirements by 2030 through the convergence of three key technologies — solar photovoltaic, wind, and battery storage.[28] Its report, *100 per cent Solar, Wind and Batteries is Just the Beginning*[29] shows that the cost of solar photovoltaic panels has dropped by 80 per cent since 2010 and will fall by a further 70 per cent by 2030. Over this same period, the cost of batteries has fallen by 90 per cent and will fall by another 80 per cent by 2030. Comparable figures for wind power are 45 per cent and 40 per cent, respectively. The report predicts that the transmission costs of conventional electricity distribution from distant power stations will continue to increase, while those from solar panels will effectively be zero — you can't put a coal-fired power plant on your roof but you can install solar panels. Their conclusion is that, not only is 100 per cent clean and affordable electricity achievable by 2030, but that the extra *Super Power* generated will be enough to support positive externalities such as water desalination, electrified transport, and waste recycling. So, renewable energy can now produce more *yield* than that from declining fossil reserves at a lower cost to the consumer, with commercial profits for investors and businesses — and can be a force *for good*.

The transformation is inevitable. Interestingly, to transform the global energy system, the RethinkX report calls not for subsidies but for political leadership.[30] According to the International Monetary Fund, in 2017 subsidies on fossil fuels were $5.2 trillion (6.5 per cent of global GDP) of which coal and petroleum accounted for 85 per cent.[31] What consumers don't pay at the pump they pay in taxes, much of which goes to the fossil-fuel industry. Why?

For the agrifood sector, as noted, there are already serious problems with the yield-for-profit model, which increasingly produces negative externalities. These are only now becoming evident to many decision makers, but they are already apparent to those who will have to live with their consequences. For example, let's take crop yield. The cheapest form of energy conversion in agriculture is in the production of calories, which is why the food system is focused on making calorie-rich products from staple crops. In recent years across much of the world, the yield of staple crops has stagnated and in regions such as parts of Africa, they have declined. The problem with using calories as a measure of yield is that it removes the contribution of all other more *expensive* constituents such as proteins, vitamins, and minerals that are essential for a healthy diet. Since nutrients contribute relatively little to yield (i.e., the calorific weight of a crop), they can be ignored in its maximization.

The consequences are that, while relatively cheap, our calorie-rich diet leads to negative externalities such as the triple burden of malnutrition — over nutrition (obesity), undernutrition (underweight), and hidden hunger (lack of nutrients).[32] The number of overweight people now exceeds those who are underweight — both of which in principle can be reversed — but the most frightening consequences are those of hidden hunger, which leads to stunting, impaired brain development, and vulnerability to disease. In a world that is calorie-rich and nutrient-poor, over 20 per cent of children under the age of five are already stunted.[33] Those who survive will all grow up to be stunted adults, which amounts to a global population of 2 billion stunted adults — equivalent to the combined populations of

India, the USA, and Indonesia. The cost of bad diets such as obesity, diabetes, hypertension, and stunting are negative externalities for which individuals and taxpayers pay everything and the food industry pays nothing. Societies across the world incur the cost of non-communicable diseases, lost workforce productivity, family expenses, and early mortality in exchange for cheap, unhealthy food.

What about profit? Again, when calculating profits, the financial models of extractive industries do not take the full cost of negative externalities into consideration. For example, the natural environment and future generations pick up the bill for the loss of biodiversity, pollution, soil degradation, and damaged health. None of these are paid out of company revenue or shareholder dividends. The introduction of ESG criteria may reduce negative externalities, but they do not by themselves affect profits other than through marginal costs for regulation.

Why We Need to Transform Food Systems for Good

Until now, the industrial agrifood system has utilized a handful of species and a single production system to ensure that the increase in global food supply has continued to exceed that of population growth. Just three cereal crops — rice, maize and wheat — constitute the staple diets of over 4 billion people.[34] In response to concerns about the wisdom of using so few staple crops for so many people, the answer has been 'food has never been cheaper, more plentiful, and more available to more people on the planet'. So why change?

Human beings are just one of about 7.77 million animal species, 298,000 plant species, and 611,000 fungal species on Earth.[35] In 2019, the Global Assessment Report on Biodiversity and Ecosystem Services[36] estimated that around one million animal and plant species are now threatened with extinction, many within decades, and that since 1990 the average number of native species in most major land-based habitats has fallen by at least 20 per cent. The report estimated that over 9 per cent of all domesticated breeds of mammals used for food and agriculture had become extinct by 2016, and at least 1,000 more breeds are still threatened — a similar trend is occurring for domesticated crops and their wild relatives. In short, the rate of loss of biodiversity during the past 50 years is unprecedented in human history and its main cause is us; our global food system is the main driver of habitat loss and species extinction mainly through the conversion of natural ecosystems to crop production or pasture.[37]

Setting aside mass extinctions, desertification, pollution, and carbon dioxide emissions, our one-size-fits-all food system served us well in benign climates when human behaviour was more-or-less predictable. The mindset of proponents of industrial agriculture is often that, *we must help poorer countries and their marginalized and indigenous communities catch up and become more like us.* In so doing, research, development, and promotion focus on increasing the yield of staple crops. A good example is the aspiration to increase maize yields in Africa from a current average of 1.5 tonnes per hectare to 7.0 tonnes per hectare and, with new biological and mechanical technologies, reach even higher yields.[38] The logic of this model places the blame for the declining performance of African agriculture on the failure of its practitioners and governments to follow the formula that is already successful in other parts of the world. Meanwhile, crop yields in Africa continue to decline, the gap between potential and actual yield of staple food crops expands, and hunger, malnutrition, and poverty increase. The medicine is not working now and will become even more unpalatable and ineffective as climate change and pandemics threaten those least able to withstand their impacts.

Instead, we might learn from those communities that for millennia have been experts in surviving on the margins of climate and society despite politicians, scientists, advisers and investors. What climate resilience skills and knowledge of agricultural methods might they teach us? And in return, how can we provide new tools and technologies that could help them become the agents of change rather than its victims? For this we must diversify agriculture beyond the species and agricultural systems that have been promoted to marginalized communities and, rather, include those that they have cultivated and cherished for millennia and on which they, not we, are the experts and primary sources of knowledge.

Even *without* climate change, the yield-for-profit paradigm is no longer fit for purpose. *With* climate change, it will have to generate ever higher yields and larger profits while feeding more people on a hotter planet. A 2017 study[39] modelled the impacts of warmer temperatures on major staple crops. It predicted that each 1°C rise of temperature would reduce the yields of maize by 7.4 per cent, wheat by 6 per cent, rice by 3.2 per cent and soybean by 3.1 per cent. If we use an estimate of 3°C temperature increase this would mean average yield losses of about 22 per cent in maize, 18 per cent in wheat, 9.6 per cent in rice and 9.3 per cent in soybean — *from temperature alone!* Of course, these averages take no account of incidences of drought and disease, local and demographic factors and extreme events. Not only will farm yields be affected by ever more hostile growing conditions, increases in atmospheric carbon dioxide are likely to reduce the nutritional content of staple crops.[40] Similarly, climate shocks, vulnerable supply chains, and increasingly stringent conditions for investment will affect food prices, input costs, and corporate profits. We need a different paradigm or as the Buzzcocks would say 'another music in a different kitchen.'[41]

The options most commonly put forward for the yield-for-profit agrifood system do not explicitly put nutritional outcomes *and* sustainable practices on top and are invariably based on permutations of business-as-usual. An alternative paradigm means transforming the food system to redefine yield and profit as positive externalities in which their contributions are seen as global public goods and not just calories for cash. For this we need to show how yield beyond calories can include tangible benefits such as healthier, more nutritious, and diverse diets as well as intangible benefits such as food culture, societal cohesion, and mental health. Similarly, we need to show how profits can go beyond financial returns to include environmental stewardship, ecosystem services, soil health, and biodiversity conservation. In brief, we need to rethink the values of the global agrifood system irrespective of climate change. It is not about putting a *price* on natural capital it is about putting *values* to it.

Sustainable Development and the Climate Crisis

In a dismal decade, the year 2015 stands out as one of hope and vision in which the governments of the world took collective leadership — twice. Within a few weeks of each other, two globally significant events took place. On 27 September, all member states of the United Nations General Assembly approved the United Nations 2030 Agenda for Sustainable

Development with its vision to *transform our world* through an action plan based on *people, planet, and prosperity*. The sheer ambition of the agenda and the unanimous commitment of all member states is astonishing, some might say foolhardy. The Agenda includes 17 Sustainable Development Goals (SDGs),[42] with timelines and deliverables to be met by 2030;

GOAL 1:   No Poverty
GOAL 2:   Zero Hunger
GOAL 3:   Good Health and Well-being
GOAL 4:   Quality Education
GOAL 5:   Gender Equality
GOAL 6:   Clean Water and Sanitation
GOAL 7:   Affordable and Clean Energy
GOAL 8:   Decent Work and Economic Growth
GOAL 9:   Industry, Innovation and Infrastructure
GOAL 10:   Reduced Inequality
GOAL 11:   Sustainable Cities and Communities
GOAL 12:   Responsible Consumption and Production
GOAL 13:   Climate Action
GOAL 14:   Life Below Water
GOAL 15:   Life on Land
GOAL 16:   Peace, Justice and Strong Institutions
GOAL 17:   Partnerships to achieve the Goals.

Achieving SDG1 (*End poverty in all its forms everywhere*) and SDG2 (*End hunger, achieve food security and improved nutrition and promote sustainable agriculture*) within 15 years must have seemed wildly optimistic in 2015 and events since then and our lack of action mean that the likelihood of achieving any of the SDGs is vanishing fast. Every year, the UN produces a Sustainable Development Report as a global assessment of the progress of individual countries towards achieving the SDGs. Its overall score measures a country's progress towards achieving all 17 SDGs. A score of 100 indicates that a country is on track to meet all of its SDG commitments. By 2020, no country was on target to achieve all the SDGs.

Even the top five ranked countries (Sweden, Denmark, Finland, France, and Germany) only achieved scores of between 80 and 85 per cent.[43] At the other end of the scale, the lowest ranked five countries (Central African Republic, South Sudan, Chad, Somalia, and Liberia) achieved scores below 50 per cent. It is hardly surprising that those countries suffering from civil conflict and that are also on the front line of the climate crisis are least equipped to achieve the SDGs. It is equally surprising that politically stable countries in benign climates with ample resources are still not on track to achieving all of the goals.

Less than 11 weeks after the launch of the 2030 Agenda for Sustainable Development, 196 UN member states met in France at the UN Framework Convention on Climate Change (UNFCCC), COP 21. The resulting Paris Agreement was adopted on 12 December 2015 and signed by 189 members of the UN in 2016. The Paris Agreement's signature goal is to *keep the increase in global average temperature to well below 2°C above pre-industrial levels; and to pursue efforts to limit the increase to 1.5°C.*[44] The agreement also aims to help individual states adapt to the impacts of climate change, and *make finance flows consistent with a pathway towards low greenhouse gas emissions and climate-resilient development.* In 2017, President Trump announced that the USA would withdraw from the Agreement in 2020 and changes in US policy that were contrary to the Paris Agreement were enacted by his Administration. Nevertheless, recognizing the seriousness of the climate crisis, an alliance of 24 US states and Puerto Rico pledged to uphold the Paris goals. In 2021 the incoming Biden administration reaffirmed its commitment to the Paris Agreement; we have yet to see how.

While the Paris Agreement ostensibly relates to SDG13 (Climate Action), its provisions and targets contribute to every SDG. However, neither the Sustainable Development Agenda nor the Paris Agreement can be met without radically transforming food systems to reduce carbon emissions, increase mitigation, and adapt to climate change to achieve zero poverty [SDG1] *and* zero hunger [SDG2]. Put simply, without transforming food systems we have no chance of *transforming our world.* And without diverse options linked with a moral purpose, we have no chance of transforming food systems for good.

The Ninth Revolution — Transforming Food Systems for Good

We Need New 'Light Bulbs' for the Food System

> *"Electric light did not come from the continuous improvement of candles."*
>
> Oren Harari[45]

Light bulbs did not simply replace candles as a source of uniform light; they provided other benefits such as being useful in remote places, under water, on mountains, inside machines, and in enclosed spaces. They could also operate for long periods of time without close attention, and they overcame the problems of candles which were smelly, created a mess and smoke, went out in a gust of wind, and could even set fire to your house! Despite the advantages of light bulbs, the transformation from candles to light bulbs wasn't instantaneous or even immediately welcomed by the candle industry. Nor was the invention of the light bulb a single intentional event — there was no 'light bulb' moment for light bulbs.

In fact, the invention of the light bulb involved a sequence of steps by different people from different backgrounds and places. Alessandro Volta, an Italian Physicist, developed the 'voltaic pile' as the first practical method to generate electricity, using a glowing copper wire. This inspired Sir Humphrey Davy, an English Chemist and aristocrat to produce the world's first incandescent electric lamp by connecting 2000 voltaic piles to create a brief and bright arc of light between two charcoal electrodes. The American inventor Thomas Edison worked on developing a longer-lasting incandescent lamp. Initially he tried using a filament made of cardboard but this proved to be unsuitable for sustained generation of light. Lewis Latimer, a draughtsman whose mother and father had escaped from slavery in Virginia, resolved this challenge by inventing a process for making carbon filaments for long lasting lamps that were subsequently used in large-scale lighting systems for New York City, Philadelphia, Montreal, and London. Volta did not set out to invent a light bulb but his technology led to another purpose for which there was a perceived need. The critical link was *innovative* thinking that transformed the existing model of generating electricity through a sequence of actions that led to a new form of lighting. More importantly, the innovation process involved *diverse* thinking of individuals with different backgrounds (one a knight and another the son of slaves) and specialisms to invent the light bulb.

Pursuing this analogy, we can think of the modern agrifood system as built around four candles (wheat, rice, maize, and soybean). By prioritizing investment and research to improve the performance of these four candles we are unlikely to meet the needs of nourishing 10 billion people by 2050. That will require innovative *and* diverse thinking to transform the existing model of the agrifood system from candles to light bulbs. The form and technologies involved in each light bulb are not as yet clear but the intellectual energies for their invention will come from diverse human and biological resources and not the usual suspects.

In his brilliant and prescient book 'Rebel Ideas: The Power of Diverse Thinking,' Matthew Syed introduces the concept of *Cognitive Diversity*.[46] He shows how *homophily* 'love of sameness' creates social groups that are made up of individuals with similar views, beliefs and experiences. Whilst such groups may quickly agree on a common position on any single issue, they develop a collective blindness to new, uncertain, and complex situations with which they are unfamiliar. Since each individual is essentially a clone of the whole group, they aspire to a form of '*clonal intelligence*' in which concepts of 'excellence' and 'innovation' are based on rapid and linear improvements of the current model rather than variations from it. The outcome is incremental or *linear innovation*. However, whilst individuals or inventions may be the 'best in class' and offer the potential of '*quick wins and silver bullets,* they are ill suited to the unexpected, unforeseen or complex, however excellent their clonal intelligence.

In contrast, cognitive diversity allows complex challenges to be tackled by social groups (teams) made up of individuals with different viewpoints, experiences, cultural backgrounds, and gender identities. Matthew Syed argues that '*Diversity isn't some optional add-on.... Rather, it is the basic ingredient of collective intelligence*'. [47] We can see this diversity in modern computer systems and mobile applications for complex situations. Diverse predictors rather than algorithms lie at the heart of artificial intelligence, machine learning and the general circulation models that are used to predict the impacts of climate change. As well as complex computer systems, this requires diverse thinking from individuals with different backgrounds, specialisms, viewpoints and even belief systems. In human societies, rather than linear innovation, cognitive diversity brings opportunities for what Matthew Syed calls *recombinant innovation*, where different approaches and ideas are brought together to address a complex challenge. In such cases we need the collective intelligence of the whole group in which each individual (whatever their status) is suited to a part but not all of the challenge.

Can we apply Matthew Syed's concepts of cognitive diversity and collective intelligence to the agrifood system? The evolution of modern agriculture is based on the search for the *best in class* crops or animals which can achieve the yield potential set by their genetics (breeding and selection) and the resources provided by their environment (natural and managed). In crops, to achieve this potential, we grow identical plants on uniform soils supplied with optimal inputs. To support this uniformity, research, education and advocacy mostly focus on high yielding crop varieties, narrow scientific disciplines and elite academic institutions. The problem is that in searching for linear innovations and rewarding clonal intelligence we have a collective blindness to the different (crops, environments, communities), the unexpected (climate change) and the unforeseen (say pandemics). Homophily reinforces the rejection of the different until the different becomes the new normal (think of Uber, Airbnb, electric vehicles and renewable energy.)

By themselves, clonal intelligence, homophily and linear innovations will not be radical enough or fast enough to meet the challenges that we and the planet face. Current and future crises simply reinforce the need for many, different 'light bulbs' for the food system, if not society in general. Greenhouse gas concentrations are at their highest levels in over 800,000 years,[48] we are in the midst of the warmest five-year period on record, and the world is engulfed in a global pandemic that will set economic development back for decades.[49] The agrifood system cannot rely on any version of business-as-usual based on linear increments in the current model but requires a fundamental and purposeful transformation of the system itself. There are no *quick wins* or *silver bullets* for this transformation. To meet complex challenges, we need to collectively embrace the messiness of diverse thought, species, systems, cultures, and foods, move outside our comfort zones and specialisms, and challenge convention. Diversification will have to be transformative, but that doesn't mean we have to start from scratch. There is enormous power in collaboration around specific actions that support diversification in which individuals, communities, institutions, foundations, and even countries have particular expertise, interest, or facilities. Such contributions should be transparent, mutually beneficial, and aimed at the global public good rather than being an excuse for short-term and narrow self-interest. Crucially, such actions must be collective and based on diverse thinking. The scale and urgency of the challenge is such that, if they are to work, collective actions must include practitioners as partners and not as *recipients* of innovation. The sum of the collective actions — the Ninth Revolution — is the contribution they make to achieving the SDGs through a partnership for the goals (SDG17).

The Light Bulb is Diversity: Nine Collective Actions to Transform Food Systems for Good

> *"Decision-makers must also recognize there are no geographical barriers to the food and agriculture disruption... Policymakers must, therefore, start planning for the modern food disruption now in order to capture the extraordinary economic, social, and environmental benefits it has to offer."*

Rethinking Food and Agriculture 2020–2030, RethinkX Report[50]

So where to start? The transformation of food systems will occur not with our current most-favoured species but with those that we have ignored, dismissed or are yet to discover. This will need *us* to reimagine agriculture beyond stereotypes that see farming as cultivating a few staple crops and rearing a few domesticated animals in prepared fields and enclosures.

Our understanding of food systems will need to include the farming of living organisms in marginal landscapes, urban spaces, rooftops, buildings, laboratories, bioreactors, ponds, and vats. It will also mean thinking beyond farmers as either traditional sons of the soil scratching a living from impoverished lands, or as grain barons traversing endless fields of maize and soybean. The next generation of farmers will include women, youth, urban professionals, retirees, community associations, and marginalized communities — all involved in raising living organisms for food or raw materials. All farmers will need to rethink the sources of living organisms that they farm, not just as species and *commodities*, but as *ingredients*, including proteins, lipids, bioactive and medicinal compounds for new foods and supplements, and raw materials for animal feeds, energy, biomaterials, phytoremediation, and discovery chemicals. In each case, next generation farms, farmers, investors, and researchers will have to overcome the five negative challenges to agricultural diversification identified in Chapter 6: lack of markets; absence of supply chains; limited quality ingredients; little research interest and lack of credible evidence. To overcome these challenges, we need a shared vision that links novel and traditional technologies and knowledge with quantitative science, along with financial and human capital, into collective actions to transform food systems. Agriculture and the research that supports it cannot aspire to simply be transactional suppliers of *better candles,* but rather an intellectual and physical resource for new light bulbs. This requires nothing less than a Ninth Revolution — to transform food systems for good.

In the sections that follow, nine opportunities and related actions are described, each of which can help to diversify the global agrifood system and all of which, together, can help to bring about a Ninth Revolution that can transform food systems for good. Here *good* is used as a shorthand for *global public good* but it is also meant in the sense that changes in the agrifood system need to be for the long term and that they should be morally right. As well as these desirable attributes, transformation of the agrifood system must be a pragmatic alternative that is better than business as usual. It has to be economically viable, desirable to end-users and just.

The collective actions identified below are not intended to be exhaustive, but rather they are a suggested starting point. They need not be sequential but they have a natural order that starts with building the global evidence-base for food system transformation and ends with measuring its impacts on the sustainable development goals by 2030. As with the energy sector, the agents of change will come from outside the modern agrifood industry. However, progress would be faster if those with the technical skills, physical facilities, and financial resources — at present focused entirely on mainstream food systems — ceased claiming that alternatives to business as usual are impossible and instead invested their energy and resources into making them a reality.

I accept that the nine proposed collective actions are a personal wish list with which many may find fault. Nevertheless, if we want to effect beneficial change, we have to be serious about doing things differently and decide how this can be done rather than continue to spend time and energy on why it can't. Whether or not the collective actions that I propose here are taken up, transformational changes in our food systems will have to occur because our current model is unsustainable and will become increasingly so with many more people on a warmer planet. Here, it is important to distinguish between *collective* actions that are based on diversity of thought, outputs and outcomes and the disastrous 'command and control' *collectivism* proposed by Trofim Lysenko and endorsed by Joseph Stalin that set Soviet agriculture back by decades. We do not need a more *collectivist* food system and consolidated research, education and investment systems to underpin it. Rather, we need diverse *collective* actions that help support agricultural communities to become the sources and owners of innovations and agents of their own transformation.

In fact, new and diverse options for food systems are already being initiated by individuals and communities as viable alternatives to business, research, and education as usual. The agents of change are diverse stakeholders, old and young; diverse systems, traditional and novel; and diverse ingredients and products; novel and rediscovered. The following collective actions would build on their leadership and furnish tools and secure resources that are needed for this critically important transformation of food systems for good.

Collective Action 1: A Global Evidence-Base for Agricultural Diversification

Challenge: *Scientific evidence is published through specialist peer-reviewed journals. How can we broaden global knowledge to include vernacular, unconventional, qualitative, and unpublished evidence to support a digital evidence base for agricultural diversification?*

Opportunity: There is a need for a credible, global, and trusted evidence-base of species, systems, and products that can inform opportunities to diversify agriculture and food systems, and that would enable these options to be compared with conventional solutions. The evidence-base needs to go beyond published and readily available data to integrate scientific (experimental), grey (unpublished), vernacular (unwritten), qualitative (observed), and sociocultural information. By linking published data, new evidence and traditional knowledge, we can provide the broadest possible evidence-base for diversification within a single evolving repository of information. We could also curate multigenerational knowledge about agriculture and food cultures before it is lost forever.

Outputs: A global evidence base for agricultural diversification that would be transparent and accessible with open standards, protocols, semantics, and data structures. These standards would be understandable not only to humans but also algorithms that could utilize them to answer complex questions, identify strengths or weaknesses of proposed solutions, and rank options. By linking *big data* that can be analyzed to reveal patterns and trends, especially relating to human behaviour, the evidence base could help to answer generic and location-specific questions, and rank the adaptability, adoptability, and profitability of diversification options. This would minimize the learning curve to develop user-friendly systems and link big data and machine learning with crowd-sourced and unconventional datasets. The evidence base would support next generation decision-support systems and dedicated applications to help guide local and regional policies. Such systems and applications would also simulate the outcomes and efficiency gains of alternative solutions now and under future climates. The evidence base would enable traceability mechanisms and mandatory ESG disclosure from across the agrifood sector. In addition, it could catalyze capital deployment towards projects that meet SDGs, and support research and innovation in agricultural diversification.

Collective Action 2: Community of Practice for Research on Agricultural Diversification

Challenge: Research training is based on disciplines and academic qualifications. How can research support a global community of practice for agricultural diversification in which communities and agribusinesses are co-innovators and agents of change?

Opportunity: Transdisciplinary and cross-boundary research could build on the strengths of existing academic institutions by linking their research activities with partner communities and the private sector through a

Community of Practice (COP) for agricultural diversification. Such a COP could develop intervention strategies that are responsive and relevant to local challenges, as well as meet the wider objectives of agricultural diversification and academic goals. This would mean that initiatives were demand-led by the communities themselves in which research and development provided a supporting function and evidence base for diversification. The COP would build a critical mass of expertise, talent, and evidence that pivots doctoral training, research leadership, and community partnerships towards a single objective: adapting to climate change through agricultural diversification.

Outputs: The agricultural diversification COP would provide a transformational and replicable model for research, training, and education that allows communities to be partners and co-owners of innovation and facilitate capacity development and knowledge exchange. To achieve this, the COP could build *research value chains* from genetics through to markets for diverse species and agricultural systems, and link research with practitioners and investors. By acting as a catalyst for novel solutions, the COP would: build global research leadership in agricultural diversification; foster south-south cooperation; strengthen rural organizations and extension services; enhance economic and social development of smallholder farmers; and provide entry points for diverse local, climate-resilient species and systems for livelihoods, especially of women and youth. As part of a global effort to meet SDGs through agricultural diversification, the COP would also drive collaboration and support capacity building at local and international academic and research institutions. Specific research and training activities would complement and strengthen other collective actions, and could be benchmarked against their overall contribution to specific SDGs.

Collective Action 3: Food and Sustainable Development Curricula for Schools

Challenge: At present, educational systems focus on teaching skills that train students to meet the perceived current needs of industry. How can we educate students to have the multiple skill sets necessary for healthy diets and sustainable lifestyles in the uncertain climates of the future?

Opportunity: In much of the world, there is a recent generation of adults who have never learned how to grow food or cook because they were neither taught at home nor in school. Agriculture and food systems need to become an intrinsic part of the educational system. As such, they would be an ideal vehicle to teach other core subjects such as biology, chemistry, physics, and mathematics. In addition, an educational focus on agriculture and food would inculcate transferable skills for healthy lifestyles that would benefit students throughout their lives. By learning how to grow and prepare food from raw ingredients, children would also learn about nutritious diets and the role of their local agrifood system in cultural traditions, sustainable development, and resilience to climate change. By knowing how to cook, source ingredients, and design menus, school students could become the vanguard for transformation of their agrifood system.

Outputs: Curricula on food and sustainable development for healthy lifestyles would link school education with global initiatives such as the Planetary Health Diet, SDGs, and climate change mitigation. Schools and national educational systems would work together to design locally applicable curricula that put food, nutrition and sustainable development at the heart of the educational system, and that use this focus as a vehicle to teach core school subjects. As well as curriculum design and content, the initiative would provide: guidelines on the preparation of school meals and menus; training and support for schools to build and maintain edible gardens; teaching aids for food cultivation and preparation; and resources to protect and celebrate local and regional food cultures. This collective action would link with others, especially Collective Action 8 and the proposed forgotten foods initiative as a driver for intergenerational change.

Collective Action 4: Decision Support for Climate-Resilient Agroecosystems

Challenge: Opportunities to diversify agroecosystems in fields, landscapes, and urban environments are limited by inadequate scientific understanding of their complexity. How can we provide expertise, guidance, and support to maximize the potential of agroecosystems?

Opportunity: As agriculture moves into marginal landscapes, urban spaces, and controlled-environments, an understanding of agroecological principles is critical to optimize the benefits of diversification. Complex agroecosystems can be more productive, stable, and resilient to climate shocks, but their complexity has discouraged wider adoption, new investment, and novel research. Many of these limitations can be overcome by the use of Industrial Revolution 4.0 (IR4.0) technologies that link physical, digital, and biological dimensions and allow monitoring through remote sensing, geospatial mapping, computational biology, and modelling.[51] Internet-of-Things (IOT) approaches can help manage agroecosystems through robotics and other technologies that reduce labour needs and optimize performance.[52] There is an opportunity to identify, test, and operationalize how innovations that are allied with diverse agroecosystems of greater productivity and resilience can outweigh their complex management to provide viable and more resilient solutions.

Outputs: A Climate-Resilient Agroecosystems Decision Support System is needed to monitor, measure, and manage complex agroecosystems. The Decision Support System would link new technologies with research, advice, and support to help agritech start-up companies and communities benefit from more efficient, environmentally stable, and profitable agroecosystems. As well as better management, research and advisory services could help practitioners to integrate: component systems, such as hydroponics (cultivation of plants in soil-free nutrient solutions), aquaculture (farming of fish, crustaceans, molluscs, aquatic plants, algae, and other organisms), aquaponics (aquaculture with hydroponics), vertical farming and other controlled-environment systems; reduce inputs (through biological control of pest and disease vectors); optimize outputs by fine-tuning the environmental requirements for component species and provide early warning of pests and diseases. The use of dedicated mobile applications would give practitioners real-time access for decision making and advice.

Collective Action 5: Decision Support System for Climate-Resilient Species

Challenge: There are thousands of underutilized species of plants, animals, insects, and microbes, many of which are useful for future agriculture. How can we identify those best suited to specific environments, their end-uses, potential returns returns on investment and societal benefits?

Opportunity: Decisions on whether to introduce diverse agricultural species into production systems need specialist guidance on their suitability at any location, the likely demand for their ingredients, efficacy of production methods, guidance on business development and returns on investment and effort. All of these factors need to be considered before starting production of any novel species, along with a risk assessment and cost/benefit analysis of different options against current practices. To support decision-making and reduce the risks of failure, researchers, sponsors, and partners need to provide the most reliable available evidence in a form that is readily accessible to the end-user. As well as suitability, such information needs to link easily accessible advice on optimal management practices, the economic or social value of outputs, and routes to market for different ingredients.

Outputs: A digital Climate-Resilient Species Decision Support System is needed to help evaluate the potential of diverse agricultural species, their suitability in different environments, nutritional content, quality, safety, and processing characteristics against conventional options. The Decision Support System could link with applications that support geospatial mapping of different species, their climate resilience, likely returns on investment, functionality, and performance under different climate change scenarios. The Decision Support System would link with the global evidence base arising from Collective Action 1, and provide selection criteria for the genetic characterization and species improvement objectives of Collective Action 7.

Collective Action 6: Digital Tools for Traceable Agrobiodiverse Value Chains

Challenge: The global food system depends on opaque supply chains that distribute commodities around the world. How can we develop traceable and agrobiodiverse supply chains to deliver diverse products to consumers and ensure benefits to grower communities?

Opportunity: In response to consumer demand, food companies and entrepreneurs are introducing more diversity into their supply chains and selling novel foods and ingredients that support biodiversity. While there is growing demand for agrobiodiverse products and new opportunities for producers, these need traceability metrics to ensure that there is diversity at each stage of the supply chain, fair compensation, and benefits shared with growers. A major limitation for food companies and producers is a lack of information about the supply sources and demand for food ingredients that support biodiversity, and the opaqueness of supply chains. Lack of transparency means that supply chains often remain "*blind*," leading to dubious practices, false claims, loss of consumer confidence, with negative consequences for the entire industry. Evidence of supply-chain transparency is critical for food companies, producers, and consumers to trust the sourcing of ingredients that enhance biodiversity and reward farmers for biodiverse practices. Technologies can now trace the exact location of a farm, the grower community, and agroecosystem involved.

Outputs: This collective action would develop a digital toolkit and chain-of-custody tracker for agrobiodiversity that enables producers and buyers to authenticate the safeguarding of biodiversity and sustainable practices in the supply chain. Traceability of biodiversity across the entire value chain would support transparency in sourcing and production. Ideally, this innovative tracker would be linked with specific SDGs and assurance and certification metrics for growers, food companies, and consumers. Evidence of biodiversity along the supply chain would encourage more sustainable practices that build on traditional or local knowledge and align with agroecological principles to improve soil health, support agroecosystems, encourage sustainable practices, and enhance climate resilience.

Collective Action 7: Genetic Resources to Breed and Improve Agricultural Species

Challenge: Most research, breeding, and investment is focused on genetic improvement of mainstream agricultural species. How can we utilize genetic and breeding technologies to rapidly evaluate and improve climate-resilient underutilized species for future agriculture?

Opportunity: The time and costs involved in the genetic characterization, breeding, and improvement of currently underutilized species have deterred research and investment for their wider use and improvement by breeding programmes. Recent advances in genetic technologies, bioinformatics, and robotic systems, and associated cost reductions, mean that we now have the genetic tools to incorporate desirable traits into crops and animals faster and more cost effectively than ever before. The use of molecular markers allows us to rapidly improve desirable attributes — such as higher and more stable yields — of a range of currently underutilized species, and to remove the limitations that have deterred breeders and investors. Molecular markers and allied techniques can also be used to ensure genotype tracing and quality control in breeding programmes. In addition to generating data within species, the use of cross-species approaches, and in particular, translating knowledge and resources from model and major species, is a pragmatic way to develop a wide range of underutilized species and the best use of limited resources for their improvement.

Outputs: An integrated breeding and improvement platform would allow systematic screening of germplasm and genetic and phenotypic characterization in association with relevant gene banks. Community-led pre-breeding lines could be developed and tested in association with seed companies, and a research and development pipeline established to screen potentially useful species and genotypes. Such screening would enable selection of those with the greatest potential for coordinated breeding and improvement programmes. The characterization of individual genomes and genotypes for promising species would allow multilocational testing and selection for yield, climate resilience, and quality characteristics, along with rapid screening of suitable material for breeding and improvement. Translation of genetic information, as well as approaches and technologies from mainstream and model species, would further allow rapid and cost-effective progress in agricultural diversification.

Collective Action 8: National Guidelines for Dietary Diversification

Challenge: Consumers are familiar with a global diet of uniform foods that are cheap, plentiful, and readily available. How can we diversify diets with nutritious, locally available ingredients that support traditional food cultures and planetary guidelines for health and environment?

Opportunity: Changing eating habits is not just about providing more nutritional information and dietary guidelines, but requires a shift in consumer attitudes about food and in the affordable availability of a wider variety of foods. The globalized Western diet is rapidly displacing unique local food cultures that cannot compete with the marketing, economic, and political power for homogenization. However, diverse diets bring with them nutritional, health, social, and cultural benefits beyond the content of their ingredients. The EAT-Lancet Planetary Health Diet[53] provides guidelines on ranges of different food groups that constitute an optimal diet for human health and environmental sustainability. It emphasizes a plant-forward diet where wholegrains, fruits, vegetables, nuts, and legumes comprise a greater proportion of foods consumed. However, it does not explicitly advocate dietary diversification to include underutilized, nutritious, and locally available foods. Using the Planetary Health Diet as a benchmark, we need to develop national dietary guidelines that also respect cultural identities and regional cuisines, reduce carbon footprint, and support livelihoods.

Outputs: National dietary guidelines could help to achieve global outcomes for human health and environmental sustainability while respecting and promoting traditional culinary cultures. The guidelines would use a common evidence-base of ingredients, foods, and food combinations that personalise individual and national needs. Digital technologies would enable individuals and societies to choose healthier diets that celebrate their own food cultures, and rediscover foods and culinary methods that contribute to dietary diversification and nutritional security. Capacity building would allow stakeholders to access generational knowledge on *forgotten foods* and their ingredients. Small-scale and subsistence farmers and food processors, supermarkets, and the general public would become the joint custodians of their own food cultures. In addition, champions, researchers, and communities could show how regional cuisines can contribute to nutritional security through dietary diversification.

Collective Action 9: Global Action Plan for Agricultural Diversification (GAPAD)

Challenge: The UN Sustainable Development Agenda includes 17 Sustainable Development Goals (SDGs) to achieve zero poverty and zero *hunger by 2030. How can we include agricultural diversification into specific SDGs, and through a global action plan, as a partnership for the goals (SDG17)?*

Opportunity: The SDGs provide a globally-agreed framework for sustainable development based on five pillars — people, planet, prosperity, peace, and partnership. Climate change will have variable impacts on food systems in different parts of the world, and reciprocally the global agrifood system is now the largest contributor to greenhouse gas emissions. Without transforming and diversifying the agrifood system, there is no realistic prospect of achieving the Paris Agreement target of reducing global warming to within 1.5°C of pre-industrial levels nor meeting the targets of specific SDGs. Agriculture provides a common link across the SDGs, but there is no plan for how the wider adoption of diverse, climate-resilient species and agricultural systems can explicitly contribute to specific SDGs in variable and volatile environments. Such an action plan is urgently needed if agricultural diversification beyond mainstream species is to make a meaningful contribution to the Sustainable Development Agenda, societal adaptation to the climate crisis, and social justice.

Outputs: A Global Action Plan for Agricultural Diversification (GAPAD)[54] would link stakeholders from communities, research institutions, and agribusiness with consumers and the global public in alliances of mutual best interest. Such alliances would explore and demonstrate how agricultural diversification can meet the needs of a warmer world and contribute to the eradication of poverty (SDG1), hunger (SDG2), and other specific SDGs (especially 7, 12, 13 and 15). Importantly, GAPAD would provide the essence of a global supporting partnership (SDG 17) to achieve these goals. Each of the other collective actions and their progress toward specific SDGs would contribute to the activities and metrics of GAPAD, which, together, would truly transform food systems for good. Development of such an action plan would draw upon the principles, and generally build upon the Declaration on Agricultural Diversification that was done alongside the Paris UNFCCC climate conference in 2015 and signed by leading researchers and international agencies.

Final Thoughts

Of the first eight agricultural revolutions described in this book, only the most recent — the Green Revolution — purposefully aimed to change the whole world. It succeeded spectacularly, but not always as intended. Those who rallied behind The Green Revolution aspired to embrace the principles of maximum crop yields, economies of scale, mechanization, control of resources and standardization of products to liberate the world from food insecurity. In large measure they have achieved this, but persistent and continued adherence to this single-minded purpose even in the face of disastrous unintended consequences has now confronted humanity with a number of unprecedented crises. Industrialized agriculture of very few mainstream crops produced in vast scale monocultures has only been made possible through: the use of fossil fuel mechanization; long-haul fossil fuel transportation; harmful herbicides and pesticides; clearing of forests; destruction of biodiversity at ecosystem, species and genetic levels; loss of top soils; depletion of soil organic carbon; desertification of drylands; diversion and pollution of fresh water; eutrophication of nearshore marine waters; diminution of the nutritional quality of diets; malnutrition and associated non-communicable diseases and dislocation of rural workers and communities. Critically, industrialized food production has led to the disruption and loss of confidence in many forgotten communities and marginalized sectors of society in their own forgotten agricultural methods, forgotten foods, forgotten food systems and forgotten knowledge honed over thousands of years and now on the verge of extinction.

Our modern agrifood system, the massive subsidies required to sustain it and its increasingly negative externalities is unsustainable for humanity and our planet. Many people have not yet seen the inevitability of change and will continue to deny and resist it. However, The Ninth Revolution in food systems is already being enabled by a new generation of consumers, activists and practitioners who see it not only as a necessity but also an opportunity. A different kind of agrifood system can utilize the Fourth Industrial Revolution and post-carbon technologies with a common purpose to reduce global atmospheric carbon, preserve biological diversity, and improve human nutrition *and* quality of life and embrace diversity.

This Is Not a Drill

If Captain Smith had listened to the six warnings issued on 14 April 1912 about sea ice, and the sightings by his own lookouts in the crow's nest, the Titanic would probably not have struck an iceberg at around 23:40 on Sunday, 14 April 1912 and sunk two hours and forty minutes later.[53] The sinking of the Titanic caused international condemnation over a lack of lifeboats, inadequate regulations, and the different treatment of the three classes of passengers during the evacuation. This led to major changes in maritime regulations and the establishment of the *International Convention for the Safety of Life at Sea* in 1914. Of course, by then it was too late for the 1500 passengers who had perished on the morning of 15 April, 1912. All of Titanic's lookouts survived the disaster.

The Titanic agrifood system on which we are all passengers will either have to change direction or hit an iceberg, the most dangerous of which is the climate crisis. This may seem alarmist and even provocative, but we have had numerous and persistent warnings from many lookouts in the crow's nest, including activists (young and old), UN agencies, national governments, leading researchers, and some of their institutions. If the agrifood system was working effectively the number of malnourished people would not exceed those who are healthy. Agriculture and its related industries would also not have been the primary cause of over 36 million square kilometres of dryland desertification and the largest contributor to biodiversity loss; nor would they be the source of 37 per cent of anthropogenic greenhouse gas emissions; nor would they be subsidized from the public purse.[56]

In a comprehensive review, the eminent research biologist Roger Leakey and his co-authors evaluated 297 reports on modern agricultural practices and found that only 64 per cent were positive.[57] They found similar patterns and consequences for food production, delivery of ecosystem services, farmer livelihoods, and regulatory processes. Since nearly two-thirds of modern agricultural practices have had positive impacts, perhaps we could conclude that on balance the global agrifood system is serving us well. On this basis, we might equally assume that if 64 out of 100 aircraft landed safely but all the others crashed, the aviation system is fit for purpose.

Of course, we would not come to this conclusion for aviation safety. We would conduct urgent investigations into why 36 aircraft had crashed and act accordingly. In comparative terms, and staying with this analogy, while we are improving the safety, fuel economy, and comfort of those parts of the food system that are still 'flying', we are not solving the problems of those that continue to crash. Those problems include declining agricultural productivity in regions where it is most needed, along with interrelated environmental, social, and economic effects that hasten land degradation, increase social deprivation, and lead to loss of biodiversity, agroecosystem functions, and soil fertility. These negative externalities of conventional modern agriculture are often described as the inevitable consequences of intensification and modernization but not usually by those who suffer them. For smallholder farmers and marginalized communities in much of the tropics and sub-tropics, the effects of our current approach are immediately existential because the increasing gap between the genetic potential and actual yield of staple food crops means more hunger, increased malnutrition, loss of livelihoods, a cycle of debt, despair and conflict, and ultimately economic migration. Indeed, the failure of agriculture in the global south is now so dire that it is already having detrimental consequences for the industrial societies of the global north. The disproportionate impacts of the current pandemic on poor and vulnerable nations and people, the '*have-nots*' will only increase these pressures on the '*haves*.' Like previous food riots and popular uprisings we know that they are inevitable but we do not know when, where and how they will be triggered, nor how we will suppress them without sacrificing our individual freedoms and societal integrity.

We are left with no choice but to rethink agriculture and food systems, not just as a means to grow and consume food, but as the fulcrum of many of the most pressing global issues confronting humanity. This requires us to diversify our thinking, approaches, species, systems, supply chains, products, markets, and partnerships. For the increasingly arid and impoverished tropics, diversification means rehabilitating degraded farmland and landscapes — areas that will never suit major staples — with crops and cropping systems that can deliver an array of highly nutritious and marketable foods. It also means rebuilding (and not destroying) agroecological functions and creating new local business opportunities to kick-start rural economies, enhance social well-being, and restore the dignity and self-respect of farming communities. Together, initiatives to diversify agriculture can promote livelihoods and social justice, mitigate and enable adaptation to climate change, and restore wildlife habitats. The Ninth Revolution proposes the transformation of food systems, based on scientific principles, the best available evidence and the 'forgotten' knowledge of farming communities that can collectively provide an adaptable model of diverse solutions that best fit specific environments, communities and their cultures.

So, how can we transform food systems? One option is to use the model set out by Wendell Berry in his *50-Year Farm Bill*. Berry proposes a ratio of 80:20 perennial to annual cropping.[58] Using this as a guideline, we would commit to an 80:20 ratio of land area between next-generation and mainstream agriculture. As well as diversifying agricultural species, next-generation agriculture includes a plethora of emerging agritech solutions in marginal landscapes, perennial systems, and urban spaces. For marginal landscapes, the incentives will not just be agricultural yields *per se* but rewards for *chain of custody* of biodiversity, carbon capture and sequestration, protection of ecosystem functions and services and diverse and nutritious products based on agrobiodiverse value chains. Next-generation food systems will include sourcing of ingredients from diverse cereals, legumes, and oilseeds, as well as fruit, vegetables, and novel bioactive compounds. Using the 80:20 ratio proposed by Berry, perennial crops can become the major sources of better nutrition and livelihoods without destroying soil ecosystems or accelerating climate change.

For urban spaces, the opportunities are virtually unlimited because various forms of urban agriculture, such as vertical farming, allow for: year-round food production; elimination of nutrient and pollutant leakage; restriction of pesticide, herbicide, fertilizer and water use; reduction in food miles through localized value chains; higher food safety standards and traceability and new employment opportunities for urban dwellers. Whether in marginal landscapes, perennial systems or urban spaces, we will need to scale up existing prototypes and use emerging technology supported by further investment, research, and development.

So, how can research support transformation of food systems? Much of the knowledge, resources, and infrastructure to transform food systems already exists. Universities and research institutes can make important contributions through in-kind expertise, studentships, facilities, and technology transfer from mainstream species and systems to their alternatives. As a metric of progress, these institutions might measure their research contributions to specific SDGs through transformation of food systems. Perhaps new categories of university ranking can include institutional contributions to the Sustainable Development Agenda.

To guide such research, there is an array of recent reports that identify mechanisms to embed an agroecological or 'nature-based' approach into mainstream policy thinking. For example, the comprehensive review by Tim Benton and colleagues[59] identifies: more plant-based diets; more land as protected natural habitat; and the adoption of more sustainable farming methods as three 'levers' to guide food policies. In 2019, a report by the High-Level Panel of Experts (HLPE) on Food Security and Nutrition of the Committee on World Food Security[60] proposed agroecological and other approaches for sustainable agriculture and food systems that support food security and nutrition. The report describes 13 agroecological principles related to: recycling; reduced inputs; soil health; animal health and welfare; biodiversity; managing interactions; economic diversification; embracing local knowledge and global science; social values and diets; fairness; connectivity; land and natural resource governance; and participation. The report also calls for greater investments in public and private research and development, and in national and international research systems to support agroecological and other innovative approaches. Critically, the report starts from the recognition of human rights (though not specifically linguistic rights) as the basis for ensuring sustainable food systems.

So, how can we fund transformation of food systems? The main resistance to change is the mindset of decision makers, vested interests and ourselves. The most powerful way to influence them is to nudge the behaviour of consumers who are yet to be convinced of the necessity, desirability, and value of transformation. Nevertheless, change will also require resource transfers and a repurposing of efforts towards new and more diverse approaches. In terms of financial resources, support can be sourced from the $20 trillion of professionally managed assets in ESG investments and green climate funds,[61] direct commercial funding, and enlightened philanthropy. The potential sums available for investment in climate action and food system transformation are staggering. Launched in 2017, Climate Action 100+[62] is now the largest ever investor engagement initiative on climate change. The initiative includes 160 global companies that are responsible for an estimated 80% of global industrial emissions. The 545 signatory investors to Climate Action 100+ have over $52 trillion of assets. As well as pledging to transition their businesses to a net-zero emissions by 2050 (almost three decades from now!), imagine if each signatory made an immediate commitment to climate actions that transform food systems to achieve specific SDGs and was measured by their related actions.

James Carville, a strategist in Bill Clinton's successful 1992 presidential campaign, is often quoted as telling campaign workers that, *"it's the economy, stupid"*.[63] As the criticality of conditions affecting the human condition unfold, modern counterparts of Carville will inevitably feel compelled to advise their political leaders *"it's the future of humanity, stupid."* There will never be a better or perhaps even another opportunity for transformation but this will require diverse thinking and diverse solutions.

Whilst 2015 will be remembered as the year of promises on climate change and sustainable development, 2021 will have to be the year of delivery. There is already a packed agenda of international summits and conferences in which food systems and biodiversity are a common theme. These include the world's first ever UN Food Systems Summit (UNFSS).[64] The UN Secretary General, Antonio Guterres, will convene this Summit that promises to launch 'bold new actions to transform the way the world produces and consumes food, delivering progress on all 17 Sustainable Development Goals'.

We now have the resources, justification, mechanisms, incentives and imperatives to transform global food systems for good. This book has provided a personal and incomplete account of how we can do this. Its main message is that to have any chance of transforming food systems we must harness rather than diminish, enhance rather than constrain, the full spectrum of human thought, knowledge and capacities and diversity of biological and natural resources for the benefit of humanity and the planet.

Food is Joy

"By adding the text by Schiller, a philosopher whom he greatly admired, and incorporating the sound and inflection of the human voice, Beethoven conveyed a broad existential philosophy that embraced his belief in unity, tolerance, peace and joy."

Conductor Marin Alsop.[65]

The year 2020 marked 250 years since the birth of Ludwig Van Beethoven. Other composers have written nine symphonies, but *the Ninth* is associated with only one composer. We can easily understand why. Its composition marks a transformative journey from darkness to light, chaos to order, fear to joy. Its uplifting message of optimism that all mankind will be brothers has inspired millions of people across generations and throughout the world. The Ninth has been used at many openings of the Olympic Games. It became the official anthem of the European Community in 1985 and then the European Union in 1993, and its concluding Ode to Joy has become a symbol of freedom, protest, and hope. Chilean demonstrators sang it under Pinochet's dictatorship, it was performed in Berlin on Christmas Day 1989, weeks after the city's reunification, and it is part of New Year festivities in Japan. Beethoven's Ninth Symphony transcends expectations even for listeners who do not speak German, who do not usually enjoy classical music or who are not familiar with other orchestral works. It is the most popularly requested track that radio interviewees have chosen to take with them on a desert island. When asked to explain why, most identify the powerful, uplifting emotion elicited by the *Ode to Joy*.

In the quarter of a millennium since Beethoven's birth, people have endured horrific conflicts and lived in peace, empires have risen and fallen, and technologies have come and gone. Each of these events has been chronicled by contemporaries and studied by historians as we endeavour to understand how human beings have affected each other. However, until relatively recently, what was less familiar was the effects that human beings and our largest enterprise, agriculture, have had on the planet. In the brief period since the birth of Beethoven, the global population has increased from around 800 million to 7.8 billion, in the last

50 years more than half of planetary biodiversity has become extinct and we have lost most of our languages and the knowledge curated within them, not least about agriculture and agricultural biodiversity. When will we say 'enough uniformity'?

Responding to the tenets of neo-classical economics and the late 20th century emergence of neoliberalism, we have seen: a transfer of educational, research and food systems from public to private ownership; a shift from national assets to global monopolies and a change in public policy focus from human rights to commodities for profit. There have been undeniable material benefits for some people and institutions in some parts of the world in this transfer: lower general taxes; a reduction of restrictive practices especially for private-sector businesses and financial-sector enterprises; a greater focus on outcomes and cost savings through 'rationalization' of labour and infrastructure; the globalization and streamlining of systems and, of course, 'quality control.' In this context, the perception of agriculture was diminished from being a global public good to becoming an extractive industry just like oil, gas and coal.

Agriculture, however, is different. An oil rig worker or coal miner can take pride in the quality of their work and its societal purpose, but they are unlikely to derive the same joy in the fruits of their labour as a farmer. A piece of coal is not the same as a stalk of wheat or a loaf of bread, a meal with friends, or a shared recipe. Whether in a garden, allotment, farm, or landscape, the first sentiment in the cultivation of food is the joy that it brings to the grower. This joy goes well beyond the pragmatic purpose of food. Similarly, cooks, chefs, families, and their communities experience a sense of joy in eating and sharing food that goes beyond sustenance or even nutrition. Indeed, the very culture of certain countries is associated with the joy of their cuisine, while others are associated with its absence. Inherent in The Ninth Revolution will be an understanding that the production and consumption of food goes well beyond immediate materialistic needs; it is the essence of life, and as we have seen through examination of past agricultural revolutions, food is the single most important factor determining the fate of civilizations.

This book is about revolutions and empires and the role of agriculture in each. The Ninth Revolution will have to look beyond previous empires which sought a monopoly over water, hegemony over people, their territories, genetic resources and knowledge and control over the food, industrial and technological revolutions that followed them. It will have to be better than our current *Empire of the Market* which at its core is a doctrine of individual free choice over regulation by an *overbearing* state. The prevailing argument has been for less government, lower taxes, deregulation, and businesses freed from the *dead hand of bureaucracy* or any wider obligations to society. The problem with this model is that it takes no wider responsibility for society or planet. For over half a century, advocates of the Empire of the Market — who now largely run the world — have consistently argued that a *free market* is the only system that works, and that it is the best way to build wealth, distribute services, and grow the economy. They are wrong now and will be more so in the future.

Faced with a global pandemic and warming planet, and with the future of our way of life in doubt, we might be wise to reflect on whether there really is no alternative to the Empire of the Market. There have already been warnings of its limitations, but so far these have been largely unheeded. When the crisis of 2008 revealed the risks of an unfettered financial services sector, it nonetheless was left to self-regulate. When extractive industries were shown to be polluting and damaging our environment and health, they were subsidized by governments to keep doing so and we carried on consuming their products. When climate change disturbed planetary ecosystems, critically necessary actions were deemed too expensive. When the global food system was consolidated into the hands of only a few major players, this was accepted as a necessary condition for cheap food. In each case, we let a *free market* — subsidized and underwritten by tax payers and unconstrained by social values — decide what was best for us *as individuals*, not for us as part of society, nor indeed as temporary custodians and guests of the planet.

The Ninth Revolution will have to transform our agrifood system and our world to an extent that many will not yet be able to anticipate or comprehend. This will not just be a revolution only concerning alternative food ingredients or diverse products, but rather will demonstrate that there is an alternative to the Empire of the Market — one that is based on shared societal values, diverse options, science, innovations, and full realization of the importance of food and its ties to human culture and, yes, joy.

ACKNOWLEDGEMENTS

The idea for this book has been in my mind for several years but it wasn't until the circumstances and enforced lockdown of 2020 that I had the time and space to write it. Political events and the global pandemic further convinced me to make the case for diversification of food systems as one of our greatest allies in the challenges that humanity and the planet face. My purpose was to approach a larger public audience than only the academic community but, for this, I realized that without a historical perspective of food systems the case for diversification could not be placed in context nor fully justified. Again, it became clear that I needed to link agricultural biodiversity, knowledge diversity and the climate crisis if I was to convince the reader that food systems must be transformed and not just tinkered with. All this led me to consider my own personal journey in the evolution of food systems and the research and education systems that underpin them. This journey has not been mine alone nor has the process of writing this book.

For the inception and completion of the book, I gratefully acknowledge the enormous help and inspiration that I got from Max Herriman and Seth

Beckerman. Both willingly gave their time and energies to discuss ideas, monitor progress and review drafts. Their timely and critical feedback helped move the manuscript from idea to first draft in under six months. I am also grateful to George Rothschild, Peter Gregory, Sean Mayes, Craig Donald and Rolf Heemskerk for reviewing early drafts and especially appreciate the suggestions, feedback, proof reading and fact checking by Sue Azam-Ali and the help from Ebrahim Jahanshiri and colleagues in sourcing initial references. Ee Von Goh and Sami Azam-Ali spent considerable time, effort and patience with me in carefully sorting the final list of references from my jumbled notes and recollections and Bilal Riazuddin and Sue Azam-Ali for proofreading the final draft. I also thank Elizabeth McKenzie for the initial design concept of the book cover and Joe Charuy for helpful notes on ESG compliance.

I am enormously indebted to the professionalism and commitment of Joy Quek at World Scientific Publishing, her willingness to incorporate changes and revisions to the original text, the sheer speed in which she turned material around and her personal courtesy to me. I also thank her colleagues in completing the cover design and format. As an author I could not have asked for a better editor or publisher.

Friends, colleagues, partners, institutions, critics and circumstances have been part of my journey. Many are mentioned in the book but, again, I gratefully acknowledge their immense contribution and that of others not named to whom I apologize. As part of the unique group led by John Monteith at Nottingham, I acknowledge Peter Gregory who has been a constant support throughout my career, Geoff Squire (with whom I shared many ideas over many years and wrote 'Principles of Tropical Agronomy'), and colleagues such as Robin Matthews, David Harris, Chin Ong, Lester Simmonds, Jorge Garcia-Huidobro and Clare Stirling for friendship and collaboration. During the years of the Tropical Crops Research Unit, I acknowledge the support of Keith Scott, the work of Ian Stronach and Sean Clifford and a whole raft of PhD and MSc students many who continued to work with me, most notably Phibion Nyamudeza, Simon Mwale, Joe Berchie, Karen Hampson and Kimpton Chavula. We could not have completed so much without early support from the University of Nottingham, the UK Department for International Development and various scholarship schemes.

I acknowledge the personal interest of Frank Begemann and Anita Linnemann that inspired me to start active research on bambara groundnut. Its continuation would not have been possible without funding from various EU programmes and the encouragement of scientific officers such as Dirk Poittier. That this support was for a crop on which there was limited knowledge, few advocates and an uncertain future brings great credit to the European Union who funded multidisciplinary activities in three continents and for more than two decades – a true model for international collaboration and long term research. During these EU projects, I was honoured to collaborate with colleagues from many countries including, amongst others, Abu Sesay, Simon Karikari, Nuhu Hatibu and MS Basu. Much of the research at Nottingham would not have been possible without the exemplary contributions of Sarah Collinson, Festo Massawe, Shravani Basu, Asha Karunaratne, countless students and support staff at the Sutton Bonington Campus, most notably the late David Hodson.

For the early life of CFFRC I am indebted to Hannah Jaenicke, Michael Hermann, George Rothschild and academic colleagues, most notably Festo Massawe, Sean Mayes and Debbie Sparkes. Its establishment would not have been possible without the support of the Malaysian Government which I gratefully acknowledge. However, my greatest debt of gratitude is to each of the wonderful staff and extended family of CFFRC that I had the honour to lead, privilege to learn from and pleasure to work with. I thank you all.

In this book, I have tried to provide as many supporting references as possible throughout the text. However, the sheer breadth of the topics covered means that there are omissions both in range and depth of sources. In my description of the life and career of John Monteith, I relied heavily on the article by Mike Unsworth that appears in Biographical Memoirs of Fellows of the Royal Society. In Chapter 4, I use the example of 'Nutritionism' and related ideas that were introduced to me through reading 'In defense of food: an eater's manifesto' by Michael Pollan. In Chapter 5 the idea to write about the East African Groundnut Scheme as a case study in hubris came from an article written by Alan Thomas in 'The Oldie' magazine in October 2019. In Chapter 8, I use the concept of 'Cognitive Diversity' that is so eloquently described by Matthew Syed in his most recent book '*Rebel ideas: the power of diverse thinking*'. The concept of cognitive diversity needs testing across all aspects of human behaviour as an essential framework for the wider biological and linguistic diversity advocated in this book. But it seems to me that it is our food systems that most urgently need to apply its principles since their success or failure affect us all and their transformation needs diverse thoughts, participants and actions. Perhaps this is the biggest challenge ahead.

ENDNOTES

1. The First Agricultural Revolution

1. Vasey DE. An ecological history of agriculture: 10,000 B.C.–A.D. 10,000. Ames: Iowa State University Press; 1992. 344 p.

2. Shelef O, Weisberg PJ, Provenza FD. The value of native plants and local production in an era of global agriculture. Frontiers in Plant Science. 2017 Dec 5;8:2069. DOI: 10.3389/fpls.2017.02069.

3. Ruiz N, Noe-Bustamante L, Saber N. Coming of age: By 2050, as birth rates continue to drop and people live longer, the world's population will change [Internet]. Washington DC: Finance & Development; 2020 [cited 2021 Feb 3]. https://www.imf.org/external/pubs/ft/fandd/2020/03/infographic-global-population-trends-picture.htm.

4. Eberhard DM, Simons GF, Fennig CD. (eds.). Ethnologue: Languages of the world. Twenty-third edition. Dallas: SIL International; 2020.

5. Shiva V. Monocultures of the mind: Perspectives on biodiversity and biotechnology. Penang: Zed Books and Third World Network; 1993. 184 p.

6. Renfrew C, Bahn P. Archaeology: Theories, methods, and practice. Eighth Edition. London: Thames & Hudson Publications; 2019. 656 p.

7. Ungar PS, Sponheimer M. The diets of early Hominins. Science. 2011 Oct 14; 334(6053): 190–193. DOI: 10.1126/science.1207701.

8. Hayden B. Subsistence and ecological adaptations of modern hunter-gatherers. In: Harding RSO, Teleki G, editors. Omnivorous primates. New York: Columbia University Press; 1981. 78 p. [xx-xx].

9. McConnell DJ, Dharmapala KA, Attanayake, SR, Upawansa GK. The forest farms of Kandy: And other gardens of complete design. Hampshire: Ashgate; 2003. 535 p.

10. Azam-Ali SN. Fitting underutilised crops within research-poor environments: Lessons and approaches. South African Journal of Plant and Soil. 2010 Jan 1;27(4):293–8. DOI: 10.1080/02571862.2010.10639997.

11. Kew. Seed Collection [Internet]. London: Kew; [date unknown] [cited 2021 Jan 19]. https://www.kew.org/science/collections-and-resources/collections/seed-collection.

12. Brockway LH. Science and colonial expansion: The role of the British Royal Botanic Gardens. American Ethnologist. 1979 Aug; 6(3): 449–65. DOI: 10.1525/ae.1979.6.3.02a00030.

13. Brockway LH. Science and colonial expansion: The role of the British Royal Botanic Gardens. American Ethnologist. 1979 Aug; 6(3): 449–65. DOI: 10.1525/ae.1979.6.3.02a00030.

14. Brockway LH. Science and colonial expansion: The role of the British Royal Botanic Gardens. American Ethnologist. 1979 Aug; 6(3): 449–65. DOI: 10.1525/ae.1979.6.3.02a00030.

15. Kew. Seed Collection [Internet]. London: Kew; [date unknown] [cited 2021 Jan 19]. https://www.kew.org/science/collections-and-resources/collections/seed-collection.

16. CGIAR. CGIAR Genebank Platform [Internet]. [place unknown]: CGIAR; [date unknown] [cited 2021 Jan 19]. https://www.cgiar.org/research/program-platform/genebank-platform/.

17. Crop Trust. Svalbard Global Seed Vault [Internet]. Bonn: Crop Trust; [date unknown] [cited 2021 Jan 19]. https://www.croptrust.org/our-work/svalbard-global-seed-vault/.

18. Deshpande S. Science Versus Ideology. Resonance. 2019 Dec; 24(12): 1355–73. DOI: 10.1007/s12045-019-0903-4.

19. Villa TC, Maxted N, Scholten M, Ford-Lloyd B. Defining and identifying crop landraces. Plant Genetic Resources. 2005 Dec;3(3):373–84. DOI: 10.1079/PGR200591.

20. Hummer KE, Hancock JF. Vavilovian centers of plant diversity: Implications and impacts. Horticultural Science. 2015 Jun 1;50(6):780–3. DOI: 10.21273/HORTSCI.50.6.780

21. Pringle P. The murder of Nikolai Vavilov: The story of Stalin's persecution of one of the greatest scientists of the 20th century. New York: Simon & Schuster; 2011. 384 p.

22. Kupzow AJ. Vavilov's law of homologous series at the fiftieth anniversary of its formulation. Economic Botany. 1975 Oct;29(4):372–9. DOI: 10.1007/BF02862184.

23. Harlan JR. Evolution of cultivated plants. In: Frankel OH, Bennett E, editors. Genetic Resources in Plants — IBP Handbook No 11. London: International Biological Programme; 1970. 554 p. [19–32].

24. Harmand S, Lewis JE, Feibel CS, Lepre CJ, Prat S, Lenoble A, et al. 3.3-million-year-old stone tools from Lomekwi 3, West Turkana, Kenya. Nature. 2015 May; 521(7552): 310-5. DOI: 10.1038/nature14464.

25. Michalowski P. Sumerian. In: Woodward RD, editor. The Cambridge Encyclopedia of the World's Ancient Languages. Cambridge: Cambridge University Press; 2004. 1162 p. [xx–xx].

26 . Mingren W. Sumerian Tablets: A deeper understanding of the oldest known written language [Internet]. [place unknown]: Ancient Origins; 2019 May [cited 2021 Jan 19]. https://www.ancient-origins.net/artifacts-ancient-writings/sumerian-tablets-0011895.

27. Eberhard DM, Simons GF, Fennig CD. (eds.). Ethnologue: Languages of the world. Twenty-third edition. Dallas, Texas: SIL International; 2020.

28. Eberhard DM, Simons GF, Fennig CD. (eds.). Ethnologue: Languages of the world. Twenty-third edition. Dallas, Texas: SIL International; 2020.

29. Krauss M. The world's languages in crisis. Language. 1992; 68(1): 4–10. DOI: 10.1075/slcs.142.01sim.

30. Endangered Languages Project. The Endangered Languages Project [Internet]. Mānoa: Endangered Languages Project; [date unknown] [cited 2021 Jan 19]. http://www.endangeredlanguages.com/.

31. Lee NH, Van Way J. Assessing levels of endangerment in the Catalogue of Endangered Languages (ELCat) using the Language Endangerment Index (LEI). Language in Society. 2016 Apr 1;45(2):271. DOI: 10.1017/S0047404515000962.

32. Campbell L, Lee NH, Okura E, Simpson S, Ueki K, Van Way J. New knowledge: Findings from the Catalogue of Endangered Languages (ELCat). Paper presented at the 3rd International Conference on Language Documentation and Conservation. 2013 February 28–March 3; University of Hawaii at Mānoa; [cited 2021 Feb 3]. https://scholarspace.manoa.hawaii.edu/bitstream/10125/26145/26145.pdf.

33. Tsunoda T. Language endangerment and language revitalization: An introduction. Berlin: Mouton de Gruyter; 2005. 307 p.

34. Evans N. Dying words: Endangered languages and what they have to tell us. Massachusetts: Wiley-Blackwell; 2010. 287 p.

35. Crystal D. Language death. Cambridge: Cambridge University Press; 2000. 198 p.

36. Skutnabb-Kangas T. Legitimating or delegitimating new forms of racism — the role of researchers. Journal of Multilingual & Multicultural Development. 1990 Jan 1;11(1–2):77-100. DOI: 10.1080/01434632.1990.9994402.

37. UN General Assembly. Convention on the Prevention and Punishment of the Crime of Genocide. Signed 9 December 1948, entered into force 12 January 1951. 78 UNTS 277.

38. Pakir A. Baba Malay as an endangered language. Paper presented at the Sixth International Conference on Austronesian Linguistics. 1991 May 20–24; Honolulu, Hawaii; [cited 2021 Feb 3]. https://indoling.com/ical/.

39. Convention on Biological Diversity. The Convention on Biological Diversity [Internet]. [place unknown]: CBD Secretariat; 2020 Apr [cited 2021 Jan 18]. https://www.cbd.int/convention/.

40. IUCN. World Conservation Strategy: Living Resource Conservation for Sustainable Development. IUCN; 1980. https://portals.iucn.org/library/efiles/documents/WCS-004.pdf.

41. UN General Assembly. World Charter for Nature, 28 October 1982, A/RES/37/7. UN General Assembly, 37th sess.: 1982–1983. https://digitallibrary.un.org/record/39295/files/A_RES_37_7-AR.pdf.

42. World Commission on Environment and Development. Our common future. Oxford: Oxford University Press; 1987. 383 p.

43. UN Conference on Environment and Development. Agenda 21, Rio Declaration, Forest Principles. New York: United Nations; 1992. https://sustainabledevelopment.un.org/content/documents/Agenda21.pdf.

44. UN General Assembly. Universal Declaration of Human Rights, 10 December 1948, 217 A (III). https://www.ohchr.org/EN/UDHR/Documents/UDHR_Translations/eng.pdf.

45. Harmon D. Sameness and silence: Language extinctions and the dawning of a biocultural approach to diversity. Global Biodiversity. 1998;8(3):2–10.

46. Skutnabb-Kangas T. Marvelous human rights rhetoric and grim realities: Language rights in education. Journal of Language, Identity, and Education. 2002 Jul 2;1(3):179–205.

47. Skutnabb-Kangas T. On biolinguistic diversity: Linking language, culture and (traditional) ecological knowledge [transcription on internet]. Invited Plenary lecture at Department of Linguistics and Philosophy, Universidad Autónoma de Madrid and Cosmocaixa; 2004 [cited 2021 Feb 3]. https://textarchive.ru/c-1894055-pall.html.

48. Mace R, Pagel M. A latitudinal gradient in the density of human languages in North America. Proceedings of the Royal Society of London. Series B: Biological Sciences. 1995 Jul 22;261(1360):117–21. DOI: 10.1098/rspb.1995.0125.

49. Skutnabb-Kangas T. When languages disappear, are bilingual education or human rights a cure? Two scenarios. In: Wei L, Dewaele JM, Housen A, editors. Opportunities and Challenges of Bilingualism. New York: Walter de Gruyter; 2002. 346 p. [45–67].

50. Sapir E. The status of linguistics as a science. Language. 1929 Dec 1:207–14. DOI: 10.2307/409588.
51. Kew. Seed Collection [Internet]. London: Kew; [date unknown] [cited 2021 Jan 19]. https://www.kew.org/science/collections-and-resources/collections/seed-collection.
52. Vavilov NI. Five continents. Love D, translator. Rome: International Plant Genetic Resources Institute; 1997. 198 p.

2. Empires of Power

1. Herrera RJ, Garcia-Bernard R. Ancestral DNA, Human Origins, and Migrations. First Edition. London: Elsevier; 2018. p. 475–509.
2. Wittfogel KA. Oriental despotism: A comparative study of total power. New Haven: Yale University Press; 1957. 556 p.
3. Liverani M. Akkad: The first world empire: Structure, ideology, traditions. Sixth Edition. Padova: Sargon; 1993. 182 p.
4. Yannopoulos SI, Lyberatos G, Theodossiou N, Li W, Valipour M, Tamburrino A, *et al*. Evolution of water lifting devices (pumps) over the centuries worldwide. Water. 2015 Sep; 7(9): 5031–60. DOI: 10.3390/w7095031.
5. Ellison R. Some thoughts on the diet of Mesopotamia from c. 3000–600 BC. Iraq. 1983 Apr 1; 45(1): 146–50. DOI: 10.2307/4200193.
6. Zettler R. Reconstructing the world of ancient Mesopotamia: Divided beginnings and holistic history. Journal of the Economic and Social History of the Orient. 2003 Jan 1; 46(1): 3–45.
7. Ricketts C. The growth of the Roman empire explained [Internet]. [place unknown]: History Hit; 2018 Aug 9 [cited 2021 Jan 18]. https://www.historyhit.com/the-growth-of-the-roman-empire-explained/.
8 . Fussell GE. Farming systems of the classical era. Technology and Culture. 1967 Jan 1; 8(1): 16–44. DOI: 10.2307/3101523.
9. Cartwright M. Food in the Roman world [Internet]. [place unknown]: Ancient History Encyclopedia Foundation; 2014 May 6 [cited 2021 Jan 19]. https://www.ancient.eu/article/684/food-in-the-roman-world/.
10. Ricketts C. Divorce and decline: the division of East and West Roman empires [Internet]. [place unknown]: History Hit; 2018 Jul 30 [cited 2021 Jan 19]. https://www.historyhit.com/divorce-and-decline-the-division-of-east-and-west-roman-empires/.

11. Idrisi Z. The Muslim agricultural revolution and its influence on Europe [Internet]. Manchester: FSTC Limited; 2005 [cited 2021 Feb 4]. 19 p. https://www.muslimheritage.com/uploads/AgricultureRevolution2.pdf

12. Imamuddin, SM. Muslim Spain 711–1492 AD: A sociological study. Leiden: Brill; 1981. 269 p.

13. Watson AM. The Arab agricultural revolution and its diffusion, 700–1100. The Journal of Economic History. 1974 Mar 1; 8–35. DOI: 10.1017/S0022050700079602.

14. Harvey JH. Gardening books and plant lists of Moorish Spain. Garden History. 1975 Apr 1; 3(2): 10–21. DOI: 10.2307/1586375.

15. Norman JM. Hulagu Khan's army threw so many books into the Tigris river that they formed a bridge that would support a man on horseback [Internet]. [place unknown]: HistoryofInformation.com; 2021 Jan 19 [cited 2021 Jan 19]. https://www.historyofinformation.com/detail.php?id=249.

16. McCarthy N. The biggest empires in human history [Internet]. [place unknown]: Statista; 2020 May 25 [cited 2021 Jan 19]. https://www.statista.com/chart/20342/peak-land-area-of-the-largestempires/#:~:text=In%201913%2C%20412%20million%20people,of%20the%20world's%20land%20area.

17. Darwin J. The end of the British empire: The Historical Debate. New Jersey: Wiley; 2006. 148 p.

18. Darwin J. The end of the British empire: The Historical Debate. New Jersey: Wiley; 2006. 148 p.

19. Mokyr J. Great famine [Internet]. [place unknown]: Encyclopædia Britannica; 2020 Feb 4 [cited 2021 Jan 19]. https://www.britannica.com/event/Great-Famine-Irish-history.

20. Akuamoa GK. Kwame, The Last Slave From West Africa. Accra: Lulu; 2011. 223 p.

21. Murdoch HA. A Legacy of Trauma: Caribbean Slavery, Race, Class, and Contemporary Identity in "Abeng". Research in African Literatures. 2009 Dec 1; 65–88. DOI: 10.1353/ral.0.0204.

22. Walvin J. The slave trade, abolition and public memory. Transactions of the Royal Historical Society. 2009 Jan 1; 139–49. DOI: 10.1017/S0080440109990077.

23. Latimer J. The apprenticeship system in the British West Indies. The Journal of Negro Education. 1964 Jan 1; 33(1): 52–7. DOI: 10.2307/2294514.

24. Roos D. How the East India Company became the world's most powerful monopoly [Internet]. [place unknown]: A&E Television Networks; 2020 Oct

23 [cited 2021 Jan 19]. https://www.history.com/news/east-india-company-england-trade.

25. Dalrymple W. The anarchy: The relentless rise of the East India Company. New York: Bloomsbury Publishing; 2019. 544 p.

26. Broadberry S, Overton M, Klein A, Van Leeuwen B. British economic growth, 1270–1870: An output-based approach. Cambridge: Cambridge University Press; 2015. 461 p.

27. Hodgson GM. 1688 and all that: Property rights, the Glorious Revolution and the rise of British capitalism. Journal of Institutional Economics. 2017 Mar; 13(1): 79–107. DOI: 10.1017/S1744137416000266.

28. Bennett RJ, Law CM, Robson B, Langton J. Urban population database, 1801–1911. SN: 7154. UK Data Service; 2012; [cited 2021 Feb 4]. https://beta.ukdataservice.ac.uk/datacatalogue/doi/?id=7154#!#1.

29. Allen R. Agriculture during the industrial revolution. In: Floud R, McCloskey DN, editors. The economic history of Britain since 1700. Second Edition. Cambridge: Cambridge University Press; 1994. p. 96–123.

30. Spring J. Education and the rise of the global economy. London: Routledge; 1998. 248 p.

31. Schreuder Y. The British Caribbean World: Barbados. In: Amsterdam's Sephardic Merchants and the Atlantic Sugar Trade in the Seventeenth Century. Cham: Palgrave Macmillan; 2019. p. 77–118.

32. Splitstoser JC, Dillehay TD, Wouters J, Claro A. Early pre-Hispanic use of indigo blue in Peru. Science Advances. 2016 Sep 1; 2(9): e1501623. DOI: 10.1126/sciadv.1501623.

33. Moulherat C, Tengberg M, Haquet JF, Mille B. First evidence of cotton at Neolithic Mehrgarh, Pakistan: Analysis of mineralized fibres from a copper bead. Journal of Archaeological Science. 2002 Dec 1; 29(12): 1393–401. DOI: 10.1006/jasc.2001.0779.

34. Maddison A. Monitoring the world economy, 1820–1992. Paris: Development Centre of the OECD; 1995. 30 p.

35. Eacott JP. Making an imperial compromise: The Calico Acts, the Atlantic Colonies, and the structure of the British Empire. The William and Mary Quarterly. 2012 Oct 1; 69(4): 731–62. DOI: 10.5309/willmaryquar.69.4.0731.

36. Marshall PJ. The Cambridge illustrated history of the British Empire. Cambridge: Cambridge University Press; 1996. 400 p.

37. Zydenbos S. Arable farming — arable crops today [Internet]. [place unknown]: Te Ara — the Encyclopedia of New Zealand; 2008 Nov 24 [cited 2021 Jan 20]. http://www.TeAra.govt.nz/en/arable-farming/page-3.

38. Australia Government Department of Agriculture, Water and the Environment. Crops [Internet]. Canberra: Department of Agriculture, Water and the Environment; 2019 Nov 4 [cited 2021 Jan 20]. https://www.agriculture.gov.au/ag-farm-food/crops.

39. The Borneo Project. Biodiversity Conservation [Internet]. Berkeley: The Borneo Project; [Date unknown] [cited 2021 Jan 20]. https://borneoproject.org/borneo/biodiversity-conservation/#:~:text=The%20Borneo%20rainforest%20is%20the,one%20of%20the%20most%20biodiverse.&text=There%20are%20about%2015%2C000%20species,of%20resident%20birds%20in%20Borneo.

40. Lange RT. Australia- animal life [Internet]. [place unknown]: Encyclopædia Britannica; 2021 Jan 18 [cited 2021 Jan 20]. https://www.britannica.com/place/Australia/Animal-life.

3. The Green Revolution

1. Khoury CK, Bjorkman AD, Dempewolf H, Ramirez-Villegas J, Guarino L, Jarvis A, *et al*. Increasing homogeneity in global food supplies. PNAS. 2014 Mar 18; 111(11): 4001–4006. DOI: 10.1073/pnas.1313490111.

2. Massawe F, Mayes S, Cheng A. Crop diversity: An unexploited treasure trove for food security. Trends in Plant Science. 2016 May; 21(5): 365–368. DOI: 10.1016/j.tplants.2016.02.006.

3. Taiz L. Agriculture, plant physiology, and human population growth: Past, present, and future. Theoretical and Experimental Plant Physiology. 2012 Dec; 25(3): 167–181. DOI: 10.1590/S2197-00252013000300001.

4. Pimm S, Joppa L. How many plant species are there, where are they, and at what rate are they going extinct? Annals of the Missouri Botanical Garden. 2015 Mar; 100(3):170–176. DOI: 10.3417/2012018.

5. FAO. Once neglected, these traditional crops are our new rising stars [Internet]. Rome: Food and Agriculture Organization of the United Nations; 2018 Oct 2 [cited 2021 Jan 21]. http://www.fao.org/fao-stories/article/en/c/1154584/.

6. FAO. Once neglected, these traditional crops are our new rising stars [Internet]. Rome: Food and Agriculture Organization of the United

Nations; 2018 Oct 2 [cited 2021 Jan 21]. http://www.fao.org/fao-stories/article/en/c/1154584/.

7. FAO. Once neglected, these traditional crops are our new rising stars [Internet]. Rome: Food and Agriculture Organization of the United Nations; 2018 Oct 2 [cited 2021 Jan 21]. http://www.fao.org/fao-stories/article/en/c/1154584/.

8. New World Encyclopedia contributors. Norman Borlaug [Internet]. [place unknown]: New World Encyclopedia; 2019 Sep 7 [cited 2021 Jan 21]. https://www.newworldencyclopedia.org/p/index.php?title=Norman_Borlaug&oldid=1024353.

9. CIMMYT. Our history [Internet]. [place unknown]: CIMMYT; [date unknown] [cited 2021 Jan 21]. https://www.cimmyt.org/about/our-history/.

10. Borlaug N. Nobel lecture [Internet]. [place unknown]: Nobel Media AB 2021; 1970 Dec 11 [cited 2021 Jan 21]. https://www.nobelprize.org/prizes/peace/1970/borlaug/lecture/.

11. Singh S. Norman Borlaug: A billion lives saved [Internet]. [place unknown]: AgBioWorld; 2011 [cited 2021 Jan 21]. http://www.agbioworld.org/biotech-info/topics/borlaug/special.html.

12. Pradhan DL. A textbook of environmental studies. Pune: Devine House Publication; 2015. p. 14.

13. Cagliarini A, Rush A. Economic development and agriculture in India. RBA Bulletin. 2011 Jun; 15–22.

14. FAO. FAOSTAT Database. Crops. "Philippines: 1961 and 2009 crop production statistics". [cited 2021 Jan 21]. http://www.fao.org/faostat/en/#data/QC.

15. Calpe C. International trade in rice: Recent developments and prospects. Manilla: International Rice Research Institute (IRRI); 2005. Report No.:492

16. Henkel M. 21st Century homestead: Sustainable agriculture II: Farming and natural resources. Raleigh: Lulu Press; 2015. 394 p.

17. Rothamsted Research. The history of Rothamsted Research [Internet]. [Place unknown]: Rothamsted Research; 2021 [cited 2021 Jan 21]. https://www.rothamsted.ac.uk/history-and-heritage.

18. Paull J. A Century of Synthetic Fertilizer: 1909–2009. Journal of Bio-Dynamics Tasmania. 2009 Jun; 94 :16–21.

19. Paull J. A Century of Synthetic Fertilizer: 1909–2009. Journal of Bio-Dynamics Tasmania. 2009 Jun; 94: 16–21.

20. The Editors of Encyclopaedia Britannica. Haber-Bosch process [Internet]. [Place unknown]: Encyclopædia Britannica; 2020 Feb 11 [cited 2021 Jan 21]. https://www.britannica.com/technology/Haber-Bosch-process.

21. Friedrich B, Hoffmann D, Renn J, Schmaltz F, Wolf M. One hundred years of chemical warfare: Research, deployment, consequences. Cham: Springer Nature; 2017. 408 p.

22. Science History Institute. Fritz Haber [Internet]. [Place unknown]: Science History Institute; 2021 [cited 2021 Jan 21]. https://www.sciencehistory.org/historical-profile/fritz-haber

23. Kudsk P, Streibig JC. Herbicides–a two-edged sword. Weed Research. 2003 Apr; 43(2): 90–102. DOI: 10.1046/j.1365-3180.2003.00328.x.

24. Green JM, Owen MD. Herbicide-resistant crops: Utilities and limitations for herbicide-resistant weed management. Journal of Agricultural and Food Chemistry. 2011 Jun 8; 59(11): 5819–29. DOI: 10.1021/jf101286h.

25. Tuzimski T. Herbicides and pesticides. In: Worsfold P, Poole C, Townshend A, Miro M, editors. Encyclopedia of Analytical Science. Third Edition. Amsterdam: Elsevier; 2019. p. 391–398.

26. Friedmann T. Heard any good Jews lately? In: Logan C. Counterbalance: Gendered perspectives for writing and language. Ontario: Broadview Press; 1997. p. 258–263.

27. Lichtfouse E, Schwarzbauer J, Robert D. Environmental chemistry for a sustainable world volume 1: Nanotechnology and health risk. New York: Springer; 2012. 410 p.

28. Poulopoulos S, Inglezakis V. Environment and development: Basic principles, human activities, and environmental implications. London: Elsevier; 2016. 594 p.

29. Nelson AR, Ravichandran K, Antony U. The impact of the green revolution on indigenous crops of India. Journal of Ethnic Foods. 2019 Dec; 6(1): 1–0. DOI: 10.1186/s42779-019-0011-9.

30. FAO. FAOSTAT Database. Crops. "Indonesia: 1966 and 2000 crop production statistics" [cited 2021 Jan 21]. http://www.fao.org/faostat/en/#data/QC.

31. Rosset P, Collins J, Moore Lappé F. Lessons from the green revolution: Do We Need New Technology to End Hunger. Tikkun. 2000 Mar 1; 15(2): 52–6.

32. Rega C. Ecological rationality in spatial planning: Concepts and tools for sustainable land-use decisions. Cham: Springer; 2020. 198 p.

33. Otero G. Food for the few: Neoliberal globalism and biotechnology in Latin America. Austin: University of Texas Press; 2008. 335 p.

34. Cohen MJ, Reeves D. Causes of hunger: 2020 vision brief 19. Washington DC: International Food Policy Research Institute; 1995. 5 p.

35. Roser M, Ritchie H. Food supply [Internet]. [place unknown]: Our World In Data; 2013 [cited 2021 Jan 21]. https://ourworldindata.org/food-supply.

36. Singh S. Norman Borlaug: A billion lives saved [Internet]. [place unknown]: AgBioWorld; 2011 [cited 2021 Jan 21]. http://www.agbioworld.org/biotech-info/topics/borlaug/special.html.

37. Borlaug N. Nobel lecture [Internet]. [place unknown]: Nobel Media AB 2021; 1970 Dec 11 [cited 2021 Jan 21]. https://www.nobelprize.org/prizes/peace/1970/borlaug/lecture/.

38. SeedQuest. New Zealand farmer breaks world record with massive 17.398 tonne per hectare wheat crop [Internet]. New Zealand: SeedQuest; 2020 Jul 8 [2021 Jan 21]. https://www.seedquest.com/news.php?type=news&id_article=119048.

39. Alexander RJ. Lázaro Cárdenas [Internet]. Britannica.com: Encyclopædia Britannica; 2020 Oct 15 [cited 2021 Jan 21]. https://www.britannica.com/biography/Lazaro-Cardenas.

40. Patel R. The Long Green Revolution. The Journal of Peasant Studies. 2013 Jan 1; 40(1): 1–63. DOI: 10.1080/03066150.2012.719224.

41. Sandhu J. Green revolution: A case study of Punjab. Proceedings of the Indian History Congress. 2014 Jan; 75: 1192–1199.

42. Berg E. Accelerated development in sub-Saharan Africa: a plan for action. Washington DC: International Bank for Reconstruction and Development/ The World Bank; 1981. 217 p.

43. Heidhues F, Obare GA. Lessons from structural adjustment programmes and their effects in Africa. Quarterly Journal of International Agriculture. 2011; 50(1): 55–64. DOI: 10.22004/ag.econ.155490.

44. Paarlberg R. Food Politics: What everyone needs to know. Oxford: Oxford University Press; 2013. 224 p.

45. Paarlberg R. Food Politics: What everyone needs to know. Oxford: Oxford University Press; 2013. 224 p.

46. Freebairn DK. Did the Green Revolution concentrate incomes? A quantitative study of research reports. World Development. 1995 Feb 1; 23(2): 265–79. DOI: 10.1016/0305-750X(94)00116-G.

47. Borlaug N. Nobel lecture [Internet]. [place unknown]: Nobel Media AB 2021; 1970 Dec 11 [cited 2021 Jan 21]. https://www.nobelprize.org/prizes/peace/1970/borlaug/lecture/.

48. Hazell PB. The Asian Green Revolution. Volume 911 of IFPRI Discussion Paper. Washington DC: Intl Food Policy Res Inst; 2009. 31 p.

49. Aktar W, Sengupta D, Chowdhury A. Impact of pesticides use in agriculture: Their benefits and hazards. Interdisciplinary Toxicology. 2009 Mar 1; 2(1): 1–2. DOI: 10.2478/v10102-009-0001-7.

50. FAO. FAOSTAT Database. Fertilisers by Nutrient. "World: 1961 and 1998 fertilisers by nutrient". [cited 2021 Jan 21]. http://www.fao.org/faostat/en/#data/RFN.

51. Nelson AR, Ravichandran K, Antony U. The impact of the green revolution on indigenous crops of India. Journal of Ethnic Foods. 2019 Dec; 6(1): 1–0. DOI: 10.1186/s42779-019-0011-9.

52. Gates B. Prepared remarks. [Speech presented at World Food Prize Symposium]. Des Moines, Iowa. 2009 Oct 15.

53. Patel R. The Long Green Revolution. The Journal of Peasant Studies. 2013 Jan 1; 40(1): 1–63. DOI: 10.1080/03066150.2012.719224.

54. Patel R. The Long Green Revolution. The Journal of Peasant Studies. 2013 Jan 1; 40(1): 1–63. DOI: 10.1080/03066150.2012.719224.

55. DeWalt BR. Mexico's second green revolution: Food for feed. Mexican Studies. 1985 Jan 1; 1(1): 29–60. DOI: 10.2307/1051979.

56. Paddock WC. How green is the green revolution? BioScience. 1970 Aug 15; 897–902. DOI: 10.2307/1295581.

57. Carson R. Silent spring. Boston: Houghton Mifflin Company; 1962. 368 p.

58. United States of America. National Environmental Policy Act of 1969, Public Law 91–190.

59. Conway G. The doubly Green Revolution. New York: Cornell University Press; 1997. 334 p.

4. The Titanic Agrifood System

1. Mbow C, Rosenzweig C, Barioni LG, Benton TG, Herrero M, Krishnapillai M, Liwenga E, Pradhan P, Rivera-Ferre MG, Sapkota T, Tubiello FN, Xu Y. Food Security. In: Shukla PR, Skea J, Calvo Buendia E, Masson-Delmotte V, Pörtner HO, Roberts DC, Zhai P, Slade R, Connors S, Van Diemen R,

Ferrat M. Climate Change and Land: An IPCC special report on climate change, desertification, land degradation, sustainable land management, food security, and greenhouse gas fluxes in terrestrial ecosystems. Geneva: United Nations' Intergovernmental Panel on Climate Change; 2019. p. 437–550. https://www.ipcc.ch/srccl/chapter/chapter-5/.

2. Holt-Giménez E, Shattuck A, Altieri M, Herren H, Gliessman S. We Already Grow Enough Food for 10 Billion People … and Still Can't End Hunger. Journal of Sustainable Agriculture. 2012 Jul 24; 36(6): 595–8. DOI: 10.1080/10440046.2012.695331.

3. Kavanagh M. Household food spending divides the world [Internet]. London: The Financial Times Limited; 2019 Jan 8 [cited 2021 Jan 23]. https://www.ft.com/content/cdd62792-0e85-11e9-acdc-4d9976f1533b.

4. Carmody R, Wrangham R. Cooking and the human commitment to a high-quality diet. Cold Spring Harbour Symposia on Quantitative Biology 2009 Oct 20; 74: 427–34. DOI: 10.1101/sqb.2009.74.019.

5. Roach J. 4,000-year-old noodles found in China [Internet]. Washington DC: National Geographic; 2005 Oct 12 [cited 2021 Jan 22]. https://www.nationalgeographic.com/news/2005/10/4-000-year-old-noodles-found-in-china/.

6. Chokshi N. Archaeologists find 3,200-year-old cheese in an Egyptian tomb [Internet]. New York: New York Times; 2018 Aug 16 [cited 2021 Jan 22]. https://www.nytimes.com/2018/08/16/science/oldest-cheese-ever-egypt-tomb.html.

7. Powell EA. The first bakers [Internet]. New York: Archaeology; 2019 Feb [cited 2021 Jan 22]. https://www.archaeology.org/issues/323-1901/features/7197-jordan-shubayqa-natufian-pita.

8. History.com Staff. Popcorn was popular in ancient Peru, discovery suggests [Internet]. [place unknown]: History.com; 2012 Jan 20 [updated 2018 Aug 29; cited 2021 Jan 22]. https://www.history.com/news/popcorn-was-popular-in-ancient-peru-discovery-suggests.

9. EUFIC. Evolution of food processing and labelling in food production [Internet]. Brussels: The European Food Information Council; 2014 Dec 8 [cited 2021 Jan 22]. https://www.eufic.org/en/food-production/article/food-production-1-3-the-evolution-of-meeting-nutritional-needs-through-proc.

10. Simpson R, Núñez H, Almonacid S. Sterilization process design. In: Ahmed J, Rahman MS, editors. Handbook of food process design, 2 volumes. First Edition. Chichester: Wiley-Blackwell; 2012. p. 362–380.

11. Simpson R, Núñez H, Almonacid S. Sterilization process design. In: Ahmed J, Rahman MS, editors. Handbook of food process design, 2 volumes. First Edition. Chichester: Wiley-Blackwell; 2012. p. 362–380.

12. FAO. 2050: A third more mouths to feed [Internet]. Rome: Food and Agriculture Organization of the United Nations; 2009 Sep 23 [cited 2021 Jan 22]. http://www.fao.org/news/story/en/item/35571/icode/#:~:text=Food%20 demand&text=Annual%20cereal%20production%20will%20have,from%20 the%2058%20percent%20today.

13. Union of Concerned Scientists. The hidden costs of industrial agriculture [Internet]. Massachusetts: Union of Concerned Scientists; 2008 Jul 11 [cited 2021 Jan 22]. https://www.ucsusa.org/resources/hidden-costs-industrial-agriculture.

14. Lusk J. The evolution of American agriculture [Internet]. West Lafayette: Jayson Lusk; 2016 Jun 27 [cited 2021 Jan 22]. http://jaysonlusk.com/ blog/2016/6/26/the-evolution-of-american-agriculture.

15. Behere PB, Bhise MC. Farmers' suicide: Across culture. Indian Journal of Psychiatry. 2009 Oct; 51(4): 242. DOI: 10.4103/0019-5545.58286.

16. Fraser CE, Smith KB, Judd F, Humphreys JS, Fragar LJ, Henderson A. Farming and mental health problems and mental illness. International Journal of Social Psychiatry. 2005 Dec; 51(4): 340–9. DOI: 10.1177/0020764005060844.

17. Gruère G, Sengupta D. Bt cotton and farmer suicides in India: An evidence-based assessment. The Journal of Development Studies. 2011 Feb 1; 47(2): 316–37. DOI: 10.1080/00220388.2010.492863.

18. Das A. Farmers' suicide in India: implications for public mental health. The International Journal of Social Psychiatry. 2011 Jan 1; 57(1): 21–9. DOI: 10.1177/0020764009103645.

19. FAO. Food security. Policy brief. Rome: Food and Agriculture Organization of the United Nations; 2006 Jun. Issue 2. http://www.fao.org/fileadmin/ templates/faoitaly/documents/pdf/pdf_Food_Security_Concept_Note.pdf.

20. Shaw DJ. World food security. London: Palgrave Macmillan UK; 2007. 510 p.

21. FAO. Food security. Policy brief. Rome: Food and Agriculture Organization of the United Nations; 2006 Jun. Issue 2. http://www.fao.org/fileadmin/ templates/faoitaly/documents/pdf/pdf_Food_Security_Concept_Note.pdf.

22. Clay E. Trade reforms and food security: Conceptualising the linkages. Rome: Food and Agriculture Organization of the United Nations; 2003. http://www.fao.org/3/a-y4671e.pdf.

23. FAO. Rome Declaration on World Food Security and World Food Summit Plan of Action. World Food Summit 1996 November 13–17. Rome: Food and Agriculture Organization of the United Nations; 1996. http://www.fao.org/wfs/.

24. FAO, IFAD, UN Children's Fund, WFP, WHO. The state of food security and nutrition in the world (SOFI) report 2019 — in brief. Rome: Food and Agriculture Organization of the United Nations; 2019. https://docs.wfp.org/api/documents/WFP-0000106760/download/?_ga=2.82667615.870359159.1612941460-968858223.1612941460.

25. FAO, IFAD, UN Children's Fund, WFP, WHO. The state of food security and nutrition in the world (SOFI) report 2019 — in brief. Rome: Food and Agriculture Organization of the United Nations; 2019. https://docs.wfp.org/api/documents/WFP-0000106760/download/?_ga=2.82667615.870359159.1612941460-968858223.1612941460.

26. FAO, IFAD, UN Children's Fund, WFP, WHO. The state of food security and nutrition in the world (SOFI) report 2019 — in brief. Rome: Food and Agriculture Organization of the United Nations; 2019. https://docs.wfp.org/api/documents/WFP-0000106760/download/?_ga=2.82667615.870359159.1612941460-968858223.1612941460.

27. FAO, IFAD, UN Children's Fund, WFP, WHO. The state of food security and nutrition in the world (SOFI) report 2019 — in brief. Rome: Food and Agriculture Organization of the United Nations; 2019. https://docs.wfp.org/api/documents/WFP-0000106760/download/?_ga=2.82667615.870359159.1612941460-968858223.1612941460.

28. FAO, IFAD, UN Children's Fund, WFP, WHO. The state of food security and nutrition in the world (SOFI) report 2019 — in brief. Rome: Food and Agriculture Organization of the United Nations; 2019. https://docs.wfp.org/api/documents/WFP-0000106760/download/?_ga=2.82667615.870359159.1612941460-968858223.1612941460.

29. FAO, IFAD, UN Children's Fund, WFP, WHO. The state of food security and nutrition in the world (SOFI) report 2019 — in brief. Rome: Food and Agriculture Organization of the United Nations; 2019. https://docs.wfp.org/api/documents/WFP-0000106760/download/?_ga=2.82667615.870359159.1612941460-968858223.1612941460.

30. The Economist. Far too many of the world's youngsters are overweight [Internet]. London: The Economist; 2019 Oct 15 [cited 2021 Jan 24].

https://www.economist.com/graphic-detail/2019/10/15/far-too-many-of-the-worlds-youngsters-are-overweight.

31. Ritchie H, Roser M. Obesity [Internet]. [Place unknown]: Our World In Data; 2017 [cited 2021 Jan 24]. https://ourworldindata.org/obesity.

32. Ritchie H, Roser M. Obesity [Internet]. [Place unknown]: Our World In Data; 2017 [cited 2021 Jan 24]. https://ourworldindata.org/obesity.

33. Ritchie H, Roser M. Obesity [Internet]. [Place unknown]: Our World In Data; 2017 [cited 2021 Jan 24]. https://ourworldindata.org/obesity.

34. Ritchie H, Roser M. Obesity [Internet]. [Place unknown]: Our World In Data; 2017 [cited 2021 Jan 24]. https://ourworldindata.org/obesity.

35. Ritchie H, Roser M. Obesity [Internet]. [Place unknown]: Our World In Data; 2017 [cited 2021 Jan 24]. https://ourworldindata.org/obesity.

36. UN Environment Programme. How to feed 10 billion people [Internet]. Nairobi: United Nations Environment Programme; 2020 Jul 13 [cited 2021 Jan 25]. https://www.unenvironment.org/news-and-stories/story/how-feed-10-billion-people.

37. Shaw DJ. World food security. London: Palgrave Macmillan UK; 2007. 510 p.

38. OECD/WHO. Health at a Glance: Asia/Pacific 2018: Measuring Progress towards Universal Health Coverage. Paris: OECD Publishing; 2018. 132 p.

39. WHO. Stunting in a nutshell [Internet]. Geneva: World Health Organization; 2015 Nov 19 [cited 2021 Jan 24]. https://www.who.int/news/item/19-11-2015-stunting-in-a-nutshell.

40. WHO. Stunting in a nutshell [Internet]. Geneva: World Health Organization; 2015 Nov 19 [cited 2021 Jan 24]. https://www.who.int/news/item/19-11-2015-stunting-in-a-nutshell.

41. UN Children's Fund. Low birthweight [Internet]. New York: United Nations Children's Fund; 2019 May [cited 2021 Jan 24]. https://data.unicef.org/topic/nutrition/low-birthweight/.

42. WHO. Malnutrition [Internet]. Geneva: World Health Organization; 2020 Apr 1 [cited 2021 Jan 24]. https://www.who.int/news-room/fact-sheets/detail/malnutrition.

43. WHO. Nutrition landscape information system (NLIS) country profile indicators interpretation guide. 2nd Edition. Geneva: World Health Organization; 2019. https://apps.who.int/iris/bitstream/handle/10665/332223/9789241516952-eng.pdf.

44. UNICEF, WHO, IBRD/The World Bank. Levels and trends in child malnutrition: Key Findings of the 2020 Edition of the Joint Child Malnutrition Estimates. Geneva: World Health Organization; 2020. Licence: CC BY-NC-SA 3.0 IGO. https://apps.who.int/iris/bitstream/handle/10665/331621/9789240003576-eng.pdf?sequence=1&isAllowed=y.

45. Centers for Disease Control and Prevention. Defining childhood obesity [Internet]. Atlanta: Centers for Disease Control and Prevention; 2018 Jul 3 [cited 2021 Jan 24]. https://www.cdc.gov/obesity/childhood/defining.html.

46. Centers for Disease Control and Prevention. Defining childhood obesity [Internet]. Atlanta: Centers for Disease Control and Prevention; 2018 Jul 3 [cited 2021 Jan 24]. https://www.cdc.gov/obesity/childhood/defining.html.

47. Centers for Disease Control and Prevention. Defining childhood obesity [Internet]. Atlanta: Centers for Disease Control and Prevention; 2018 Jul 3 [cited 2021 Jan 24]. https://www.cdc.gov/obesity/childhood/defining.html.

48. Ritchie H, Roser M. CO2 and Greenhouse Gas Emissions [Internet]. Oxford: Our World In Data; 2017 May [cited 2021 Jan 24]. https://ourworldindata.org/co2-and-other-greenhouse-gas-emissions.

49. Ritchie H, Roser M. CO2 and Greenhouse Gas Emissions [Internet]. Oxford: Our World In Data; 2017 May [cited 2021 Jan 24]. https://ourworldindata.org/co2-and-other-greenhouse-gas-emissions.

50. Ritchie H, Roser M. CO2 and Greenhouse Gas Emissions [Internet]. Oxford: Our World In Data; 2017 May [cited 2021 Jan 24]. https://ourworldindata.org/co2-and-other-greenhouse-gas-emissions.

51. Ritchie H, Roser M. CO2 and Greenhouse Gas Emissions [Internet]. Oxford: Our World In Data; 2017 May [cited 2021 Jan 24]. https://ourworldindata.org/co2-and-other-greenhouse-gas-emissions.

52. Ritchie H, Roser M. CO2 and Greenhouse Gas Emissions [Internet]. Oxford: Our World In Data; 2017 May [cited 2021 Jan 24]. https://ourworldindata.org/co2-and-other-greenhouse-gas-emissions.

53. Salawitch RJ, Bennett BF, Canty TP, Hope AP, Tribett WR. Paris Climate Agreement: Beacon of Hope. Cham: Springer International Publishing; 2017. 186 p.

54. TheWorldCounts. World average temperature (°C) [Internet]. Copenhagen: TheWorldCounts; 2021 [cited 2021 Jan 24]. https://www.theworldcounts.com/challenges/climate-change/global-warming/average-global-temperature/story.

55. Ritchie H, Roser M. CO2 and Greenhouse Gas Emissions [Internet]. Oxford: Our World In Data; 2017 May [cited 2021 Jan 24]. https://ourworldindata.org/co2-and-other-greenhouse-gas-emissions.

56. Cama T, Henry D. Trump: We are getting out of Paris climate deal [Internet]. Washington DC: The Hill; 2017 Jun 1 [cited 2021 Jan 25]. https://thehill.com/policy/energy-environment/335955-trump-pulls-us-out-of-paris-climate-deal.

57. Biden Jr. Paris Climate Agreement [Internet]. Washington DC: The White House; 2021 Jan 20 [cited 20201 Jan 25]. https://www.whitehouse.gov/briefing-room/statements-releases/2021/01/20/paris-climate-agreement/.

58. European Commission. Causes of climate change [Internet]. Brussels: European Commission; 2021 [cited 2021 Jan 25]. https://ec.europa.eu/clima/change/causes_en.

59. Mbow C, Rosenzweig C, Barioni LG, Benton TG, Herrero M, Krishnapillai M, Liwenga E, Pradhan P, Rivera-Ferre MG, Sapkota T, Tubiello FN, Xu Y. Food Security. In: Shukla PR, Skea J, Calvo Buendia E, Masson-Delmotte V, Pörtner HO, Roberts DC, Zhai P, Slade R, Connors S, Van Diemen R, Ferrat M. Climate Change and Land: An IPCC special report on climate change, desertification, land degradation, sustainable land management, food security, and greenhouse gas fluxes in terrestrial ecosystems. Geneva: United Nations' Intergovernmental Panel on Climate Change; 2019. p. 437–550. https://www.ipcc.ch/srccl/chapter/chapter-5/.

60. Haldar I. Global warming: The causes and consequences. New Delhi: Readworthy Press Corporation; 2011. 200 p.

61. Tschakert P, Zimmerer K, King B, Baum S, Wang C. Livestock's long shadow [Internet]. Pennsylvania: Penn State's College of Earth and Mineral Sciences; 2020 [cited 2021 Jan 25]. https://www.e-education.psu.edu/geog30/node/364.

62. IPCC. Climate Change 2014: Impacts, Adaptation, and Vulnerability. Part A: Global and Sectoral Aspects. Contribution of Working Group II to the Fifth Assessment Report of the Intergovernmental Panel on Climate Change [Field CB, Barros VR, Dokken DJ, Mach KJ, Mastrandrea, Bilir TE *et al.*, eds. Cambridge and New York: Cambridge University Press; 2014. 1132 p.

63. Kornher L. Maize markets in Eastern and Southern Africa (ESA) in the context of climate change. The state of agricultural commodity markets (SOCO) 2018. Rome: Food and Agriculture Organization of the United Nations; 2018. http://www.fao.org/3/ca2155en/CA2155EN.pdf

64. Peng S, Huang J, Sheehy JE, Laza RC, Visperas RM, Zhong X, *et al.* Rice yields decline with higher night temperature from global warming. Proceedings of the National Academy of Sciences. 2004 Jul 6; 101(27): 9971–5. DOI: 10.1073/pnas.0403720101.

65. Rao P, Patil Y. Reconsidering the impact of climate change on global water supply, use, and management. Hershey: IGI Global; 2017. 430 p.

66. Sarkar MS, Begum RA, Pereira JJ. Impacts of climate change on oil palm production in Malaysia. Environmental Science and Pollution Research. 2020 Mar; 27(9): 9760–70. DOI: 10.1007/s11356-020-07601-1.

67. Loladze I. Rising atmospheric CO2 and human nutrition: Toward globally imbalanced plant stoichiometry?. Trends in Ecology & Evolution. 2002 Oct 1; 17(10): 457–61. DOI: 10.1016/S0169-5347(02)02587-9.

68. Bottemiller Evich H. The great nutrient collapse [Internet]. Arlington: Politico; 2017 Sep 13 [cited 2021 Jan 25]. https://www.politico.com/agenda/story/2017/09/13/food-nutrients-carbon-dioxide-000511/.

69. Smith MR, Myers SS. Impact of anthropogenic CO2 emissions on global human nutrition. Nature Climate Change. 2018 Sep; 8(9): 834–9. DOI: 10.1038/s41558-018-0253-3.

70. Mbow C, Rosenzweig C, Barioni LG, Benton TG, Herrero M, Krishnapillai M, Liwenga E, Pradhan P, Rivera-Ferre MG, Sapkota T, Tubiello FN, Xu Y. Food Security. In: Shukla PR, Skea J, Calvo Buendia E, Masson-Delmotte V, Pörtner HO, Roberts DC, Zhai P, Slade R, Connors S, Van Diemen R, Ferrat M. Climate Change and Land: An IPCC special report on climate change, desertification, land degradation, sustainable land management, food security, and greenhouse gas fluxes in terrestrial ecosystems. Geneva: United Nations' Intergovernmental Panel on Climate Change; 2019. p. 437–550. https://www.ipcc.ch/srccl/chapter/chapter-5/.

71. Beddington J. Food, energy, water and the climate: A perfect storm of global events? London: Government Office for Science; 2009. https://citeseerx.ist.psu.edu/viewdoc/download?doi=10.1.1.522.3978&rep=rep1&type=pdf.

72. Azam-Ali SN. Production systems and agronomy | Multicropping. In: Thomas B, editor. Encyclopedia of Applied Plant Sciences. Amsterdam: Elsevier; 2003. p. 978–984.

73. No-Till Farmer. Intercropping increases yields while reducing the use of fertilizers [Internet]. Brookfield: No-Till Farmer; 2020 Jun 12 [cited 2021 Jan 25]. https://www.no-tillfarmer.com/articles/9758-intercropping-increases-yields-while-reducing-the-use-of-fertilizers.

74. No-Till Farmer. 'Intercropping increases yields while reducing the use of fertilizers [Internet]. Brookfield: No-Till Farmer; 2020 Jun 12 [cited 2021 Jan 25]. https://www.no-tillfarmer.com/articles/9758-intercropping-increases-yields-while-reducing-the-use-of-fertilizers.

75. Azam-Ali SN. Production systems and agronomy | Multicropping. In: Thomas B, editor. Encyclopedia of Applied Plant Sciences. Amsterdam: Elsevier; 2003. p. 978–984.

76. International Development Research Centre. Facts & figures on food and biodiversity. Ottawa: International Development Research Centre; 2010 Dec 23 [cited 2021 Jan 25]. https://www.idrc.ca/en/research-in-action/facts-figures-food-and-biodiversity#:~:text=Of%20those%20150%2C%20just%2012,%2C%20medicine%2C%20shelter%2C%20transportation.

77. United States Department of Agriculture. Pigeon peas (red gram), mature seeds, raw [Internet]. Washington DC: United States Department of Agriculture; 2019 Apr 1 [cited 2021 Jan 25]. https://fdc.nal.usda.gov/fdc-app.html#/food-details/172436/nutrients.

78. Pollan M. In defense of food: An eater's manifesto. New York: Penguin Press; 2008. 256 p.

79. Monteiro C. The big issue is ultra-processing. 'Carbs': The answer. World Nutrition. 2011 Feb; 2(2): 86–97.

80. Paull J. A century of synthetic fertilizer: 1909–2009. Journal of Bio-Dynamics Tasmania. 2009 Jun; 94:16–21.

81. McDowell LR. Vitamins in animal and human nutrition. Second edition. Ames: Iowa State University Press; 2000. 793 p.

82. Pollan M. Unhappy meals [Internet]. New York: The New York Times Magazine; 2007 Jan 28 [cited 2021 Jan 25]. https://michaelpollan.com/articles-archive/unhappy-meals/.

83. Remans R, Wood SA, Saha N, Anderman TL, DeFries RS. Measuring nutritional diversity of national food supplies. Global Food Security. 2014 Nov 1; 3(3–4): 174–82. DOI: 10.1016/j.gfs.2014.07.001.

84. IDF. IDF Diabetes Atlas. 9th edition. Brussels: International Diabetes Federation; 2019. 148 p.

85. Movono L. 'Our diet is killing us quietly': Fiji's diabetes crisis [Internet]. London: The Guardian; 2020 Jun 26 [cited 2021 Jan 25]. https://www.theguardian.com/world/2020/jun/27/our-diet-is-killing-us-quietly-fijis-diabetes-crisis.

86. Butler DA. Unsinkable: The full story of the RMS Titanic. Mechanicsburg: Stackpole Books; 1998. 336 p.

5. Global Agricultural Research and Education

1. Ranking Web of Universities. Methodology [Internet]. Madrid: Ranking Web of Universities; 2020 July [cited 2021 Jan 26]. http://www.webometrics.info/en/Methodology.

2. Kowarski I. 10 universities with the biggest endowments [Internet]. New York: U.S. News; 2020 Sep 22 [cited 2021 Jan 26]. https://www.usnews.com/education/best-colleges/the-short-list-college/articles/10-universities-with-the-biggest-endowments.

3. The World Bank Data. GDP (current US$) [Internet]. Washington, D.C.: The World Bank Data; 2021 [cited 2021 Jan 26]. https://data.worldbank.org/indicator/NY.GDP.MKTP.CD

4. WeAreFreeMovers. Harvard University — Faculty of Arts And Sciences (Master Level) [Internet]. [place unknown]: WeAreFreeMovers; 2021 [cited 2021 Jan 26]. https://www.wearefreemovers.com/universities/harvard-university-faculty-of-arts-and-sciences-master-level/.

5. Fan Y. The University of al-Qarawiyyin: The oldest university in the world [Internet]. Washington DC: Arab America; 2020 Nov 25 [cited 2021 Jan 26]. https://www.arabamerica.com/the-university-of-al-qarawiyyin-the-oldest-university-in-the-world/.

6. Università di Bologna. The numbers of history [Internet]. Bologna: Università di Bologna; 2021 [cited 2021 Jan 26]. https://www.unibo.it/en/university/who-we-are/our-history/the-numbers-of-history.

7. Università degli Studi di Napoli Federico II. The University of Naples Federico II is the oldest public university in the world [Internet]. Naples: Università degli Studi di Napoli Federico II; 2021 [cited 2021 Jan 26]. http://www.international.unina.it/.

8. The Library of Congress. Primary documents in American history: Morrill Act [Internet]. Washington DC: The Library of Congress; 2017 Oct 26 [cited 2021 Jan 26]. https://www.loc.gov/rr/program/bib/ourdocs/morrill.html.

9. Gombrich RF. British Higher Education Policy in the Last Twenty Years. The Murder of a Profession. Synthesis philosophica. 2013; 28(1–2): 7–29.

10. Academic Ranking of World Universities. About academic ranking of world universities [Internet]. Shanghai: ShanghaiRanking Consultancy; 2020 [cited 2021 Jan 26]. http://www.shanghairanking.com/aboutarwu.html.

11. Crew B. World university rankings: Explained [Internet]. London: Springer Nature; 2019 Aug 22 [cited 2021 Jan 26]. https://www.natureindex.com/news-blog/world-university-rankings-explainer-times-higher-education-arwu-shanghai-qs-quacquarelli-symonds

12. Editorial Team. The history and development of higher education ranking systems [Internet]. London: Quacquarelli Symonds Limited; 2018 Mar 2 [cited 2021 Jan 26]. https://qswownews.com/history-development-higher-education-ranking-systems/.

13. International Ranking Expert Group (IREG). Who we are [Internet]. Brussels: International Ranking Expert Group; 2019 [cited 2021 Jan 26]. https://ireg-observatory.org/en/about-us/.

14. International Ranking Expert Group (IREG). Berlin Principles [Internet]. Brussels: International Ranking Expert Group; 2019 [cited 2021 Jan 26]. https://ireg-observatory.org/en/about-us/.

15. Brown S. Research assessment in higher education: The impact on institutions, staff and educational research in Scotland. ACCESS: Critical Perspectives on Communication, Cultural & Policy Studies. 2008 Jan; 27(1/2): 141–51.

16. Saha S, Saint S, Christakis DA. Impact factor: A valid measure of journal quality? Journal of the Medical Library Association. 2003 Jan; 91(1): 42.

17. Royal Agricultural Society of England. Home [Internet]. Warwickshire: Royal Agricultural Society of England; [date unknown] [cited 2021 Jan 26]. https://www.rase.org.uk/home.

18. Roxburgh CW, Pratley JE. The future of food production research in the rangelands: Challenges and prospects for research investment, organisation and human resources. The Rangeland Journal. 2015 Apr 17; 37(2): 125–38. DOI: 10.1071/RJ14090.

19. Stirling CM, Harris D, Witcombe JR. Managing an agricultural research programme for poverty alleviation in developing countries: An institute without walls. Experimental Agriculture. 2006 Apr 1; 42(2): 127.

20. Alston JM, Wyatt TJ, Pardey PG, Marra MC, Chan-Kang C. A meta-analysis of rates of return to agricultural R & D. Washington DC: International Food Policy Research Institute (IFPRI); 2000. Report No.: 113.

21. Pardey PG, Beintema NM. Slow magic; Agricultural R&D a century after Mendel. Food Policy Report. Washington DC: International Food Policy Research Institute; 2001. http://ebrary.ifpri.org/utils/getfile/collection/p15738coll2/id/59951/filename/59952.pdf.

22. FAO. Our priorities — The Strategic Objectives of FAO. Rome: Food and Agriculture Organization of the United Nations; 2019. http://www.fao.org/3/i8580en/I8580EN.pdf.

23. Staples LS, Sayward AL. The birth of development: How the World Bank, Food and Agriculture Organization, and World Health Organization changed the world, 1945–1965. Ohio: Kent State University Press; 2006. 349 p.

24. O'Broin S. FAO in seven decades. In: Small A, O'Broin S, editors. 70 Years of FAO (1945–2015). Rome: Food and Agriculture Organization of the United Nations; 2015. p. 14–81.

25. UN. What are UN specialized agencies, and how many are there? [Internet]. New York: United Nations; 2020 Oct 21 [cited 2021 Jan 26]. https://ask.un.org/faq/140935.

26. WHO. International food standards (FAO/WHO Codex Alimentarius) [Internet]. Geneva: World Health Organization; 2021 [cited 2021 Jan 26]. https://www.who.int/foodsafety/areas_work/food-standard/en/.

27. FAO. FAO: Challenges and Opportunities in a Global World. Rome: Food and Agriculture Organization of the United Nations; 2019. 325 p.

28. O'Broin S. FAO in seven decades. In: Small A, O'Broin S, editors. 70 Years of FAO (1945–2015). Rome: Food and Agriculture Organization of the United Nations; 2015. p. 14–81.

29. Phillips RW. FAO: Its origins, formation and evolution 1945–1981. Rome: Food and Agriculture Organization of the United Nations; 1981. 200 p.

30. Gaia@gaia.net. UN criticized for supporting corporate agribusiness instead of small farmers & the poor [Internet]. Minnesota: Organic Consumers Association; 2021 cited [2021 Jan 26]. https://www.organicconsumers.org/news/un-criticized-supporting-corporate-agribusiness-instead-small-farmers-poor.

31. Leavitt C. All Roads Lead to Rome: New acquisitions relating to the Eternal City [Internet]. Indiana: University of Notre Dam; 2011 Sep 14 [cited 2021 Jan 26]. https://italianstudies.nd.edu/news-events/news/all-roads-lead-to-rome-new-acquisitions-relating-to-the-eternal-city/.

32. World Bank. The CGIAR at 31: Celebrating its Achievements, Facing its Challenges. Washington DC: World Bank; 2003. Report No.: 232.

33. Harrington L, Cook SE, Lemoalle J, Kirby M, Taylor C, Woolley J. Cross-basin comparisons of water use, water scarcity and their impact on livelihoods: present and future. Water International. 2009 Feb 18; 34(1): 144–54. DOI: 10.1080/02508060802661584.

34. Bouis HE, Saltzman A. Improving nutrition through biofortification: A review of evidence from HarvestPlus, 2003 through 2016. Global Food Security. 2017 Mar 1; 12: 49–58. DOI: 10.1016/j.gfs.2017.01.009.

35. Bruskiewich R, Senger M, Davenport G, Ruiz M, Rouard M, Hazekamp T, Takeya M, *et al.* The Generation Challenge Programme Platform: Semantic Standards and Workbench for Crop Science. International Journal of Plant Genomics. 2008 Dec 1; 2008: 70–5. DOI: 10.1155/2008/369601

36. CGIAR. Research centers [Internet]. Montpellier: Consultative Group on International Agricultural Research; 2018 [cited 2021 Jan 26]. https://www.cgiar.org/research/research-centers/.

37. Ricepedia. Rice productivity [Internet]. Manilla: Ricepedia; [date unknown] [cited 2021 Jan 26]. http://ricepedia.org/rice-as-a-crop/rice-productivity.

38. The Crop Trust. Our mission [Internet]. Bonn: The Crop Trust; [date unknown] [cited 2021 Jan 26]. https://www.croptrust.org/our-mission/.

39. CGIAR. One CGIAR [Internet]. Montpellier: Consultative Group on International Agricultural Research; 2018 [cited 2021 Jan 26]. https://www.cgiar.org/food-security-impact/one-cgiar/.

40. International Panel of Experts on Sustainable Food Systems. Open letter | 'One CGIAR' with two tiers of influence? [Internet]. Brussels: International Panel of Experts on Sustainable Food Systems; 2020 Jul 21 [cited 2021 Jan 26]. http://www.ipes-food.org/pages/OneGGIAR#.

41. AIRCA. AIRCA members [Internet]. Nairobi: Association of International Research and Development Centers for Agriculture; 2016 [cited 2021 Jan 26]. http://www.airca.org/index.php/airca-members.

42. AIRCA. AIRCA members [Internet]. Nairobi: Association of International Research and Development Centers for Agriculture; 2016 [cited 2021 Jan 26]. http://www.airca.org/index.php/airca-members.

43. AIRCA. AIRCA members [Internet]. Nairobi: Association of International Research and Development Centers for Agriculture; 2016 [cited 2021 Jan 26]. http://www.airca.org/index.php/airca-members.

44. Millar R. On this day 1953… Discovering DNA [Internet]. London: Union Press Ltd; 2017 Feb 28 [cited 2021 Jan 26]. https://www.thedrinksbusiness. com/2017/02/on-this-day-1953-discovering-dna/#:~:text=On%2028%20 February%201953%20staff,found%20the%20secret%20of%20 life.%E2%80%9D.

45. Jump P. Crick and Watson rejected? [Internet]. Washington DC: Inside Higher Ed; 2015 Aug 6 [cited 2021 Jan 26]. https://www.insidehighered. com/news/2015/08/06/journal-article-speculates-crick-and-watson-would- have-difficulty-today-publishing.

46. Watts J. James Lovelock: 'The biosphere and I are both in the last 1% of our lives' [Internet]. London: Guardian News & Media Ltd; 2020 Jul 18 [cited 2021 Jan 26]. https://www.theguardian.com/environment/2020/jul/18/james- lovelock-the-biosphere-and-i-are-both-in-the-last-1-per-cent-of-our- lives?CMP=share_btn_fb.

6. Underutilized Crops

1. Azam-Ali SN. Fitting underutilised crops within research-poor environ- ments: Lessons and approaches. South African Journal of Plant and Soil. 2010 Jan 1; 27(4): 293–8. DOI: 10.1080/02571862.2010.10639997.

2. Monteith JL. Solar radiation and productivity in tropical ecosystems. Journal of Applied Ecology. 1972 Dec 1; 9(3): 747–66. DOI: 10.2307/2401901.

3. Unsworth MH. John Lennox Monteith. 3 September 1929–20 July 2012. Biographical Memoirs of Fellows of the Royal Society. 2014; 60: 299–329. DOI: 10.1098/rsbm.2014.0005.

4. Monteith JL. Climatic variation and the growth of crops. Quarterly Journal of the Royal Meteorological Society. 2007 Jul; 107(454): 749–774. DOI: 10.1002/qj.49710745402.

5. Penman HL. Natural evaporation from open water, bare soil and grass. Proceedings. Mathematical, physical, and engineering sciences. 1948 Apr 22; 193(1032): 120–145. DOI: 10.2135/cropsci2019.05.0292.

6. Monteith JL. Evaporation and surface temperature. Quarterly Journal of the Royal Meteorological Society. 1981; 107(451): 1–27. DOI: 10.1002/ qj.49710745102.

7. Monteith JL. Conservative behaviour in the response of crops to water and light. In: Rabbinge R, Goudriaan J, van Keulen H, Penning de Vries FWT, van Laar HH, editors. Theoretical production ecology: Reflections and prospects. Wageningen: Pudoc, 1990. p. 3–16.

8. Howell TA, Evett SR. The Penman-Monteith method. Texas: USDA Agricultural Research Service; 2004. 14 p.

9. Howell TA, Evett SR. The Penman-Monteith method. Texas: USDA Agricultural Research Service; 2004. 14 p.

10. Unsworth MH. John Lennox Monteith. 3 September 1929–20 July 2012. Biographical Memoirs of Fellows of the Royal Society. 2014; 60: 299–329. DOI: 10.1098/rsbm.2014.0005.

11. Unsworth MH. John Lennox Monteith. 3 September 1929–20 July 2012. Biographical Memoirs of Fellows of the Royal Society. 2014; 60: 299–329. DOI: 10.1098/rsbm.2014.0005.

12. Unsworth MH. John Lennox Monteith. 3 September 1929–20 July 2012. Biographical Memoirs of Fellows of the Royal Society. 2014; 60: 299–329. DOI: 10.1098/rsbm.2014.0005.

13. Monteith JL. Drying patterns on roads in the Scottish Highlands. Weather. 1996 Mar; 51(3): 95–100. DOI: 10.1002/j.1477-8696.1996.tb03954.x.

14. Monteith JL. Crepuscular rays formed by the Western Ghats. Weather 1986 Sep; 41(9): 292–299. DOI: 10.1002/j.1477-8696.1986.tb03862.x.

15. Quotepark. "If I have seen further than others, it is by standing upon the shoulders of giants" [Internet]. [place unknown]: Quotepark; 2020 May 24 [cited 2021 Jan 26]. https://quotepark.com/quotes/1214856-isaac-newton-if-i-have-seen-further-than-others-it-is-by-stand/.

16. Monteith JL. Principles of environmental physics. New York: American Elsevier Publishing Company; 1973. 241 p.

17. Punia S, Siroha AK, Sandhu KS, Gahlawat SK, Kaur M, editors. Pearl Millet: Properties, Functionality and Its Applications. CRC Press; 2020. 174 p.

18. Pasupuleti J, Nigam SN, Pandey MK, Nagesh P, Varshney RK. Groundnut improvement: Use of genetic and genomic tools. Frontiers in Plant Science. 2013 Feb 25; 4:23. DOI: 10.3389/fpls.2013.00023.

19. Heller J, Begemann F, Mushonga J, editors. Bambara groundnut, Vigna subterranea (L.) Verdc. Proceedings of the workshop on Conservation and

Improvement of Bambara Groundnut (Vigna subterranea (L.) Verdc.); 1995 Nov 14–16; Harare, Zimbabwe. Rome, Italy: International Plant Genetic Resources Institute; 1997.

20. Linneman AR. Phenological development in bambara groundnut (Vigna subterranea) at constant exposure to photoperiods of 10 to 16 h. Annals of Botany. 1993 May 1; 71(5): 445–52. DOI: 10.1006/anbo.1993.1058.

21. Mungoma C. Photoperiod sensitivity in tropical maize accessions, early inbreds, and their crosses [Dissertation]. Ames: Iowa State University; 1988.

22. Harris D, Azam-Ali SN. Implications of daylength sensitivity in bambara groundnut (Vigna subterranea) for production in Botswana. The Journal of Agricultural Science. 1993 Feb; 120(1): 75–8. DOI: 10.1017/S0021859600073615.

23. Linneman AR. Phenological development in bambara groundnut (Vigna subterranea) at constant exposure to photoperiods of 10 to 16 h. Annals of Botany. 1993 May 1; 71(5): 445–52. DOI: 10.1006/anbo.1993.1058.

24. Harris D, Azam-Ali SN. Implications of daylength sensitivity in bambara groundnut (Vigna subterranea) for production in Botswana. The Journal of Agricultural Science. 1993 Feb; 120(1): 75–8. DOI: 10.1017/S0021859600073615.

25. Harris D, Azam-Ali SN. Implications of daylength sensitivity in bambara groundnut (Vigna subterranea) for production in Botswana. The Journal of Agricultural Science. 1993 Feb; 120(1): 75–8. DOI: 10.1017/S0021859600073615.

26. Harris D, Azam-Ali SN. Implications of daylength sensitivity in bambara groundnut (Vigna subterranea) for production in Botswana. The Journal of Agricultural Science. 1993 Feb; 120(1): 75–8. DOI: 10.1017/S0021859600073615.

27. Linneman AR. Phenological development in bambara groundnut (Vigna subterranea) at constant exposure to photoperiods of 10 to 16 h. Annals of Botany. 1993 May 1; 71(5): 445–52. DOI: 10.1006/anbo.1993.1058.

28. Azam-Ali SN, Aquilar-Manjarrez P. Bannayan M. A Global Mapping System for Bambara Groundnut Production. Rome: Food and Agriculture Organization of the United Nations; 2001. 56 p.

29. Azam-Ali SN, Aquilar-Manjarrez P. Bannayan M. A Global Mapping System for Bambara Groundnut Production. Rome: Food and Agriculture Organization of the United Nations; 2001. 56 p.

30. Hampson K, Azam-Ali SN, Sesay A, Mukwaya S. Assessing opportunities for increased utilization of bambara groundnut in Southern Africa. Loughborough: University of Nottingham; 2001. 67 p.

31. BAMLINK. Second Annual Report: Molecular, Environmental and Nutritional Evaluation of Bambara Groundnut (Vigna subterranea L.Verdc.) for Food Production in Semi-Arid Africa and India. [place unknown]: BAMLINK; 2007.

32. BAMLINK. Second Annual Report: Molecular, Environmental and Nutritional Evaluation of Bambara Groundnut (Vigna subterranea L.Verdc.) for Food Production in Semi-Arid Africa and India. [place unknown]: BAMLINK; 2007.

33. BAMLINK. Second Annual Report: Molecular, Environmental and Nutritional Evaluation of Bambara Groundnut (Vigna subterranea L.Verdc.) for Food Production in Semi-Arid Africa and India. [place unknown]: BAMLINK; 2007.

34. BAMLINK. Second Annual Report: Molecular, Environmental and Nutritional Evaluation of Bambara Groundnut (Vigna subterranea L.Verdc.) for Food Production in Semi-Arid Africa and India. [place unknown]: BAMLINK; 2007.

35. Kole C. Pulses, sugar and tuber crops. Heidelberg: Springer Verlag; 2007. 306 p.

36. Mayes S, Ho WK, Kendabie P, Chai HH, Aliyu S, Feldman A, *et al.* Applying molecular genetics to underutilised species — problems and opportunities. Malaysian Applied Biology. 2015; 44(4): 1–9.

37. BAMLINK. Final report and project summary: Molecular, Environmental and Nutritional Evaluation of Bambara Groundnut (Vigna subterranea L. Verdc.) for Food Production in Semi-Arid Africa and India. [place unknown]: BAMLINK; 2011.

38. Chaitanya PK, Priya BV, Babu NV, Avinash P, Asha S, Kumar RB. In-vitro plant regeneration studies of Arachis hypogaea L.(groundnut) for its heavy metal tolerance (K2Cr2O7). Research Journal of Pharmaceutical, Biological and Chemical Sciences. 2014; 5(6): 394–406.

39. The Nibble. Food 101: The history of peanuts for National Peanut Day [Internet]. New York: The Nibble; 2018 Sep 13 [cited 2021 Jan 27]. https://blog.thenibble.com/2018/09/13/food-101-the-history-of-peanuts-for-national-peanut-day/.

40. Nautiyal PC, Mejia D, Lewis B, editors. Groundnut: Post-harvest operations. Sagdividi: National Research Centre for Groundnut (ICAR); 2002. 127 p.

41. Virginia Carolinas Peanuts. The history of peanuts [Internet]. North Carolina: Virginia Carolinas Peanut Promotions; 2019 [cited 2021 Jan 27]. https://www.aboutpeanuts.com/peanut-facts/origin-history-of-peanuts.

42. Cherfas J. Peanuts and world history [Internet]. [place unknown]: Agricultural Biodiversity Weblog; 2017 Jan 12 [cited 2021 Jan 27]. https://agro.biodiver.se/2017/01/peanuts-and-world-history/.

43. National Peanut Board. G.W. Carver [Internet]. Georgia: National Peanut Board; 2021 [cited 2021 Jan 27]. https://www.nationalpeanutboard.org/more/gw-carver/.

44. National Peanut Board. Who invented peanut butter [Internet]. Georgia: National Peanut Board; 2021 [cited 2021 Jan 27]. https://www.nationalpeanutboard.org/peanut-info/who-invented-peanut-butter.html.

45. European Patent Office. Michel Lescanne (France) [Internet]. Munich: Eiropean Patent Office; 2015 May 13 [cited 2021 Jan 27]. https://www.epo.org/news-events/events/european-inventor/finalists/2015/lescanne.html.

46. Cavendish R. Britain abandons the Groundnuts Scheme [Internet]. London: History Today; 2001 Jan 9 [cited 2021 Jan 27]. https://www.historytoday.com/archive/britain-abandons-groundnuts-scheme.

47. Davie E. Public enterprises, guided by "mean-wellers", are prone to disaster [Internet]. South Africa: The Gremin; 2017 Nov 16 [cited 2021 Jan 27]. https://www.thegremlin.co.za/2017/11/16/public-enterprises-guided-by-mean-wellers-are-prone-to-disaster/.

48. Westcott N. Imperialism and development: The East African groundnut scheme and its legacy. Suffolk: James Curry; 2020. 260 p.

49. UC Berkeley. Journey to Mali: 1350–1351 [Internet]. Valifornia: UC Berkeley; 2021 [cited 2021 Jan 27]. https://orias.berkeley.edu/resources-teachers/travels-ibn-battuta/journey/journey-mali-1350–1351.

50. Harris JB. The Africa cookbook: Tastes of a continent. New York: Simon and Schuster; 1998. 382 p.

51. Forsythe L, Nyamanda M, Mbachi Mwangwela A, Bennett B. Beliefs, taboos and minor crop value chains: the case of Bambara Groundnut in Malawi. Food, Culture & Society. 2015 Jul 3; 18(3): 501–17. DOI: 10.1080/15528014.2015.1043112.

7. Crops for the Future

1. SeedQuest. New Zealand farmer breaks world record with massive 17.398 tonne per hectare wheat crop [Internet]. New Zealand: SeedQuest; 2020 Jul 8 [2021 Jan 21]. https://www.seedquest.com/news.php?type=news&id_article=119048.

2. Garg M, Sharma GM, Sharma N, Kapoor S, Kumar P, Chunduri A, *et al.* Biofortified crops generated by breeding, agronomy, and transgenic approaches are improving lives of millions of people around the world. Frontiers in Nutrition. 2018 Feb 14; 5: 12. DOI: 10.3389/fnut.2018.00012.

3. Phillips T. Genetically modified organisms (GMOs): Transgenic crops and recombinant DNA technology. Nature Education. 2008; 1(1): 213. DOI: 10.21931/RB/2018.03.01.1.

4. Phillips T. Genetically modified organisms (GMOs): Transgenic crops and recombinant DNA technology. Nature Education. 2008; 1(1): 213. DOI: 10.21931/RB/2018.03.01.1.

5. FAO. Once neglected, these traditional crops are our new rising stars [Internet]. Rome: Food and Agriculture Organization of the United Nations; 2018 Oct 2 [cited 2021 Jan 21]. http://www.fao.org/fao-stories/article/en/c/1154584/.

6. Matthews PJ, Ghanem ME. Perception gaps that may explain the status of taro (Colocasia esculenta) as an "orphan crop". Plants, People, Planet. 2020; 00: 1–14. DOI: 10.1002/ppp3.10155.

7. Mabhaudi T, Chimonyo VGP, Hlahla S, Massawe F, Mayes S, Nhamo L, *et al.* Prospects of orphan crops in climate change. Planta. 2019 Sep; 250(3): 695–708. DOI: 10.3390/su9020291.

8. Bioversity International and the International Center for Tropical Agriculture. An alliance for accelerated change [Internet]. Cali: International Center for Tropical Agriculture; 2019 [cited 2021 Jan 27]. https://ciat.cgiar.org/alliance/.

9. Gregory PJ, Mayes S, Hui CH, Jahanshiri E, Julkifle A, Kuppusamy G, *et al.* Crops For the Future (CFF): An overview of research efforts in the adoption of underutilised species. Planta. 2019 Sep; 250(3): 979–88. DOI: 10.1007/s00425-019-03179-2.

10. Gregory PJ, Mayes S, Hui CH, Jahanshiri E, Julkifle A, Kuppusamy G, *et al.* Crops For the Future (CFF): An overview of research efforts in the adoption

of underutilised species. Planta. 2019 Sep; 250(3): 979–88. DOI: 10.1007/s00425-019-03179-2.

11. FAO. The State of World Fisheries and Aquaculture 2020. Sustainability in action. Rome: Food and Agriculture Organization of the United Nations; 2020. http://www.fao.org/3/ca9229en/CA9229EN.pdf.

12. Tran G, Heuzé V, Makkar HPS. Insects in fish diets. Animal Frontiers. 2015 Apr 1; 5(2): 37–44. DOI: 10.2527/af.2015-0018.

13. van Huis A. Potential of insects as food and feed in assuring food security. Annual Review of Entomology. 2013 Jan 7; 58: 563–83. DOI:10.1146/annurev-ento-120811-153704.

14. Palma L, Fernandez-Bayo J, Niemeier D, Pitesky M, VanderGheynst JS. Managing high fiber food waste for the cultivation of black soldier fly larvae. NPJ Science of Food. 2019 Sep 2; 3(1): 1–7. DOI: 10.1038/s41538-019-0047-7.

15. Association of International Research and Development Centers for Agriculture. Launch of the Forgotten Foods Network [Internet]. Nairobi: Association of International Research and Development Centers for Agriculture; 2017 Nov 3 [cited 2021 Jan 27]. http://www.airca.org/index.php/partner-news/54-launch-of-the-forgotten-foods-network.

16. Association of International Research and Development Centers for Agriculture. Launch of the Forgotten Foods Network [Internet]. Nairobi: Association of International Research and Development Centers for Agriculture; 2017 Nov 3 [cited 2021 Jan 27]. http://www.airca.org/index.php/partner-news/54-launch-of-the-forgotten-foods-network.

17. Padulosi S, Hodgkin T, Williams JT, Haq N. Underutilized crops: Trends, challenges and opportunities in the 21st Century. Rome: International Plant Genetic Resources Institute (IPGRI); 2002. https://cgspace.cgiar.org/rest/bitstreams/177929/retrieve#page=355.

18. Gregory PJ, Mayes S, Hui CH, Jahanshiri E, Julkifle A, Kuppusamy G, *et al.* Crops For the Future (CFF): An overview of research efforts in the adoption of underutilised species. Planta. 2019 Sep; 250(3): 979–88. DOI: 10.1007/s00425-019-03179-2.

19. Nohd Nizar NM, Jahanshiri E, Mohd Sinin SS, Wimalasiri EM, Mohd Suhairi TA, Gregory PJ *et al.* Open data to support agricultural diversification. Data in Brief. 2021 Jan 21: 106781. DOI: 10.1016/j.dib.2021.106781.

20. Gregory PJ, Mayes S, Hui CH, Jahanshiri E, Julkifle A, Kuppusamy G, *et al.* Crops For the Future (CFF): An overview of research efforts in the adoption of underutilised species. Planta. 2019 Sep; 250(3): 979–88. DOI: 10.1007/s00425-019-03179-2.

21. NamZ. Innovation [Internet]. Singapore: NamZ; 2019 [cited 2021 Jan 27]. https://www.namz.tech/.

22. The Lexicon. About [Internet]. California: The Lexicon; 2020 [cited 2021 Jan 27]. https://www.thelexicon.org/about/.

23. UN Educational, Scientific and Cultural Organization. Mediterranean diet [Internet]. Paris: United Nations Educational, Scientific and Cultural Organization; 2013 [cited 2021 Jan 27]. https://ich.unesco.org/en/RL/mediterranean-diet-00884.

24. Gregory PJ, Mayes S, Hui CH, Jahanshiri E, Julkifle A, Kuppusamy G, *et al.* Crops For the Future (CFF): An overview of research efforts in the adoption of underutilised species. Planta. 2019 Sep; 250(3): 979–88. DOI: 10.1007/s00425-019-03179-2.

25. LGC Biosearch Technologies. The African Orphan Crops Consortium (AOCC) [Internet]. Middlesex: LGC Biosearch Technologies; 2021 [cited 2021 Jan 27]. https://www.biosearchtech.com/sectors/agrigenomics/african-orphan-crops-consortium#:~:text=The%20African%20Orphan%20Crops%20Consortium%20(AOCC)%20is%20an%20international%20effort,among%20the%20continent's%20rural%20children.

26. LandSupport. Project [Internet]. Portici: LandSupport; 2018 [cited 2021 Jan 27]. https://www.landsupport.eu/project/.

27. Food Systems Dashboard. About the dashboard [Internet]. Maryland: Johns Hopkins University; 2020 [cited 2021 Jan 27]. https://foodsystemsdashboard.org/.

28. Thought For Food. Our approach [Internet]. Alabama: Thought For Food; 2020 [cited 2021 Jan 27]. https://thoughtforfood.org/approach/.

8. Revolution Number 9

1. African Commission on Human and Peoples' Rights. Extractive industries, land rights and indigenous populations'/communities' rights. Copenhagen: International Work Group for Indigenous Affairs; 2017. 144 p.

2. onValues Investment Strategies and Research Ltd. Investing for long-term value. Zurich: OnValues Investment Strategies and Research Ltd; 2005. 32 p.

3. Kell G. The remarkable rise of ESG [Internet]. New Jersey: Forbes; 2018 Jul 11 [cited 2021 Jan 29]. https://www.forbes.com/sites/georgkell/2018/07/11/the-remarkable-rise-of-esg/#2d3122541695.

4. Kell G. The remarkable rise of ESG [Internet]. New Jersey: Forbes; 2018 Jul 11 [cited 2021 Jan 29]. https://www.forbes.com/sites/georgkell/2018/07/11/the-remarkable-rise-of-esg/#2d3122541695.

5. GEM Report. Education increases awareness and concern for the environment [Internet]. Paris: Global Education Monitoring Report Team (GEM Report); 2015 Dec 8 [cited 2021 Jan 27]. https://gemreportunesco.wordpress.com/2015/12/08/education-increases-awareness-and-concern-for-the-environment/.

6. United States Environmental Protection Agency. Global greenhouse gas emissions data [Internet]. Washington DC: United States Environmental Protection Agency; 2020 Sep 10 [cited 2021 Jan 29]. https://www.epa.gov/ghgemissions/global-greenhouse-gas-emissions-data.

7. United States Environmental Protection Agency. Global greenhouse gas emissions data [Internet]. Washington DC: United States Environmental Protection Agency; 2020 Sep 10 [cited 2021 Jan 29]. https://www.epa.gov/ghgemissions/global-greenhouse-gas-emissions-data

8. UN Environment Programme. Renewable energy investment in 2018 hit USD 288.9 billion, far exceeding fossil fuel investment [Internet]. Nairobi: United Nations Environment Programme; 2019 Jun 18 [cited 2021 Jan 29]. https://www.unenvironment.org/news-and-stories/press-release/renewable-energy-investment-2018-hit-usd-2889-billion-far-exceeding.

9. LeBeau P. Tesla repays $465 million loan from Federal program [Internet]. Englewood Cliffs, NJ: CNBC; 2013 May 22 [cited 2021 Jan 29]. https://www.cnbc.com/id/100759230#:~:text=Less%20than%20four%20years%20after,Advanced%20Technology%20Vehicle%20Manufacturing%20program.

10. Parker S. What's Tesla's secret? [Internet]. London: World Finance; 2014 Apr 8 [cited 2021 Jan 29]. https://www.worldfinance.com/markets/whats-teslas-secret.

11. Root A. Think the Model S is fast? Tesla raced to $500 billion in value in less than a year [Internet]. New York: Dow Jones & Company; 2020 Nov 23

[2021 Jan 29]. https://www.barrons.com/articles/think-model-s-is-fast-tesla-raced-to-400-billion-in-value-in-less-than-a-year-51606170785.

12. Greene JL, Cowan T. Table egg production and hen welfare: Agreement and legislative proposals. Washington DC: Congressional Research Service; 2013. 27 p.

13. Andrews J. European Union bans battery cages for egg-laying hens [Internet]. Seattle: Food Safety News; 2012 Jan 19 [cited 2021 Jan 29]. https://www.foodsafetynews.com/2012/01/european-union-bans-battery-cages-for-egg-laying-hens/.

14. FAO. Food wastage footprint: Impacts on natural resources. Rome: Food and Agriculture Organization of the United Nations; 2013. 63 p.

15. Lobell DB, Cassman KG, Field CB. Crop yield gaps: Their importance, magnitudes, and causes. Annual Review of Environment and Resources. 2009 Oct 15; 34. DOI: 10.1146/annurev.environ.041008.093740.

16. Renfrew C, Bahn PG. Archaeology: Theories, methods, and practice. Eighth Edition. London: Thames & Hudson Publications; 2004. 656 p.

17. Berry W. The unsettling of America: Culture & agriculture. Berkeley: Counterpoint; 2015. 240 p.

18. Leiserowitz A, Maibach E, Rosenthal S, Kotcher J, Ballew M, Goldberg M, Gustafson A. Climate change in the American mind. New Haven: Yale Program on Climate Change Communication; 2018. 35 p.

19. UN Department of Economic and Social Affairs. Growing at a slower pace, world population is expected to reach 9.7 billion in 2050 and could peak at nearly 11 billion around 2100 [Internet]. New York: United Nations Department of Economic and Social Affairs; 2019 Jun 17 [cited 2021 Jan 29]. https://www.un.org/development/desa/en/news/population/world-population-prospects-2019.html.

20. Ranganathan J, Waite R, Searchinger T, Hanson C. How to sustainably feed 10 billion people by 2050, in 21 charts [Internet]. Washington DC: World Resources Institute; 2018 Dec 5 [cited 2021 Jan 29]. https://www.wri.org/blog/2018/12/how-sustainably-feed-10-billion-people-2050-21-charts.

21. Ranganathan J, Waite R, Searchinger T, Hanson C. How to sustainably feed 10 billion people by 2050, in 21 charts [Internet]. Washington DC: World Resources Institute; 2018 Dec 5 [cited 2021 Jan 29]. https://www.wri.org/blog/2018/12/how-sustainably-feed-10-billion-people-2050-21-charts.

22. Tollefson J. How hot will Earth get by 2100? [Internet]. London: Springer Nature; 2020 Apr 22 [cited 2021 Jan 29]. https://www.nature.com/articles/d41586-020-01125-x.

23. Hoegh-Guldberg O, Jacob D, Taylor M, Bindi M, Brown S, Camilloni I, Diedhiou A, Djalante R, Ebi KL, Engelbrecht F, Guiot J, Hijioka Y, Mehrotra S, Payne A, Seneviratne SI, Thomas A, Warren R, Zhou G. 2018: Impacts of 1.5°C global warming on natural and human systems. In: Masson-Delmotte V, Zhai P, Pörtner H-O, Roberts D, Skea J, Shukla PR, Pirani A, Moufouma-Okia W, Péan C, Pidcock R, Connors S, Matthews JBR, Chen Y, Zhou X, Gomis MI, Lonnoy E, Maycock T, Tignor M, Waterfield T, editors. Global Warming of 1.5°C. An IPCC Special Report on the impacts of global warming of 1.5°C above pre-industrial levels and related global greenhouse gas emission pathways, in the context of strengthening the global response to the threat of climate change, sustainable development, and efforts to eradicate poverty. Geneva: United Nations Intergovernmental Panel on Climate Change; 2018. https://www.ipcc.ch/site/assets/uploads/sites/2/2019/06/SR15_Full_Report_High_Res.pdf.

24. Global Agriculture. Industrial agriculture and small-scale farming [Internet]. Berlin: Foundation on Future Farming; [date unknown] [cited 2021 Jan 29]. https://www.globalagriculture.org/report-topics/industrial-agriculture-and-small-scale-farming.html.

25. Lowder SK, Skoet J, Raney T. The number, size, and distribution of farms, smallholder farms, and family farms worldwide. World Development. 2016 Nov 1; 87: 16–29. DOI: 10.1016/j.worlddev.2015.10.041.

26. Lagi M, Bertrand KZ, Bar-Yam Y. The food crises and political instability in North Africa and the Middle East. SSRN Electronic Journal. 2011 Aug 15; 1910031. DOI: 10.2139/ssrn.1910031.

27. Pimentel D. Handbook of Energy Utilization In Agriculture. Ohio: CRC Press; 2019. 488 p.

28. RethinkX. About [Internet]. California: RethinkX; [date unknown] [cited 2021 Jan 29]. https://www.rethinkx.com/.

29. RethinkX. Rethinking energy 2020–2030: 100% solar, wind, and batteries is just the beginning. California: RethinkX; 2020 Oct. 62 p.

30. RethinkX. Rethinking energy 2020–2030: 100% solar, wind, and batteries is just the beginning. San Francisco, CA: RethinkX; 2020 Oct. 62 p.

31. Coady D, Parry I, Le NP, Shang B. Global fossil fuel subsidies remain large: An update based on country-level estimates. Washington DC: International Monetary Fund; 2019. 39 p.

32. BBC News. More obese people in the world than underweight, says study [Internet]. London: British Broadcasting Corporation; 2016 Mar 31 [cited 2021 Jan 29]. https://www.bbc.com/news/health-35933691.

33. Roser M, Ritchie H. Hunger and undernourishment [Internet]. Oxford: Our World in Data; 2013 [cited 2021 Jan 29]. https://ourworldindata.org/hunger-and-undernourishment.

34. FAO. Staple foods: What do people eat? [Internet]. Rome: Food and Agriculture Organization of the United Nations; [date unknown] [cited 2021 Jan 29]. http://www.fao.org/3/u8480e/u8480e07.html.

35. Mora C, Tittensor DP, Adl S, Simpson AG, Worm B. How many species are there on Earth and in the ocean? PLoS Biol. 2011 Aug 23; 9(8): e1001127. DOI: 10.1371/journal.pbio.1001127.

36. Brondizio ES, Settele J, Díaz S, Ngo HT. Global assessment report on biodiversity and ecosystem services of the Intergovernmental Science-Policy Platform on Biodiversity and Ecosystem Services. Bonn: IPBES secretariat; 2019 Jul. https://ipbes.net/global-assessment.

37. Brondizio ES, Settele J, Díaz S, Ngo HT. Global assessment report on biodiversity and ecosystem services of the Intergovernmental Science-Policy Platform on Biodiversity and Ecosystem Services. Bonn: IPBES secretariat; 2019 Jul. https://ipbes.net/global-assessment.

38. van Dijk M, Morley T, van Loon M, Reidsma P, Tesfaye K, van Ittersum MK. Reducing the maize yield gap in Ethiopia: Decomposition and policy simulation. Agricultural Systems. 2020 Aug 1; 183: 1–11. DOI: 10.1016/j.agsy.2020.102828.

39. Zhao C, Liu B, Piao S, Wang X, Lobell DB, Huang Y, *et al.* Temperature increase reduces global yields of major crops in four independent estimates. Proceedings of the National Academy of Sciences. 2017 Aug 29; 114(35): 9326–31. DOI: 10.1073/pnas.1701762114.

40. Loladze I. Rising atmospheric CO_2 and human nutrition: Toward globally imbalanced plant stoichiometry? Trends in Ecology & Evolution. 2002 Oct 1; 17(10): 457–61. DOI: 10.1016/S0169-5347(02)02587-9.

41. McGartland T. Buzzcocks: The Complete History. London: Bonnier Zaffre; 2017 May 4. 300 p.

42. UN. The Sustainable Development Agenda [Internet]. New York: United Nations;2021 [cited 2021 Jan 29]. https://www.un.org/sustainabledevelopment/development-agenda/.

43. Sachs J, Schmidt-Traub G, Kroll C, Lafortune G, Fuller G, Woelm F. The Sustainable Development Goals and COVID-19. Sustainable Development Report 2020. Cambridge: Cambridge University Press; 2020. 520 p.

44. Salawitch RJ, Canty TP, Hope AP, Tribett WR, Bennett BF. Paris Climate Agreement: Beacon of Hope. Cham: Springer Climate. 2017 Jan 5. 186 p.

45. Peuc G. Debunking Bad Design Memes, Part 2: "Candles and Electric Light" quote [Internet]. California: Medium; 2017 Jan 12 [cited 2021 Jan 29]. https://medium.com/@gpeuc/debunking-bad-design-memes-part-2-candles-and-electric-light-quote-3f9990784cfe.

46. Syed M. Rebel ideas: The power of diverse thinking. London: John Murray Press; 2020. 312 p.

47. Syed M. Rebel ideas: The power of diverse thinking. London: John Murray Press; 2020. 312 p.

48. World Meteorological Organization. Greenhouse gas concentrations in atmosphere reach yet another high [Internet]. Geneva: World Meteorological Organization; 2019 Nov 25 [cited 2021 Jan 29]. https://public.wmo.int/en/media/press-release/greenhouse-gas-concentrations-atmosphere-reach-yet-another-high.

49. Stiglitz JE, Shiller RJ, Gopinath G, Reinhart CM, Posen A, Prasad E, Tooze A, D'Andrea Tyson L, Mahbubani K. How the economy will look after the Coronavirus pandemic [Internet]. Washington DC: Foreign Policy; 2020 Apr 15 [cited 2021 Jan 29]. https://foreignpolicy.com/2020/04/15/how-the-economy-will-look-after-the-coronavirus-pandemic/.

50. RethinkX. Rethinking food and agriculture 2020–2030: 100% solar, wind, and batteries is just the beginning. California: RethinkX; 2019 Sep. 76 p.

51. McGinnis D. What is the Fourth Industrial Revolution? [Internet]. California: Salesforce; 2020 Oct 27 [cited 2021 Jan 29]. https://www.salesforce.com/blog/what-is-the-fourth-industrial-revolution-4ir/.

52. McGinnis D. What is the Fourth Industrial Revolution? [Internet]. California: Salesforce; 2020 Oct 27 [cited 2021 Jan 29]. https://www.salesforce.com/blog/what-is-the-fourth-industrial-revolution-4ir/.

53. Willett W, Rockström J, Loken B, Springmann M, Lang T, Vermeulen S, et al. Food in the Anthropocene: The EAT–Lancet Commission on healthy

diets from sustainable food systems. The Lancet. 2019 Feb 2; 393(10170): 447–92. DOI: 10.1016/S0140-6736(18)31788-4.

54. Association of International Research and Development Centers for Agriculture. GAPAD — Global Action Plan for Agricultural Diversification [Internet]. Nairobi: Association of International Research and Development Centers for Agriculture; 2016 [cited 2021 Jan 26]. http://www.airca.org/index.php/airca-resources/gapad.

55. Butler DA. Unsinkable: The full story of the RMS Titanic. Pennsylvania: Stackpole Books; 1998. 292 p.

56. Rafferty JP, Pimm SL. Desertification [Internet]. [place unknown]: Encyclopedia Britannica; 2020 Jan 29 [cited 2021 Jan 29]. https://www.britannica.com/science/desertification.

57. Leakey RR. A re-boot of tropical agriculture benefits food production, rural economies, health, social justice and the environment. Nature Food. 2020 May; 1(5): 260–5. DOI: 10.1038/s43016-020-0076-z.

58. Jackson W, Berry W. A 50-Year Farm Bill [Internet]. New York: New York Times; 2009 Jan 4 [cited 2021 Jan 29]. https://www.nytimes.com/2009/01/05/opinion/05berry.html.

59. Benton T, Bieg C, Harwatt H, Pudasaini R, Wellesley L. Food system impacts on biodiversity loss. Three levers for food system transformation in support of nature. London: Chatham House; 2021 Feb. https://www.chathamhouse.org/sites/default/files/2021-02/2021-02-03-food-system-biodiversity-loss-benton-et-al_0.pdf.

60. HLPE. Agroecological and other innovative approaches for sustainable agriculture and food systems that enhance food security and nutrition. A report by the High Level Panel of Experts on Food Security and Nutrition of the Committee on World Food Security. Rome: The High Level Panel of Experts (HLPE); 2019. http://www.fao.org/fileadmin/user_upload/hlpe/hlpe_documents/HLPE_S_and_R/HLPE_2019_Agroecological-and-Other-Innovative-Approaches_S-R_EN.pdf.

61. Esg Kell G. The remarkable rise of ESG [Internet]. New Jersey: Forbes; 2018 Jul 11 [cited 2021 Jan 29]. https://www.forbes.com/sites/georg-kell/2018/07/11/the-remarkable-rise-of-esg/#2d3122541695.

62. Climate Action 100+. 2020 progress report. [place unknown]: Climate Action 100+; 2020. https://www.climateaction100.org/wp-content/uploads/2020/12/CA100-Progress-Report.pdf.

63. Riley JL. What Trump can learn from James Carville [Internet]. New York: Wall Street Journal; 2019 Jun 11 [cited 2021 Jan 29]. https://www.wsj.com/articles/what-trump-can-learn-from-james-carville-11560294122.

64. UN. The 2021 Food Systems Summit [Internet]. New York: United Nations; 2021 [cited 2021 Feb 10]. https://www.un.org/en/food-systems-summit.

65. Alsop M. Ode to Beethoven: Why we must find joy in times of crisis [Internet]. Cologny: World Economic Forum; 2020 Apr 7 [cited 2021 Jan 29]. https://www.weforum.org/agenda/2020/04/covid-19-and-joy-in-times-of-crisis/.

INDEX

Academic Ranking of World
  Universities (ARWU), 138
agricultural research, 134, 135, 143,
  145–153, 160–179, 183, 195, 198
agrisilvopastoral, 5
agroecology, 34, 115, 116, 118
agroecosystem, 4, 5, 60, 83, 116, 289,
  292, 293, 312, 313, 315, 321
agroforestry, 5, 117
AIRCA, 170–172, 249, 260
Akkadian Empire, 30, 32, 36, 52, 53
ASSESSCROP, 263, 264, 269

bambara groundnut, 182, 196–211,
  220–229, 245, 251–255, 257, 266,
  271
BAMBREED, 209
BAMFOOD, 207
BAMGROW, 207

BAMLINK, 207, 208
BamYield, 254–256
Borlaug, 61–63, 65, 69, 74, 75, 78,
  81, 86
British Empire, 7, 8, 24, 40–42, 48,
  49, 51, 52, 54–56
business as usual, 286–288, 291, 308

Calico Act, 50
centres of diversity, 12, 13
centres of origin, 9, 11
CFFPLUS, 261, 262, 271
CFFRC, 242–246, 248–253,
  256–262, 264–268, 271–273
CGIAR, 8, 9, 62, 82, 160–171, 173,
  178, 243, 258
  Challenge Programmes, 163, 164
Cheaper Food Paradigm, 91, 92, 96,
  97, 114

climate change, 3, 8, 9, 20, 32, 73, 97,
    106–113, 120, 151, 152, 164, 168,
    170, 172, 174, 183, 186, 189, 193,
    194, 280, 288–291, 295, 300, 301,
    303, 305, 306, 311, 312, 314, 318,
    320, 322, 324, 327
climate crisis, 97, 106, 289, 295
cognitive diversity, 305, 306
Collective Action on Forgotten Foods,
    268
CONNECT, 252, 253
Consultative Group on International
    Agricultural Research, 160
CropBASE, 262, 263, 271, 272
Crops For the Future (CFF), 210, 211,
    231, 233, 242, 258, 259, 265, 273
Crop Trust, 166, 251

diet, 4, 53, 82, 84, 110, 112, 125–128,
    130, 131, 163, 167, 250, 269, 274,
    299, 301, 311, 312, 316, 317, 319,
    323

East Africa Groundnut Scheme,
    215–217, 220, 229
East India Company, 42, 43, 50
ecosystem, 4, 5, 9, 19, 20
empirical, 182, 183, 185, 186
enclosures, 44–46
Enclosures Acts, 54
English Agricultural Revolution, 43,
    44, 47, 50
ESG, 279, 280, 299, 310, 323
EU LANDSUPPORT, 271

farmer suicides, 99
FishPLUS, 247, 248, 258
Food and Agriculture Organization (FAO),
    61, 101, 154–161, 165, 173, 249

food insecurity, 96, 99–101, 103
FoodPLUS, 246
food security, 91, 99–101, 125, 134,
    157, 159, 161, 163, 164, 171, 174
forgotten foods, 126–128, 130,
    249–251, 267–269, 273, 312, 317,
    319
Forgotten Foods Network, 250, 251,
    267
fossil fuel, 106, 108, 277, 278, 281,
    282, 285, 290, 296, 298

gene banks, 6, 8, 9, 165
genetic erosion, 13, 14, 25
genetic resources, 6–9, 11, 14, 19,
    20, 26
genetic traits, 64, 65
germplasm, 6, 25
Global Action Plan for Agricultural
    Diversification (GAPAD), 172,
    260, 317, 318
Global Forum for Agricultural
    Research and Innovation (GFAR),
    158, 268
global public good, 26, 281, 284, 307,
    308, 326
great nutrient collapse, 112
Green Revolution, 60, 61, 63–66, 70,
    72–75, 78–88, 91, 92, 98–101, 161,
    167, 170, 182, 227, 234, 237, 242,
    274, 314, 319
GROWCROP, 263

hidden hunger, 103, 125
high fructose corn syrup (HFCS), 96,
    118
Homologous Series in Variation, 13
hunter-gatherer, 3, 4, 5, 6, 28, 57, 58
hydraulic empire, 29, 30

Industrial Revolution, 43, 45, 49, 50,
   93, 94, 106, 146, 278, 293, 319
intercrops, 5, 116
International Centre for Underutilized
   Crops, 242
International Ranking Expert
   Group [IREG], 139
irrigation, 29–33, 37–39, 52, 53, 58,
   61, 63–66, 72, 77–79, 82, 85, 86,
   98, 110, 114, 116, 117, 121, 130

Kew, 6–8, 24

landrace, 12–14, 25
LANDSUPPORT, 271, 272
language, 2, 15–23, 25, 26
Law of homologous series, 13
Lexicon, 267, 268
linguicism, 17–19
linguistic and cultural diversity, 21
linguistic diversity, 18, 20–22
Lysenko, 10, 11, 25, 309

malnutrition, 101–105, 130, 214, 223
mechanistic, 182, 183, 186, 202,
   205
modularization, 139–141
monoculture, 2, 5, 23, 26, 52, 61, 66,
   73, 79, 82, 96, 98, 114–117, 120,
   129, 149, 173, 174, 179, 182, 288,
   293, 319
Monteith, John, 183–188, 190–193
moringa, 252, 266
multicropping, 115–120
Muslim Empire, 37, 52–54

National Agricultural Research Systems
   (NARS), 150, 151, 160, 162
Navigation Acts, 49

negative externalities, 249, 277, 278,
   280, 284–286, 298, 299, 319, 321
neoliberal, 2
Nobel Peace Prize, 63, 69
Nobel prize, 63, 68, 134, 175
Nutritionism, 121–123, 125, 126

obesity, 102, 105
ODA Tropical Crop
   Microclimatology Unit, 188, 189,
   193, 194, 201
Official Development Assistance
   (ODA), 151, 152

Paris Agreement, 291, 303
Pasteur, 94
Pasteurization, 93, 94
Penman-Monteith, 185, 186
perfect storm, 113, 120, 130
plantation, 39, 41, 48, 51, 52, 55, 148,
   149, 173, 179
Plumpy'nut, 214
positive externalities, 280, 281, 297,
   301

Quacqarelli-Symonds (QS), 138

REF, 142
Research Assessment Exercise
   (RAE), 137, 142
Research Value Chain (RVC),
   243–245, 252–256, 265
Roman Empire, 32, 34–37, 40, 53
Rothamsted, 147

sedentary agriculture, 2, 3, 5, 14, 31
sedentary farming, 28
sedentism, 2, 3, 29
seed banks, 6, 9, 13, 23

SELECTCROP, 263, 264, 271, 272
slavery, 41, 42, 49, 51–53
staple (crops), 197, 210, 321
    food crops, 2, 35, 91, 92, 96, 97,
        100, 110, 120, 121, 125, 131,
        148, 163, 166–168, 178, 195,
        198, 215, 298–301, 307
stunting, 104, 105, 130
sustainable development, 301, 324
Sustainable Development Agenda,
    178, 301, 324
Sustainable Development Goals, 324
    SDG, 178, 260, 302, 303, 307,
        310–312, 317, 318, 323, 324
    SDG1, 302, 303, 318
    SDG2, 260, 303, 318
    SDG7, 260
    SDG13, 303
    SDG17, 307, 318
Svalbard Global Seed Vault, 9, 166

Teaching Quality Assurance (TQA),
    137
The Great Nutrient Collapse, 112

Thought For Food (TFF), 273
Times Higher Education (THE) world
    university rankings, 138
Tropical Crops Research Unit
    (TCRU), 192–197, 199, 205

UK Overseas Development
    Administration (ODA), 188
underutilized crop, 170, 171, 178,
    181, 182, 194–196, 200, 204–207,
    209–211, 220, 221, 223, 226, 229,
    236, 238, 240–249, 251, 256, 257,
    259–267, 270–272
USECROP, 263
US Morrill Act, 136

Vavilov, Nikolai, 9–14, 24, 25

wasting, 104, 105
WhatIF, 266, 267

yield-for-profit, 228, 280, 296, 298,
    301

Printed in the United States
by Baker & Taylor Publisher Services

Printed in the United States
by Baker & Taylor Publisher Services